Evolutionary Concepts in End User Productivity and Performance:
Applications for Organizational Progress

Steve Clarke
University of Hull, UK

Information Science
REFERENCE

INFORMATION SCIENCE REFERENCE
Hershey · New York

Director of Editorial Content: Kristin Klinger
Assistant Development Editor: Deborah Yahnke
Director of Production: Jennifer Neidig
Managing Editor: Jamie Snavely
Assistant Managing Editor: Carole Coulson
Typesetter: Carole Coulson
Cover Design: Lisa Tosheff
Printed at: Yurchak Printing Inc.

Published in the United States of America by
Information Science Reference (an imprint of IGI Global)
701 E. Chocolate Avenue, Suite 200
Hershey PA 17033
Tel: 717-533-8845
Fax: 717-533-8661
E-mail: cust@igi-global.com
Web site: http://www.igi-global.com

and in the United Kingdom by
Information Science Reference (an imprint of IGI Global)
3 Henrietta Street
Covent Garden
London WC2E 8LU
Tel: 44 20 7240 0856
Fax: 44 20 7379 0609
Web site: http://www.eurospanbookstore.com

Library of Congress Cataloging-in-Publication Data

Evolutionary concepts in end user productivity and performance : applications for organizational progress / Steve Clarke, editor.

p. cm.

Includes bibliographical references and index.

Summary: "This book aims to represent some of the most current investigations into a wide range of end-user computing issues, enhancing understanding of recent developments"--Provided by publisher.

ISBN 978-1-60566-136-0 (hardcover) -- ISBN 978-1-60566-137-7 (ebook)

1. End-user computing. 2. Labor productivity. I. Clarke, Steve, 1950-

QA76.9.E53E96 2009

004.01'9--dc22

2008022546

British Cataloguing in Publication Data
A Cataloguing in Publication record for this book is available from the British Library.

Evolutionary Concepts in End User Productivity and Performance: Applications for Organizational Progress is part of the IGI Global series named *Advances in End User Computing Series (AEUC)* Series, ISBN: 1537-9310

Table of Contents

Selected Readings

Detailed Table of Contents

Although the discipline of information systems (IS) development is well established, IS failure and abandonment remains widespread. As a result, a considerable amount of IS research literature has investigated, among other things, the factors associated with IS success and failure. However, little attention has been given to any possible relationships that exist among the uncovered factors. In an attempt to address this, Chapter I examines the development of a successful IS, and compares the factors associated with its success against the factors most reported in our review of the literature as being associated with IS failure.

Chapter II reports on a case study performed in cooperation with a telecommunication provider. The rapidly changing business environment demands that the company has supportive, sustainable information systems to stay on the front line of the business area. The company's continuous evolution of the IT-infrastructure makes it necessary to tailor the interaction between different applications. The objective of the case study was to explore what is required to allow end users to tailor the interaction between flexible applications in an evolving IT-infrastructure to provide for software sustainability. The case study followed a design research paradigm where a prototype was created and evaluated from a use perspective. The overall result shows that allowing end users to tailor the interaction between flexible applications in an evolving IT infrastructure relies on, among other things, an organization that allows cooperation between users and developers that supports both evolution and tailoring.

Chapter III uses some special usability and ethical issues that arise from experience with what can be called captive end-user systems (CEUS). These are systems required to gain access to or participate in a private or privileged organization, or for an employee or member of another organization wishing to gain such access and participation. We focus on a few systems we list, but our discussion is relevant to many others, and not necessarily Web-based ones. The specific usability aimed at in this chapter is usability testing (UT), which we use in its usually accepted definition.

Spreadsheets are commonly used and commonly flawed, but it is not clear how often spreadsheet errors lead to bad decisions. In Chapter IV we interviewed 45 executives and senior managers/analysts in the private, public, and non-profit sectors about their experiences with spreadsheet quality control and with errors affecting decision making. Almost all of them said spreadsheet errors are common. Quality control was usually informal and applied to the analysis and/or decision, not just the spreadsheet per se. Most respondents could cite instances of errors directly leading to bad decisions, but opinions differ as to whether the consequences of spreadsheet errors are severe. Some thought any big errors would be so obvious as to be caught by even informal review. Others suggest that spreadsheets inform but do not make decisions, so errors do not necessarily lead one for one to bad decisions. Still, many respondents believed spreadsheet errors were a significant problem and that more formal spreadsheet quality control could be beneficial.

The means by which the U.S. justice system attempts to control illegal hacking are practiced under the assumption that hacking is like any other illegal crime. Chapter V evaluates this assumption by comparing illegal hacking to shoplifting. Three inhibitors of two illegal behaviors are examined: informal sanction, punishment severity, and punishment certainty. A survey of 136 undergraduate students attending a university and 54 illegal hackers attending the DefCon conference in 2003 was conducted. The results show that both groups perceive a higher level of punishment severity but a lower level of informal sanction for hacking than for shoplifting. Our findings show that hackers perceive a lower level of punishment certainty for hacking than for shoplifting, but students perceive a higher level of punishment certainty for hacking than for shoplifting. The results add to the stream of information security research and provide significant implications for law makers and educators aiming to combat hacking.

The last decade has seen the proliferation of business-to-employee (B2E) portals as integrated, efficient, and user-friendly technology platform to assist employees to increase their productivity, as well as for organizations to reduce their operating costs. To date, very few studies have focused on determining the extent to which the portal implementations have been successful. Such a study is crucial, considering that organizations have committed large investments to implementing the portals and they would certainly like to see the return on their investments. Our study in Chapter VI aims to develop a scale for measuring user satisfaction with B2E portals. The four steps of scale development: conceptual model development, item generation, content validation, and an exploratory study, are reported in this chapter. Evidence about reliability, content validity, criterion-related validity, convergent validity, and discriminant validity is presented.

Improving how knowledge is leveraged in organizations for improved business performance is currently considered as a major organizational change. Knowledge management (KM) projects are stigmatized as demanding, fuzzy, and complex, with questionable outcomes—more than 70% of them do not deliver what they promised. While most organizations have deployed knowledge management systems (KMSs), only a handful have been able to leverage these investments. The goal of Chapter VII is to propose theoretical background for design of KMS that successfully support and enable new knowledge creation and existing knowledge utilization. By using principles of the design science, design profiles proposed build upon works from organization and IS sciences, primarily the Evolutionary Information-Processing Theory of Knowledge Creation (Li & Kettinger, 2006) and the Task Technology Fit Theory (Zigurs & Buckland, 1998), the latter being amended for particularities of the KM environment.

In Chapter VIII we report the findings of a field study of the enactment of ICT supported knowledge work in a Human Resources contact centre, illustrating the negotiable boundary between what constitutes the developer and user. Drawing upon ideas from the social shaping of technology, we examine how discussions regarding producer-user relations require a degree of greater sophistication as we show how users develop technologies and work practices in-situ. In this case different forms of knowledge are practised to create and maintain a knowledge sharing system. We show how as staff simultaneously distance themselves from, and ally with, ICT supported encoded knowledge scripts, the system becomes materially important to the project of constructing the knowledge characteristic of professional identity. Our work implies that although much has been made of contextualising the user, as a user, further work is required to contextualise users as developers and moreover, developers as users.

Chapter IX describes two experiments that examined overconfidence in spreadsheet development. Overconfidence has been seen widely in spreadsheet development and could account for the rarity of testing by end-user spreadsheet developers. The first experiment studied a new way of measuring overconfidence. It demonstrated that overconfidence really is strong among spreadsheet developers. The second experiment attempted to reduce overconfidence by telling subjects in the treatment group the percentage of students who made errors on the task in the past. This warning did reduce overconfidence, and it reduced errors somewhat, although not enough to make spreadsheet development safe.

Voice recognition technology-enabled devices possess extraordinary growth potential, yet some research indicates that organizations and consumers are resisting their adoption. The study in Chapter X investigates the implementation of a voice recognition device in the United States Navy. Grounded in the social psychology and information systems literature, the researchers adapted instruments and developed a tool to explain technology adoption in this environment. Using factor analysis and structural equation modeling, analysis of data from the 270 participants explained almost 90% of the variance in the model. This research adapts the technology acceptance model by adding elements of the theory of planned behavior, providing researchers and practitioners with a valuable instrument to predict technology adoption.

Colleges of business have dealt with teaching computer literacy and advanced computer application concepts for many years, often with much difficulty. Traditional approaches to provide this type of instruction, that is, teaching tool-related features in a lecture in a computer lab, may not be the best medium for this type of material. Indeed, textbook publishers struggle as they attempt to compile and organize

The study's objective is to arrive at a theoretical model and framework to guide research into the implementation of KMS, while also seeking to inform practice. In order to achieve this, Chapter XIV applies the critical success factors (CSF) method in a field study of successful KMS implementations across 12 large multinational organisations operating in a range of sectors. The chapter first generates a 'collective set' of CSFs from extant research to construct an a priori model and framework: this is then empirically validated and extended using the field study findings to arrive at a 'collective set' of CSFs for all 12 organisations. These are then employed to refine and extend the theoretical model using insights from the literature on capability theory. It is hoped that the model and framework will aid theory building and future empirical research on this highly important and relevant topic.

Chapter XV develops a model of adoption and continued use of knowledge management systems (KMSs), which is primarily built on Rogers' (1995) innovation stages model along with two very important social psychology theories—Ajzen and Fishbein's (1980) theory of reasoned action (TRA) and Davis's (1986) technology acceptance model (TAM). It presents various factors and variables in detail. Hypotheses are developed which can be tested via empirical study. The proposed model has both theoretical and practical implications. It can be adapted for application in various organizations in national and international arena.

Selected Readings

In today's global environment, a myriad of communication mechanisms enable cultures around the world to interact with one another and form complex interrelationships. The goal of Chapter XVI is to illustrate an individual-based approach to understanding cultural similarities and differences in the borderless world. Within the context of Web communication, a typology of individual cultural value orientations is proposed. This conceptualization emphasizes the need for making distinctions first at the individual level, before group-level comparisons are meaningful, in order to grasp the complexity of today's global culture. The empirical study reported here further demonstrates the usefulness of this approach by successfully identifying 16 groups among American Web users as postulated in the proposed typology. Future research should follow the implications provided in this chapter in order to broaden our thinking about the role of culture in a world of global communication.

ChapterXVII presents the mCity Project, a project owned by the City of Stockholm, aiming at creating user-friendly mobile services in collaboration with businesses. Starting from the end-users' perspective, mCity focuses on how to satisfy existing needs in the community, initiating test pilots within a wide range of areas, from health care and education, to tourism and business. The lesson learned is that user focus creates involvement among end users and leads to the development of sustainable systems that are actually used after they have been implemented. This is naturally vital input not only to municipalities and governments but also for the IT/telecom industry at large. Using the knowledge from mCity, the authors suggest a new, broader definition of "m-government" which focuses on mobile people rather than mobile technology.

Lately, user quality of experience (QoE) during their interaction with a system is a significant factor in the assessment of most systems. However, user QoE is dependent not only on the content served to the users, but also on the performance of the service provided. Chapter XVIII describes a novel QoE layer that extends the features of classic adaptive e-learning systems in order to consider delivery performance in the adaptation process and help in providing good user perceived QoE during the learning process. An experimental study compared a classic adaptive e-learning system with one enhanced with the proposed QoE layer. The result analysis compares learner outcome, learning performance, visual quality and usability of the two systems and shows how the QoE layer brings significant benefits to user satisfaction improving the overall learning process.

One challenge within the high-tech sector is to develop products that end users will actually need and will be able to use. One way of trying to match the design of high-tech products to the needs of end users, is to let researchers and designers interact with them via a human-centred design (HCD) approach. One HCD project, in which the author of Chapter XIX works, is studied. It is shown that the relation between interacting with end users and making design decision is not straightforward or "logical." Gathering knowledge about end users is like making a grasping gesture and reduces their otherness. Making design decisions is not based on rationally applying rules. It is argued that doing HCD is a social process with ethical qualities. A role for management is suggested to organize HCD alternatively to stimulate researchers and designers to explicitly discuss such ethical qualities and to work more reflectively.

Preface

Welcome to the latest annual volume of Advances in End-User Computing (EUC). EUC research and practice continues to provide new insights into the domain, and this 2008 volume aims to represent some of the most current investigations into a wide range of End-User Computing issues. We hope that you, as researchers, educators, and professionals in the domain, find something to enhance your understanding of these most recent developments, and, not least, that you enjoy reading about them. A summary of the contents of the text is given below.

Chapter I, "Information Systems Success and Failure–Two Sides of One Coin, or Different in Nature? An Exploratory Study", by Jeremy Fowler and Pat Horan, La Trobe University, Australia, argues that, although the discipline of information systems (IS) development is well established, IS failure and abandonment remains widespread. They further suggest that little attention has been given to any possible relationships that exist among "uncovered" factors, and seek to address this by examining the development of a successful IS, and comparing the factors associated with its success against the factors most reported in our review of the literature as being associated with IS failure. The results of the study show that four of the six factors associated with the success of the investigated IS were related to the IS failure factors identified from the literature.

Chapter II, "Achieving Sustainable Tailorable Software Systems by Collaboration Between End-Users and Developers", is by Jeanette Eriksson of the Blekinge Institute of Technology, Sweden, and Yvonne Dittrich, IT-University of Copenhagen, Denmark.

The chapter looks at a case study to show how the sustainability of information systems as a way of gaining advantage in rapidly changing environments. They argue that the fast pace of change makes flexibility in software an essential part of this process, and that one way to provide this is end-user tailoring (enabling the end user to modify the software while it is being used, as opposed to modifying it during the initial development process). This has the added advantage that end users already possess domain knowledge, so by providing support for end-user tailoring alterations can be made more immediately. Their results support the claim that end-users can even tailor the interaction between business applications. Three different categories of issues emerge as important when providing end-users with the possibility to manage interactions between applications in an evolving IT-infrastructure.

Chapter III, "Usability, Testing, and Ethical Issues in Captive End-User Systems", by Marvin D. Troutt and William Acar, Kent State University, USA, and Douglas A. Druckenmiller, Western Illinois University – Quad Cities, USA, addresses some usability and ethical issues that arise from experience with captive end-user systems (CEUS). These are systems required to gain access to or participate in a private or privileged organization, or for an employee or member of another organization wishing to gain such access and participation. It is argued that the discussion is relevant to other systems than the one investigated, and has particular relevance to the domain of usability testing.

Chapter IV, "Do Spreadsheet Errors Lead to Bad Decisions? Perspectives of Executives and Senior Managers", is by Jonathan P. Caulkins, Carnegie Mellon University, Erica Layne Morrison, IBM Global Services, and Timothy Weidemann, Fairweather Consulting, all in the USA. Whilst they accept the common argument that spreadsheets are frequently flawed, they contend that it is not clear how often spreadsheet errors lead to bad decisions. The findings are based on interviews with forty-five executives and senior managers / analysts in the private, public, and non-profit sectors about their experiences with spreadsheet quality control and with errors affecting decision making. Spreadsheet errors emerged as commonplace, and quality control informal. Instances of errors directly leading to bad decisions were widely cited, but opinions differ as to whether the consequences of spreadsheet errors are severe. Overall, spreadsheet errors were seen to be a significant problem, and more formal spreadsheet quality control was widely recommended.

Chapter V, "A Comparison of the Inhibitors of Hacking vs. Shoplifting" by Lixuan Zhang, College of Charleston, USA, and Randall Young and Victor Prybutok from the University of North Texas, USA. In this chapter, grounded in information security research, the authors argue that the means by which the United States justice system attempts to control illegal hacking assumes that hacking is like any other illegal crime. This concept is evaluated by comparing illegal hacking to shoplifting. From a survey of 136 undergraduate students attending a university and 54 illegal hackers attending the DefCon conference in 2003, it emerged that both groups perceive a higher level of punishment severity but a lower level of informal sanction for hacking than for shoplifting. The results add to the stream of information security research and provide significant implications for law makers and educators aiming to combat hacking.

Chapter VI, "Developing Success Measures for Staff Portal Implementation", by Dewi Rooslani Tojib and Ly Fie Sugianto from Monash University, Australia, looks at the proliferation of Business-to-Employee (B2E) portals. The study aims to develop a scale for measuring user satisfaction with B2E portals, arguing that, to date, very few studies have focused on determining the extent to which the portal implementations have been successful.

Chapter VII, "Contingencies in the KMS Design: A Tentative Design Model", by Peter Baloh, University of Ljubljana, Slovenia, discusses the leveraging of knowledge to improve business performance. Grounded in the domain of Knowledge management (KM), the aim of this chapter is to propose theoretical background for design of KMS that successfully supports and enables new knowledge creation and existing knowledge utilization. Proposed fit profiles suggest that one-size-fits-all approaches do not work and that organizations must take, in contrast with extant literature, a segmented approach to KM activities and technological support.

Chapter VIII, "Users as Developers: A Field Study of Call Centre Knowledge Work", by Beryl Burns and Ben Light from the University of Salford, UK, reports the findings of a field study of the enactment of ICT supported knowledge work in a Human Resources contact centre, illustrating the negotiable boundary between what constitutes the developer and user. The authors examine how discussions regarding producer-user relations require a degree of greater sophistication. The research reaches the valuable conclusion that although much has been made of contextualising the user, as a user, further work is required to contextualise users as developers and moreover, developers as users.

Chapter IX, "Two Experiments in Reducing Overconfidence in Spreadsheet Development", by Raymond R. Panko of the University of Hawai`i, USA, describes two experiments that examined overconfidence in spreadsheet development. The first experiment studied a new way of measuring overconfidence, whilst the second experiment attempted to reduce overconfidence by telling subjects in the treatment group the percentage of students who made errors on the task in the past.

Chapter X, "User Acceptance of Voice Recognition Technology: An Empirical Extension of the Technology Acceptance Model", by Steven John Simon, Mercer University, USA, and David Paper, Utah State University, USA, investigates the implementation of a voice recognition device in the United States Navy. Grounded in the social psychology and information systems literature, the researchers adapted instruments and developed a tool to explain technology adoption in this environment. Using factor analysis and structural equation modeling, analysis of data from the 270 participants explained almost 90% of the variance in the model. This research adapts the technology acceptance model by adding elements of the theory of planned behavior, providing researchers and practitioners with a valuable instrument to predict technology adoption.

Chapter XI, "Educating Our Students in Computer Application Concepts: A Case for Problem-Based Learning", is by Peter P. Mykytyn, Southern Illinois University, USA. The subject of the chapter is the difficulty of teaching computer literacy and advanced computer application concepts. Traditional approaches, it is argued, are open to question, and textbooks struggle as they attempt to compile and organize appropriate material. This research has taken a problem-based learning (PBL) approach to teaching computer application concepts (in this case, Microsoft Excel and Access).

Chapter XII, "Covert End User Development: A Study of Success", by Elaine H. Ferneley, University of Salford, UK, asserts that End User Development (EUD) of system applications is typically undertaken by end users for their own, or closely aligned colleagues, business needs. EUD studies have focused on activity that is small scale, is undertaken with management consent and will ultimately be brought into alignment with the organisation's software development strategy. However, owing to the increase pace of today's organisations, EUD activity increasingly takes place without the full knowledge or consent of management, and such developments can be defined as "covert". The authors report on a covert EUD project where a wide group of internal and external stakeholders worked collaboratively to drive an organisation's software development strategy. The research highlights the future inevitability of external stakeholders engaging in end user development as, with the emergence of wiki and blog-like environments, the boundaries of organisations' technological artifacts become increasingly hard to define.

Chapter XIII, "When Technology Does Not Support Learning: Conflicts Between Epistemological Beliefs and Technology Support in Virtual Learning Environments", is by Steven Hornik and Yu Wu, University of Central Florida, USA, and Richard D. Johnson, University of South Florida, USA. Using data from 307 individuals, this research study investigated the process and outcome losses that occur when friction exists between individuals' epistemological beliefs and their perceptions of how the technology supports learning. Specifically, the results indicated that when there was friction between the technology support of learning and an individual's epistemological beliefs, course communication, course satisfaction, and course performance were reduced. Implications for design of virtual learning environments and future research are discussed.

Chapter XIV, "A Theoretical Model and Framework for Understanding Knowledge Management System Implementation", by Tom Butler, Ciara Heavin, and Finbarr O'Donovan, University College Cork, Ireland, aims to arrive at a theoretical model and framework to guide research into the implementation of KMS, while also seeking to inform practice. The chapter applies the critical success factors (CSF) method in a field study of successful KMS implementations across 12 large multinational organisations operating in a range of sectors. It is hoped that the model and framework will aid theory building and future empirical research on this highly important and relevant topic.

Chapter XV, "Exploring the Factors Influencing End Users' Acceptance of Knowledge Management Systems: Development of a Research Model of Adoption and Continued Use", is by Jun Xu, Southern Cross University, Australia, and Mohammed Quaddus, Curtin University of Technology, Australia. The chapter develops a model of adoption and continued use of knowledge management systems (KMSs), which is primarily built on Rogers' innovation stages model along with Ajzen and Fishbein's (1980) theory of reasoned action (TRA) and Davis's (1986) technology acceptance model (TAM). It presents various factors and variables in detail. Hypotheses are developed which can be tested via empirical study. The proposed model has both theoretical and practical implications. It can be adapted for application in various organizations in national and international arena.

Following on from the above fifteen chapters, we have also included a series of four selected readings which we hope you will agree enhance the quality of the text. The coverage of these, in summary, is given below.

Chapter XVI, "Classifying Web Users: A Cultural Value-Based Approach", looks at the problems inherent in today's communication mechanisms, which enable global interaction between different cultural groups, and aims to understand some of the cultural similarities and differences in this seemingly borderless world. A typology of individual cultural value orientations is proposed, emphasizing the need for making distinctions at the level of the individual, before group-level comparisons can become meaningful

Chapter XVII, "mCity: User Focused Development of Mobile Services Within the City of Stockholm", presents the mCity project in the city of Stockholm, which aims at creating user-friendly mobile services in collaboration with businesses. The project takes an end-user perspective, focusing on how to satisfy existing needs in the community, and as a result creates involvement among end users which, it is claimed, leads to the development of sustainable systems that are actually used after they have been implemented.

Chapter XVIII, "End-User Quality of Experience-Aware Personalized E-Learning", looks at how user quality of experience may be seen as dependent not only on the content served to the users, but also on the performance of the service provided. Using an experimental study, the authors compared a classic adaptive e-learning system with one enhanced with their 'quality of experience layer', to arrive at some novel and interesting findings.

Chapter XIX, "High-Tech Meets End-User", reviews matching the design of high-tech products to the needs of end users via a human-centred design (HCD) approach. A HCD project is studied, with the outcome that the relation between interacting with end users and making design decision is seen as not straightforward or "logical." It is argued that HCD is a social process with ethical qualities, and that researchers and designers need to explicitly address these qualities and to work more reflectively.

CONCLUSION: CONTRIBUTION TO THE FIELD

The chapters and readings presented above provide a wide variety of perspectives on the domain of End User Computing. The range of topics in this subject is, of course, vast, but I have been particularly pleased with the coverage of these chapters, and I hope you agree that they offer a valuable contemporary insight into EUC.

As always, I hope you enjoy reading them.

Steve Clarke
Editor-in-Chief
Advances in End User Computing, Volume 2008

Chapter I
Information Systems Success and Failure—Two Sides of One Coin, or Different in Nature?
An Exploratory Study

Jeremy Fowler
La Trobe University, Australia

Pat Horan
La Trobe University, Australia

ABSTRACT

Although the discipline of information systems (IS) development is well established, IS failure and abandonment remains widespread. As a result, a considerable amount of IS research literature has investigated, among other things, the factors associated with IS success and failure. However, little attention has been given to any possible relationships that exist among the uncovered factors. In an attempt to address this, we examine the development of a successful IS, and compare the factors associated with its success against the factors most reported in our review of the literature as being associated with IS failure. This may be an important area of study given, for example, project management practices may be affected by knowing whether success and failure are two sides of one coin, or different in nature. The results of our exploratory study showed that four of the six factors associated with the success of the investigated IS were related to the factors identified from our review of the literature as being associated with IS failure.

INTRODUCTION

The information systems (IS) profession has long been plagued by the failure and abandonment of a large number of IS projects. Recent research figures suggest somewhere in the range of 70 to 80 percent of IS projects are delivered late and over budget, often with missing functionality, while approximately 20 to 30 percent are considered outright failures (Stamati, Kanellis, Stamati, & Martakos, 2005; Standish Group, 2001). In an attempt to improve this situation a large body of research has, among other things, looked at the factors most influential in the success and failure of IS developments (e.g., Beynon-Davies, 1995; DeLone & McLean, 1992; DeLone & McLean, 2003; Ketchell, 2003; Law & Perez, 2005; Montealegre & Keil, 2000). However, little attention has been given to any possible relationships that exist among the uncovered factors. This leads us to ask, what factors are associated with a successful IS, and how do they relate to the factors identified in the literature as being associated with IS failure? Are IS success and failure two sides of one coin, or are they different in nature? This chapter, therefore, reports on an exploratory study of one organizational case of stakeholders' experiences of a successful IS, and compares factors identified as being associated with the success of the IS against a set of factors identified in the research literature as being associated with IS failure.

It is hoped this research may provide important insight into the relative importance of IS development factors in the success and failure of IS. For example, negative levels of factor 'x' might be a very important factor in the failure of IS, while positive levels of factor 'x' might only be moderately important in the success of IS.

LITERATURE REVIEW

In the first section of this literature review a brief account of the problems surrounding IS development and evaluation is presented. This includes a brief discussion of the difficulties faced when defining IS success and failure, the high failure rate of IS developments, and the question of *when* and *how* IS development outcomes should be measured. The second section then presents a brief overview of the six factors found to be the most regularly associated with IS failure during our review of the literature

IS Success and Failure: Definitions and Evaluation

Currently, no universally accepted definition of IS failure exists. Over the years researchers have developed a multitude of positions in regards to the term "failure" within the IS context. Sauer (1993), for example, defined an IS to have failed if "development of operation ceases, leaving supporters dissatisfied with the extent to which the system has served their interests" (Sauer, 1993, p. 4). Sauer described this definition as being more forgiving than most, given that many authors consider factors such as user-resistance or missed targets etc. to be sufficient grounds for describing an IS as a failure. Alternatively, the Standish Group (1994) defines an IS to have failed if it is cancelled or does not meet its budget, delivery, and business objectives. Wilson and Howcroft (2002) showed that given the multitude of descriptions developed by researchers relating to IS failure, almost any project could potentially be considered a failure of some description.

Conversely, IS success can also be viewed in a number of ways. Taylor (2000) defined an IS to be successful if it delivered to the sponsor "everything specified to the quality agreed on or within the time and costs laid out at the start" (p. 24), whilst the Standish Group (1994) view an IS to be a success if it meets its budget, delivery and business objectives.

Many authors (Dix, Finlay, Abword, & Beale, 2004; Garrity & Sanders, 1998; Seddon, Staples, Patnayakuni, & Bowtell, 1999; Wilson & Howcroft, 2002) have shown that one of the key reasons for the difficulty in defining IS success and failure is that different stakeholders view IS outcomes in highly varying ways, and thus, "different measures are likely to be needed to assess the impact of effectiveness of a system for different groups of stakeholders" (Seddon et al., 1999, p. 19). Further, it has been shown that a single stakeholder's perspective of success or failure can vary tremendously over time (Wilson and Howcroft 2002).

A good example of the highly differing views among stakeholders is described in Linberg's (1999) review of a failed medical procedure based IS. Although management had condemned the IS, it was found that five of the eight team members involved in the project deemed it to be the most successful project they had ever been involved in (the remaining three members nominated it as their second most successful). The reasons reported for this were that (1) the project was a theoretical challenge; (2) the product worked in the way it was intended to work; and (3) the team was small and high performing (Linberg, 1999). This is a good example of how an IS that one stakeholder group viewed as a failure could still be viewed by others as highly successful.

When determining IS outcomes another important question to be considered is *when* should the IS be measured? Studies have shown that IS success and failure can be measured in terms of the short-term or immediate impact, as well as the long-term or indirect impact (Garrity & Sanders, 1998). A study of an IS at the short-term stage may give a different view of its success when compared to a study of the same IS conducted at the long-term stage. In conjunction with this, the level at which IS outcomes are measured is yet another important consideration. Garrity and Sanders (1998) defined three levels at which an IS can be measured:

1. Firm or organizational level measures of success
2. Function or process level measures of success
3. Individual measures of success

At the organizational level, IS success can be measured primarily using measures related to organizational performance. This includes increased market share and/or profitability, operating efficiency, operating cost, and return on equity and stock. At the function or process level the IS can be measured in terms of the efficient use of resources and by the reduction of process cycle times. This measure includes the operating efficiency of functional areas, reduced costs, and processes that are well integrated. Finally, at the individual (or user) level the IS can be measured in terms of each user's perception of utility and satisfaction. This stage is defined by user satisfaction, user IS satisfaction, and utility of system (Garrity & Sanders, 1998).

Contrary to Garrity and Sanders' (1998) scheme, Sauer (1993) believes that measuring the performance of an IS against a set of metrics such as these will "generate useful evaluations but they do not constitute the very essence of failure" (p. 18). The premise being that although we may have a certain set of measures relating to some of the factors that contributed to an IS's outcome, we still do not have a fuller, deeper understanding about the underlying phenomena that caused the outcome. This is because a set of measured facts and figures on paper can never give a true account of the intricate web of social, political, and technical phenomena that occur during IS developments.

As demonstrated in this section, the problems associated with determining IS success and failure are complex and many. There exists a complex intertwining web of relationships between social, political, and technical factors that needs to be addressed in order to fully understand the phenomenon of IS success and failure. Questions such as how should we measure an IS, when should we measure an IS, and from whose perspective should we measure the IS, are just a few of the often subtle questions that confront researchers interested in investigating the reasons behind IS success and failure.

Failure Factor Matrix

Based on the review of the literature, we identified six factors that were the most regularly associated with IS failure (refer to Table 1). These factors will act as the framework against which the factors found to be the most influential in the success of the investigated IS are later compared.

These factors were derived using a slightly modified version of Schmidt, Lyytinen, Keil, and Cule's (2001) 29 ranked risk factors list. Using this list a review of the relevant literature was conducted and any reference to the 29 ranked risk factors was recorded in a spreadsheet. Once the literature review was complete the citations were added and the six risk factors with the highest number of recorded citations were selected (as shown in Table 1. In total, approximately 100 relevant articles were surveyed during this process.

It is important to note, however, that this is by no means a definitive list of the failure factors most associated with IS failure. The list is not based on an exhaustive review of the literature and, therefore, these six factors should not be seen as generalizable across all IS literature.

The following six sections give a brief overview of each identified factor.

Lack of Effective Project Management Skills/Involvement

A lack of effective project management skills/involvement was found to be the most commonly cited reason for IS failure during our review of the literature. One of the key reasons for its continued appearance could be that, unlike many of the other failure factors, project management is a process that lasts the full development lifecycle (Standish Group, 1999). Liebowitz (1999) found that IS managers often neglect to become involved in the implementation and training stages of a project's development, suggesting that if managers made it a priority to follow through the entire process, some of the often-cited managerial problems might be alleviated.

Table 1. The six factors found to be the most associated with IS failure during our review of the literature

Ranking:	Factor:	Example supporting literature:
1	Lack of effective project management skills/involvement	Beynon-Davies, 1999; Doherty & King, 2001; Jiang, Klein, & Discenza, 2002; McGrath 2002; Wallace, Keil, & Rai, 2004.
2	Lack of adequate user involvement	Al-Mashari & Al-Mudimigh, 2003; Chung & Peterson, 2000,2001; Wallace et al., 2004.
3	Lack of top-management commitment to the project	Irani, Sharif, & Love, 2001; Koenig, 2003; Standish Group, 1999.
4	Lack of required knowledge/skills in the project personnel	Jiang et al., 2002; Oz & Sozik, 2000.
5	Poor/inadequate user training	Das, 1999; Taylor, 2000.
6	User resistance	Baskerville, Pawlowski, & McLean, 2000; Roberts, Leigh, & Purvis, 2000.

In order to be effective, project managers need to have good leadership, communication, and administrative skills, be technically competent, and hold a position that is senior enough to command respect (Avison & Fitzgerald, 2003). They need to be able to get people working together, getting them committed to the change and building confidence (Avison & Fitzgerald, 2003).

Specifically developed management techniques such as Gantt charts, work breakdown structures, PERT analysis, and specially developed management support methodologies like PRINCE (Avison & Fitzgerald, 2003) can also be useful in aiding effective project management. Project managers also have the option of using practices such as pre-project partnering, which has been shown to improve management performance during IS development (Jiang et al., 2002). A project team in a pre-project partnering environment has a single set of jointly developed goals and procedures for carrying out project activities, and will have a set of practices in place aimed at controlling conflict and system quality (Jiang et al., 2002).

Lack of Adequate User Involvement

A lack of adequate user involvement was found to be the second most commonly cited factor in the failure of IS developments during our review of the literature. Having a lack of adequate user involvement has been found to lead to decreased system use (Choe, 1996), increased project development cycles (LaPlante, 1991), and low levels of user satisfaction and commitment (Avison & Fitzgerald, 2003).

In order to avoid these problems, it is vital IS practitioners involve all the relevant user groups in the development of IS, particularly if this involvement will result in them being involved in the decision making process (Avison & Fitzgerald, 2003). Through being involved in the development process, users are more likely to be committed to the IS, while gaining a sense of system ownership which increases the likelihood of success (Hunton & Beeler, 1997).

Through user participation, IS practitioners can gain a deeper understanding of how each individual's views impact the current IS. This will allow them to create an IS that will satisfy at least the majority of users' needs, while giving the users an opportunity to gain greater ownership of the system. This should lead to increased user job satisfaction and greater system usage, which is vital to creating successful IS.

Lack of Top-Management Commitment to the Project

Having a lack of top-management commitment to a project was found to be the third most commonly cited factor in the failure of IS development during our review of the literature. When top-management are committed to a project they will do whatever is necessary throughout all stages of the IS's development and implementation to ensure it succeeds (Ginzberg, 1981). Also, having top-management support has been identified as a key factor that can contribute to the escalation of commitment to troubled projects (Keil & Robey, 1999).

One way to assist in maximizing top-management support is to incorporate an executive sponsor into the project's development team (Oz & Sosik, 2000). This executive sponsor will act as the project champion and have a vested interest in the project's successful outcome (Standish Group, 1999).

However, too much top-management involvement can also be detrimental to project success. The American Airlines CONFIRM system case (Oz, 1994; Standish Group, 1994) is an example of top-management becoming too actively involved as project managers. This resulted in the project's termination because there were seen to be "too many cooks in the kitchen and the soup spoiled" (Standish Group, 1994, p. 3). Top-level managers who are too committed to a project are just as detrimental as those who are uncommitted (Neumann, 1997).

Lack of Required Knowledge/Skills in the Project Personnel

Having a lack of the required knowledge/skills in the project personnel was found to be the fourth most commonly cited factor in the failure of IS developments during our review of the literature. A lack of the required knowledge/skills in the project personnel can result in schedule overruns because of the need for the team to master new skills, and time and budget overruns because of inexperience with the undocumented idiosyncrasies of each new piece of hardware and software (Laudon & Laudon, 1998).

The major challenges faced by management when selecting project team members are inexperienced staff members, budget constraints, and the continual change in IS related job requirements (Cash & Fox, 1992). These problems are driven by constantly changing technologies, an ever changing business environment, and the continually changing role of IS in organisations (Lee, Trauth, & Farwell, 1995).

Combining team members from different departments or organisations can also cause special challenges (Cash & Fox, 1992). Each organisation has different goals and cultures, and special care must be taken when creating mixed departmental or organizational teams to make certain these goals and cultures are compatible (Cash & Fox, 1992).

Poor/Inadequate User Training

Poor/inadequate user training was the fifth most commonly cited factor in the failure of IS projects during our review of the literature. The goal of user training is to produce motivated users who possess the skills necessary to effectively use all the relevant features the new IS has to offer (Compeau, Olfman, Sei, & Webster, 1995). It is an important aspect of IS development as users sometimes find it difficult to adapt to the often-rapid introduction of new technologies into their working environments (Shaw, DeLone, & Niederman, 2002).

The problem with most training, however, seems to be that it is either ineffective, poorly structured, or limited in its content (Rifkin, 1991). These problems can be escalated through budget pressures, poorly qualified IS trainers, and a general lack of interest in the training process by all parties involved (Rifkin, 1991). User training is a process that should be given appropriate prior thought, and not something that is completed as an afterthought. Without this critical step, all of the previous hard work and planning may be made redundant by users who are dissatisfied with the IS, simply because they do not know how to use it properly.

User Resistance

Finally, user resistance was found to be the sixth most commonly cited factor in the failure of IS projects during our review of the literature. In order to counteract user resistance it is vital that developers make it a priority to involve users in the development process and ask them what they want from the new IS (Roberts et al., 2000). If users become involved in the IS's development they will generally accept the new system and gain a sense of ownership of it (Avison & Fitzgerald, 2003). Users also need to be given a legitimate business rationale for using the new IS (Heichler, 1995), as well as appropriate user training (Laudon & Laudon, 1998). If a new system is given to users with little or no training they are unlikely to be able to use it properly, and in turn may begin to resist it (Gaudin, 1998).

However, participation by users in implementation activities may not be enough to alleviate user resistance. Users may not always be involved in projects in a productive fashion, and may be using their position to further personal interests and gain power (Laudon & Laudon, 1998). In this situation, user participation may actually intensify resentment and resistance towards the IS, with users potentially implementing strategies that deliberately hinder the progress of a new IS (Buchanan & Badham, 1999). These strategies may be overt acts of sabotage, or more subtle acts such as 'losing' documents (Avison & Fitzgerald, 2003).

Summary

Although this is not meant to be a definitive list of the factors most associated with IS failure, it does discuss some important development aspects that, if controlled, could increase a project's chances of success. It is reasonable to assume that each of these six factors will generally play an important role in the success or otherwise of an IS project, and therefore should at least be considered at some level by those in charge of IS development efforts.

METHOD

Research Questions

As discussed earlier, little or no research has explored how the factors influential in IS failure are associated with those influential in IS success. Therefore, the first stage of this study assesses the factors considered by the participants to have been influential in a successful IS developed by a leading regional Australian-based organization. These success factors were then analyzed to see how they related to the previously defined matrix of IS failure factors (refer to Table 1). In order to achieve this, the following research questions were used:

- **Research question 1:** What factors are associated with a successful IS within a leading regional Australian based organization?
- **Research question 2:** How do the success factors from research question 1 relate to the factors identified in the literature as being associated with IS failure?

Study Design

A descriptive qualitative research approach was taken in this exploratory study. In doing so, a convenience sample of five members of a leading regional Australian based organization was used. These people were:

- One member of the project's management
- Two of the project's senior analyst/programmers
- One current user who was not involved in the IS's initial development cycle but was later involved as a business consultant over a period of approximately six months
- One current user who was involved in the IS's development as a business project manager

Data Collection

Data for the study was collected via one questionnaire and a series of five in-depth semi-structured interviews. The questionnaire's purpose was to collect background information about the IS prior to conducting the interviews. This information included details about the IS's objectives, requirement changes, organizational departments involved, previous management and personnel experience, the methodology used, and the project's planned and actual schedule and budget figures.

The second, and main stage of the data collection process involved a series of five in-depth semi-structured interviews. These interviews explored the interviewees' personal experiences of their involvement in the IS's development, as well as how, if at all, they believed the pre-determined matrix of IS failure factors related to the success of the IS. Each of the participants was asked a different series of questions based upon their role within the IS, with the exception of the two senior analyst/programmers, who were both asked the same questions.

Data Analysis

The emphasis of the analysis in this study was on drawing out the main themes associated with the successful IS, and how these themes related to the previously defined matrix of failure factors (refer to Table 1. Therefore, transcripts of interviews were first analyzed to determine perceived success factors for the investigated IS, and then analyzed again against the failure factor matrix.

RESULTS AND DISCUSSION

The first two sub-sections ('About the system' and 'Why did the participants consider the system a success?') of this section give an overview of the investigated IS as well as examining why each of the study's participants regarded the system as a success. This provides a framework on which to base the results and discussion of our research questions.

In the following sections the participants will be known by the IDs outlined in Table 2.

About the System

The IS investigated during this study is a large Internet based financial transaction service implemented in May of 2003 by a leading regional Australian based organization. The IS is used by approximately 5,000 to 6,000 users each day, generating somewhere in the order of 30,000 logons. The IS's users cover a broad spectrum of skill levels and frequency of use.

The new IS replaced an existing system used by the organization. The objectives of the new IS were to (1) replace the existing IS with technology aligned with the organization's strategic direction; (2) cater for future growth for a minimum of five years; (3) develop an architecture that provided failover and redundancy; (4) provide a re-use code base that would lower the cost and reduce delivery times of related future development efforts; and (5) train staff in the organization's new application platforms. Of these five objectives, one, three and five have been fully achieved, with objective four being partially achieved. Objective two, cater for future growth for a minimum of five years, is not yet assessable but at this stage looks likely to be fully achieved.

The project's development was planned to take a total of five months and two weeks to complete, at a significant monetary cost to the organization. In reality the project took approximately 17 months (an increase of approximately 325%) to complete (refer to Table 3) at a budget increase of approximately 84%.

Table 2. Participant information

ID:	Title:	Role during the IS's development:
PM	Project manager	One of the two project managers
AP1	Senior analyst/programmer	Senior analyst/programmer
AP2	Senior analyst/programmer	Senior analyst/programmer
IU1	Involved user	Business consultant (involved for approximately six months)
IU2	Involved user	Business project manager

Table 3. The project's planned versus actual individual development stage time frames

Number of months/weeks (elapsed time)		
Project Stage:	**Planned:**	**Actual:**
Analysis/planning stage(s)	1 month	3 months 2 weeks
Development stage(s)	4 months	12 months
Implementation stage(s)	2 weeks	6 weeks
Total:	5 months 2 weeks	17 months (approx.)

The project's development team comprised 30 people, with two in managerial roles. The current project manager (PM) took over approximately eight months into the development of the IS. A business project manager/sponsor was also appointed at that time to co-manage the project. Prior to this, a single project manager was in charge of development. The project's personnel had previously been involved in a combined total of approximately 550 projects, with management being involved in a combined total of approximately 170. A number of organizational departments were involved in the project's development including IT, online solutions, operational risk, Audit and business service.

During the IS's development a number of significant requirement changes were made including (1) giving the interface a totally different look and feel; (2) making several functional changes relating to business service; (3) creating a totally different organizational hardware and network infrastructure; and (4) implementing a number of system fixes.

The IS remains fully operational, and is universally regarded within the organization as being highly successful. The organization has also received extensive positive feedback from customers regarding the new IS.

Why Did the Participants Consider the IS a Success?

As previously discussed in the literature review, an explicit definition of what comprises a successful IS can be very hard to reach. This is particularly the case given the various stakeholder perspectives from which IS success and failure can be assessed. The following will, therefore, briefly outline why each of the study's participants regarded the IS as a success.

AP1 thought the IS was a success because it has very little downtime and when there are problems they are easy to fix. AP1 also referred to the amount of positive customer feedback received by the organization in relation to the IS. PM reported that it was successful because it was received well by the customers and although the project's delivery date had slipped the IS provided the organization with a lot more functionality than they had expected at the outset. PM also thought the IS was successful because the client was extremely happy with it and once the system was turned on there were no system failures within the first two months. AP2 thought the IS was successful because it worked well upon installation and he personally learnt a lot during development. IU2 regarded the IS as a success because it had been great for the organization and the take-up rate was excellent.

Research Question 1: What Factors are Associated with a Successful IS Within a Leading Regional Australian Based Organization?

In this section we examine the factors reported by the participants as being associated with the

Table 4. Factors participants perceived to contribute to the success of the IS

	Perceived success factors:	AP1:	PM:	AP2:	IU1:	IU2:	Total:
1	Top-management commitment:	✔	✔	✔	✔	✔	5
2	Project team commitment:	✔	✔	✔			3
3	Effective project management:	✔	✔	✔			3
4	Project personnel knowledge/skills:	✔	✔	✔			3
5	Enlisting of external contractors:		✔	✔		✔	3
6	Good working environment:	✔					1
7	Working for a good business unit:		✔				1
8	Support from vendor companies:		✔				1
9	Project personnel training:	✔					1
10	The appointment of a full-time business and project manager:		✔				1

success of the IS. Analysis of interview transcripts produced ten factors (refer to Table 4).

Of the ten factors cited by the participants as being influential in the success of the IS, five were mentioned by multiple participants. These five factors can therefore be seen as those viewed by the participants as being the most influential in the success of the IS. These factors are, therefore, now discussed in greater detail.

Top-Management Commitment

This was reportedly the most influential factor in the success of this IS, being cited by all interview participants. This factor may have played a particularly important role in this case given the project's significant cost and schedule overruns, which, in many cases, would have led to the cancellation or failure of the project. However, because the project was viewed as strategic by the organization, the philosophy was that the IS was to be successfully completed regardless of the financial cost ("because it's a strategic path we were going to put it in no matter how much it cost" - AP1), and regardless of the initial time estimates ("everyone's perspective, including the business, was we simply can't put it out until it's right and we don't care how long it takes" - PM).

PM felt one reason for the strong level of commitment from top-management might have been the project management techniques he used to keep the members of top-management informed. These techniques included sending out weekly status reports and regularly talking to the different managers who had a stake in the project in an effort to foster and maintain their commitment.

Project Team Commitment

Project team commitment was another factor that reportedly had high importance in the success of the IS. It could be argued that the high level of commitment displayed by the project personnel was most likely the result of the management techniques used (high level of communication etc.), the level of commitment from top-management, and the overall professionalism of the members of the project team. This project team commitment, coupled with the top-management commitment, appears to have had considerable influence in the success of the IS. The combination of these two factors appears to have negated many of the problems that may have otherwise led to the failure or cancellation of the project.

Effective Project Management

This case provided a persuasive example of how project management can have a considerable effect on project outcomes. The project manager who was appointed at the beginning of the project's lifecycle already had many other operational responsibilities. This meant that he could not give the project his undivided attention, creating a situation where he was essentially working on the project part-time. Although it is not entirely clear from the interviews, it would be reasonable to assume that this situation may have contributed, at some level, to the significant cost and schedule increases the project experienced during early development. This situation was rectified by the appointment of a fulltime project manager approximately eight months into development. PM felt that the point where he became involved was a lift given there was now someone dedicated to the project fulltime. PM also requested that a business project manager/sponsor be appointed to co-manage the development of the IS. PM commented, "between the business project manager and myself I think it had a very positive impact [on the project's outcome]".

Project Personnel Knowledge/Skills

It is interesting that project personnel knowledge/skills was a key contributor to the success of the IS given that at the beginning of development many members of the project team had little or no knowledge of the technologies to be used. These team members had to be trained in many areas including J2EE and Web development. The main portion of this training occurred during the first 12 months of the project, and generally consisted of on the job training as well as some more formal training for approximately three to four weeks prior to the project's commencement.

However, it was noted that despite the initial lack of skill and knowledge in some areas, on the whole, the team comprised some of the best people who were working in IT at that time. This, together with the team member training resulted in a project team that became extremely fluent in all aspects of the technology involved. This allowed the turnaround times on system functions to dramatically reduce as the project reached maturity.

Enlisting of External Contractors

Although not regularly cited during our review of the literature as a key factor in the success or failure of IS, enlisting the expertise of external contractors was one of the most highly reported reasons for this IS's success. One reason for this could be that, as previously discussed, the project was developed using technologies unfamiliar to many members of the project team. By enlisting external contractors the project team was able to not only be trained, but also have the assistance of the contractors in the development of the project.

External contractors were also employed to assist in the redevelopment of the IS's user interface. This was done in order to incorporate into the project's development people with a unique, but temporarily required skill set that the organization could not justify employing on a fulltime basis. IU2 thought "engaging the external Web designers and psychologists was definitely a very wise move for us". It helped the organization to develop a user-interface that required little or no user training, which was very important given the number of people with varied skill levels who would use the IS.

Without the assistance of these external contractors it is still likely that the project would have been completed given the commitment of all those involved. However, this would have most likely occurred at a much higher time and budget cost to the organization.

Research Question 2: How do the Success Factors from Research Question 1 Relate to the Factors Identified in the Literature as Being Associated with IS Failure?

In the following section, the six previously defined failure factors (refer to Table 1) are examined in an attempt to provide insight into how, if at all, they relate to the factors determined as being the most important in the success of the investigated IS (refer to Research Question 1).

Lack of Effective Project Management Skills/Involvement

Our review of the literature revealed that having a lack of effective project management skills/involvement was the factor that most frequently contributes to the failure of IS. Conversely, effective project management in this case was regarded by those interviewed as the third most influential factor in the success of the investigated IS.

One reason for this could be that the members of project management were involved for the entire duration of development. The failure of project management to become adequately involved for the entirety of development was one of the key contributors to the appearance of project management in lists of IS failure factors as outlined in the literature review (Liebowitz, 1999; Standish Group, 1999).

PM also appeared to possess many of the skills identified in the literature as being important in the establishment of effective project management. For example, Schmidt et al. (2001) found that having an improper definition of roles and responsibilities within the project team was one of the key predictors of poor project management. The project manager in this case made it a priority to ensure that all those involved in the development had a clear idea of their roles and responsibilities.

Having good communication skills was another key skill that was cited in the literature as being important in effective project management (Avison & Fitzgerald, 2003). This was a skill that was used frequently by PM:

So basically almost my entire role was communicating with people, going around and seeing where they were up to and feeding that information to other people.

Avison and Fitzgerald (2003) also found that effective project managers should be capable of getting people working together, getting them committed to the change and building confidence. This was another skill PM found to be a particularly important aspect of the job.

Lack of Adequate User Involvement

Given the Internet based nature of the IS, coupled with the fact that the IS would have many thousands of users each day with highly varying skill levels, it would be reasonable to assume that user involvement would play a particularly important role during development. However, user involvement was not mentioned by any of the participants as being a factor in the success of the IS.

The success of the IS, despite the absence of user involvement, does however have a logical but somewhat unusual explanation. The basis of this explanation is the IS's main objective, which was to replace an existing IS with technology aligned with the organization's strategic direction. Therefore, the mandate for the new IS was essentially to copy the old IS into the new programming and infrastructure platforms. This meant that a large percentage of the required user involvement had already been conducted during the previous IS's development. This resulted in user involvement being of much lower importance in the development of the replacement IS.

However, users were still involved during the development of the new IS. The organization ran focus groups with selected customers where they were asked to perform a series of predetermined tasks whilst their actions were recorded and analyzed. This prompted the organization to make some very slight changes to the interface before it was implemented. A series of prototypes was also tested on members of the organization's call center staff. The organization reasoned that if the call center staff were unable to understand various aspects of the IS then the customers would be unable to also.

Lack of Top-Management Commitment to the Project

Top-management commitment was found to be the third highest ranked failure factor during our review of the literature. Inversely, this factor was regarded as the most influential in the success of the investigated IS. When top-management are committed to a project they will do whatever is necessary throughout all stages of the system's development and implementation to ensure the IS solves all the required problem areas (Ginzberg, 1981). This was clearly an important facet of the investigated IS's development. Top-management was committed for the entire duration of the project and was forthright in their allocation of additional financial resources. Top-management was also prepared to allow significant extra time in which to complete the project to the standard required. This is indicative of the de-escalation in commitment to failing courses of action top-management can generate (Keil & Robey, 1999).

As mentioned earlier, one way to assist in maximizing top-management support is to incorporate an executive sponsor into the project's development team (Oz & Sosik, 2000). In this case, one of the PM's first steps was the appointment of a business sponsor who also acted as a second project manager. PM regarded this step as being critical in the development of the IS.

Finally, as in the CONFIRM car rental and hotel reservation system case (Beynon-Davies, 1995; Oz, 1994; Standish Group, 1994), PM also noted that too much top-management involvement can have a detrimental effect on project success and cited several techniques that can be used to effectively manage this problem. These techniques included collectively developing a project development and team charter that details what each person or stakeholder group's boundaries and accountabilities are. PM felt this was critical to a project's success.

Lack of Required Knowledge/Skills in the Project Personnel

A lack of the required knowledge/skills in the project personnel was found to be the fourth most influential factor in the failure of IS projects during the review of the literature. Conversely, the high level of knowledge and skill ultimately possessed by the investigated IS's project team was reported as the fourth most influential factor in the IS's success. However, the project team still suffered from many of the challenges inherent in the process of selecting project team members. Some of these challenges included:

1. The initial problem of inexperienced staff (Cash & Fox, 1992)
2. The combining of team members from different departments and organizations (Cash & Fox, 1992).

The first of these problems (i.e., inexperienced staff) was one of the driving forces in the initial schedule and budget overruns. This problem was overcome through training in the new technologies prior to, and during, the IS's development, as well as the integration of external contractors into the project team. These contractors assisted the IS's development and provided additional training to project team members.

The problem of handling the integration of people from different departments and organizations (i.e. the second of the aforementioned problems) was largely overcome via processes implemented by the project manager. These included gaining the support of a business sponsor, generating a constant flow of communication between the departments, making sure people had a clear idea of their roles and responsibilities, and the publication of key project dates.

Poor/Inadequate User Training

Although poor/inadequate training was cited as being one of the key causes of IS failure within the literature, it did not play a role in the success of the investigated IS. This is to be expected given that the IS is Internet based making it impossible to formally train every user.

However, the organization did put in place processes to assist users prior to, and after, the new IS was implemented. Prior to the IS's implementation the organization issued a demonstration version of the new IS linked from the current system so that users could become familiar with the new IS's interface and features. They also used the organization's Website to keep users informed about the benefits and features of the new IS, as well as details about when it would be implemented.

Users also had the option of using the IS's online help facility or contacting the organization's call-center staff if they had any difficulties with the IS. Also, whenever the organization releases a new version of the system they include a "What's New?" section to acclimatize users to any changes made.

It is reasonable to assume that these efforts to counteract the absence of any formal user training have, at least to some degree, been successful given the extremely small amount of negative feedback the organization has received regarding the new IS.

User Resistance

User resistance was the sixth factor identified during our review of the literature as being associated with IS failure. Although user resistance was not explicitly mentioned by any of the study's participants, frequent references were made to the positive feedback received from a large number of satisfied users. It would not be unreasonable to conclude that this tends to confirm the absence of any user resistance towards the project.

CONCLUSION

The aim of the research was to survey one organizational case of stakeholders' experiences of a successful IS, and evaluate the factors identified as being associated with the success of the IS against a set of factors identified in the research literature as being associated with IS failure. It was hoped that some parallels between these two sets of factors would be developed to give some insight into whether failed and successful IS's are affected by the same factors or are different in nature.

Results for the first research question showed that the factors reported by the study's participants as being associated with the success of the investigated IS were (1) top-management commitment; (2) project team commitment; (3) effective project management; (4) project personnel knowledge/skills; and (5) enlisting of external contractors. This set of factors would appear to be generally consistent with those regularly reported in the literature as being associated with IS success.

It is also worth noting how the combination of top-management commitment and project team commitment appears to have had a considerable influence in the successful development of the investigated IS. These stakeholders had significant buy-in to the project and were not prepared to see it fail. This combined commitment appears to have rendered irrelevant many of the problems that affected the development of the IS; problems that in many cases may have led to the failure or cancellation of the project. This high level of project commitment by both parties appears to have been influenced by the communication techniques used by the project manager. These techniques included regularly issuing status reports as well as personally going and talking to various people involved in the IS's development. This helped promote and then maintain the high level of project commitment displayed by each of these stakeholder groups.

The results for the second research question suggest that the factors most influential in IS success are closely related to the factors most influential in IS failure. Of the five factors mentioned explicitly as being critical to the success of the investigated IS, three directly correspond to the factors identified in the literature as being associated with IS failure. Add to this 'user acceptance', which was reported implicitly by many of the study's participants, and the number of direct matches increases to four of six (refer to Figure 1).

This is particularly promising, especially given the Internet based nature of the IS which, as previously discussed, largely rendered irrelevant each of the user related factors in the identified failure factor list (i.e., lack of adequate user involvement and poor/inadequate training). Only item two, project team commitment, and item five, enlisting of external contractors, from the success factor list failed to correspond in some capacity to any of the previously identified failure factors.

The absence of any correlation between these two success factors and the set of failure factors can be explained given their unusual nature. Project team commitment and the enlisting of external contractors are factors that seldom appeared during our review of the literature in regard to being influential in the success or otherwise of IS. In particular, the importance of "enlisting of external contractors" appears to have been somewhat unusual. As previously discussed, this factor was important because of the initial lack of skills possessed by many members of the project personnel.

This research, although relatively small in scope, has provided some useful insights into some areas of IS research that have received little previous attention. It is hoped that this research may inspire, and provide a basis for, future study within the presently discussed areas of IS success and failure.

The results of this area of research may be useful for managers and other IS professionals

Figure 1. How the identified failure factors corresponded to the identified success factors

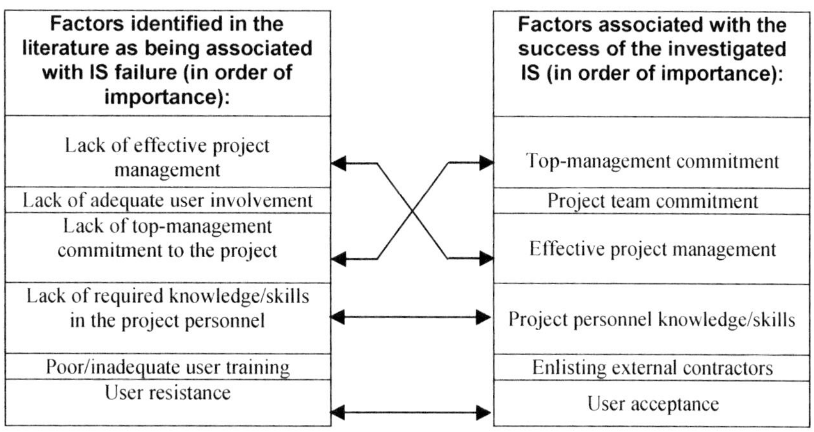

through the identification of the factors that carry a similar high level of importance in both the success and failure of IS development projects. These identified factors can then be more closely monitored and managed in an attempt to foster success and avoid failure. For example, if factor 'x' is found to be extremely influential in both the success and failure of IS development then it may be more closely monitored and managed than factor 'y' which is generally only moderately influential in IS failure, but hardly influential at all in IS success.

Further Research

There are several areas where further research might be conducted based on this study. Primarily, this would include a similar investigation conducted with a greater sample size in order to achieve more authoritative results. It might also be interesting for future research to investigate whether or not specific combinations of factors such as top-management commitment and project team commitment can overcome sets of negative factors that are present in the development of IS. The combination of certain positive factors appears to have been very influential in the investigated IS.

REFERENCES

Al-Mashari, M., & Al-Mudimigh, A. (2003). ERP Implementation: Lessons from a case study. *Information Technology & People, 16*(1), 21-33.

Avison, D., & Fitzgerald, G. (2003). *Information systems development: Methodologies, techniques, and tools* (3rd ed.). Berkshire: Mc-Graw-Hill.

Baskerville, R., Pawlowski, S., & McLean, E. (2000). Enterprise resource planning and organizational knowledge: Patterns of convergence and divergence. In S. Ang, H. Krcmar, W. Orlikowski & P. Weill (Eds.), In *Proceedings of the Twenty First International Conference on Information Systems* (pp. 396-406). Atlanta, GA: Association for Information Systems.

Beynon-Davies, P. (1995). Information systems "failure": The case of the London ambulance service's computer-aided dispatch project. *European Journal of Information Systems, 4*(3), 171-184.

Beynon-Davies, P. (1999). Human error and information systems failure: The case of the London ambulance service computer-aided dispatch system project. *Interacting with Computers, 11*(6), 699-720.

Buchanan, D., & Badham, R. (1999). Politics and organizational change: The lived experience. *Human Relations, 52*(5), 609-629.

Cash, C.H., & Fox, R. (1992). Elements of successful project management. *Journal of Systems Management, 43*(9), 10-12.

Choe, J. (1996). The relationships among performance of accounting information systems, influence factors, and evolution level of information systems. *Journal of Management Information Systems, 12*(4), 215-240.

Chung, K.S., & Peterson, D.K. (2000,2001). Developers' perceptions of information systems success factors. *Journal of Computer Information Systems, 41*(2), 29-35.

Compeau, D., Olfman, L., Sei, M., & Webster, J. (1995). End-user training and learning. *Communications of the ACM, 38*(7), 24-26.

Das, S. (1999, May 6). E-tag chaos hits tollway opening. *The Age*, p. 1-2.

DeLone, W.H., & McLean, E.R. (1992). Information systems success: The quest for the dependant variable. *Information Systems Research, 3*(1), 60-95.

DeLone, W.H., & McLean, E.R. (2003). The DeLone and McLean model of information systems success: A ten-year update. *Journal of Management Information Systems, 19*(4).

Dix, A., Finlay, J., Abword, D., & Beale, R. (2004). *Human-computer interaction* (3rd. ed.). Harlow, UK: Pearson Education Limited.

Doherty, N.F., & King, M. (2001). An investigation of the factors affecting the successful treatment of organizational issues in systems development projects. *European Journal of Information Systems, 10*(3), 147.

Garrity, E.J., & Sanders, L.G. (1998). *Information systems success measurement.* Hershey, PA: Idea Group Publishing.

Gaudin, S. (1998). Migration plans? Remember: Talk to end users up front. *Computerworld, 32*(43), 1-2.

Ginzberg, M.J. (1981). Key recurrent issues in the MIS implementation process. *MIS Quarterly, 5*(2), 47-59.

Heichler, E. (1995). Move to groupware sparks user resistance. *Computerworld, 29*(11), 12.

Hunton, J.E., & Beeler, J.D. (1997). Effects of user participation in systems development: A longitudinal field experiment. *MIS Quarterly, 21*(4), 359-388.

Irani, Z., Sharif, A.M., & Love, P.E.D. (2001). Transforming failure into success through organizational learning: An analysis of a manufacturing information system. *European Journal of Information Systems, 10*, 55-66.

Jiang, J.J., Klein, G., & Discenza, R. (2002). Pre-project partnering impact on an information system project, project team and project manager. *European Journal of Information Systems, 11*(2), 86-97.

Keil, M., & Robey, D. (1999) Turning around troubled software projects: An exploratory study of the de-escalation of commitment to failing courses of action, *Journal of Management Information Systems, 15*(4), 63-88.

Ketchell, M. (2003, February 28). RMIT to scrap $47m software system. *The Age.* Retrieved 23 April, 2004, from http://www.theage.com.au/articles/2003/02/27/1046064164879.html

Klein, G., Jiang, J.J., Shelor, R., & Ballon, J.L. (1999). Skill coverage in project teams. *Journal of Computer Information Systems, 40*(1), 69-75.

Koenig, M. (2003). Knowledge management, user education and librarianship. *Library Review, 52*(1-2), 10-17.

LaPlante, A. (1991). The ups and downs of end-user involvement. *InfoWorld, 13*(47), 45-47.

Laudon, K., & Laudon, J. (1998). *Management Information Systems: New Approaches to Organisation and Technology.* Englewood Cliffs: Prentice-Hall.

Law, W.K., & Perez, K. (2005). Crosscultural implementation of information system. *Journal of Cases on Information Technology, 7*(2), 121-130.

Lee, D., Trauth, E., & Farwell, D. (1995). Critical skills and knowledge requirements of IS professionals: A joint academic/industry investigation. *MIS Quarterly, 19*(3), 313-340.

Liebowitz, J. (1999). Information systems: Success or failure? *Journal of Computer Information Systems, 40*(1), 17-27.

Linberg, K.R. (1999). Software developer perceptions about software project failure: A case study. *Journal of Systems and Software, 49*(2-3), 177-192.

McGrath, K. (2002). The golden circle: A way of arguing and acting about technology in the London ambulance service. *European Journal of Information Systems, 11*(4), 17-26.

Montealegre, R., & Keil, M. (2000). De-escalating information technology projects: lessons from the Denver International Airport. *MIS Quarterly, 24*(3), 417-447.

Neumann, P.G. (1997). Integrity in software development. *Communications of the ACM, 40*(10), 144.

Oz, E. (1994). When professional standards are lax: The CONFIRM failure and its lessons. *Association for Computing Machinery. Communications of the ACM, 37*(10), 29-36.

Oz, E., & Sosik, J.J. (2000). Why information systems projects are abandoned: A leadership and communication theory and exploratory study. *Journal of Computer Information Systems, 41*(1), 66-78.

Rifkin, G. (1991). End-user training: Needs improvement. *Computerworld, 25*(15), 73-74.

Roberts, T.L., Leigh, W., & Purvis, R.L. (2000). Perceptions on stakeholder involvement in the implementation of system development methodologies. *The Journal of Computer Information Systems, 10*(1), 78-83.

Sauer, C. (1993). *Why information systems fail: A case study approach.* Oxfordshire: Alfred Waller Ltd.

Schmidt, R., Lyytinen, K., Keil, M., & Cule, P. (2001). Identifying software project risks: An international Delphi study. *Journal of Management Information Systems, 17*(4), 5-37.

Seddon, P.B., Staples, S., Patnayakuni, R., & Bowtell, M. (1999). Dimensions of information systems success. *Communications of AIS, 2*(3), 2-24.

Shaw, N., DeLone, W., & Niederman, F. (2002). Sources of dissatisfaction in end-user support: An empirical study. *Database for Advances in Information Systems, 33*(2), 41-56.

Stamati, T., Kanellis, P. Stamati, K. & Martaakos, D. (2005). Migration of legacy information systems. *Encyclopedia of information science and technology,* Second Edition. Hershey, PA: IGI Global Publishing.

Standish Group (1994). *The CHAOS report (1994).* Retrieved November 30, 2006, from http://www.standishgroup.com/sample_research/chaos_1994_1.php

Standish Group (1999). *Project resolution:* The 5-year view. Retrieved November 30, 2006, from *http://www.standishgroup.com/sample_research/PDFpages/chaos1999.pdf*

Standish Group (2001) *Extreme chaos.* Retrieved November 30, 2006, from http://www.standishgroup.com/sample_research/PDFpages/extreme_chaos.pdf

Taylor, P. (2000). IT projects: Sink or swim. *Computer Bulletin, 42*(1), 24-26.

Wallace, L., Keil, M., & Rai, A. (2004). How software project risk affects project performance: An investigation of the dimensions of risk and an exploratory model. *Decision Sciences, 35*(2), 289-321.

Wilson, M., & Howcroft, D. (2002). Reconstructing failure: Social shaping meets IS research. *European Journal of Information Systems, 11*(4), 236.

Chapter II
Achieving Sustainable Tailorable Software Systems by Collaboration Between End-Users and Developers

Jeanette Eriksson
Blekinge Institute of Technology, Sweden

Yvonne Dittrich
IT-University of Copenhagen, Denmark

ABSTRACT

This chapter reports on a case study performed in cooperation with a telecommunication provider. The telecom business changes rapidly as new services are continuously introduced. The rapidly changing business environment demands that the company has supportive, sustainable information systems to stay on the front line of the business area. The company's continuous evolution of the IT-infrastructure makes it necessary to tailor the interaction between different applications. The objective of the case study was to explore what is required to allow end users to tailor the interaction between flexible applications in an evolving IT-infrastructure to provide for software sustainability. The case study followed a design research paradigm where a prototype was created and evaluated from a use perspective. The overall result shows that allowing end users to tailor the interaction between flexible applications in an evolving IT infrastructure relies on, among other things, an organization that allows cooperation between users and developers that supports both evolution and tailoring.

INTRODUCTION

In most business areas today, competition is hard. It is a matter of company survival to interpret and follow up changes within the business market. The margin between success and failure is small. Possessing suitable, sustainable information systems is an advantage when attempting to stay in the front line of the business area. In order to be and remain competitive, these information systems must adapt to changes in the business environment.

Keeping business systems up to date in a rapidly and continuously changing business environment such as, in this case, the telecom business, takes a lot of effort. Owing to the fast pace of change, flexibility in software is necessary to prevent software obsolescence and to keep the software useful. This inevitably means that the system has to evolve (Lehman, 1980). One way to provide the necessary kind of flexibility is end-user tailoring. End-user tailoring enables the end user to modify the software while it is being used, as opposed to modifying it during the initial development process (Henderson & Kyng, 1991). Software development, which is mostly done by professional software developers, involves transferring some domain knowledge from users to developers (Bennett & Rajlich, 2000) which may take some time and effort. End users, however, already possess the domain knowledge, so by providing support for end-user tailoring, enabling end users to make task related changes, alterations can be made immediately, as needed. Since time is money, a company can gain advantageous competitiveness if the business software can be at the forefront of the market changes. Can tailorable software support developing business practices over a long time? In our research project we had the possibility to address and evaluate the sustainability of tailorable software: The tailoring possibilities themselves have to evolve. Tailoring has to be supported by cooperation between users and developers to allow for the evolution of the tailoring functionality.

Tailoring research so far has focused on flexible stand-alone systems. In earlier projects, we too focused on the design of flexible and end-user tailorable applications (Lindeberg et al., 2002). However, interaction with other systems turned out to be a bottleneck, since business systems in telecommunication are part of an IT-infrastructure consisting of heterogeneous data sources. Other research also indicates that software and IT-infrastructures pose new challenges for software engineering (Bleek, 2004). Normally, the data exchange between different systems is the realm of the software developers, but in this article we use the evaluation of a prototype to explore what is necessary to allow *end-users* to tailor the interaction between flexible applications in an evolving IT-infrastructure. Our results support the claim that end-users can even tailor the interaction between business applications. The analysis of a user evaluation of a case-based prototype results in a number of issues to be addressed regarding the technical design, the know-how demanded of the users, and the organizational setting, particularly the cooperation between users and developers. These issues both confirm and extend existing research on end-user development and tailoring.

We start by presenting how our research relates to others' work. In the following section (The Case Study), we describe our research approach in detail, the relevant work practices and business systems of our industrial partner is briefly described and the design of the prototype is presented to provide a basis for the evaluations and discussions. Thereafter, we present the outcome of the evaluation, which points out three different categories of issues that are important when providing end-users with the possibility to manage interactions between applications in an evolving IT-infrastructure. The discussion relates these results to the state of the art.

RELATED WORK

The research on end-user tailoring addresses mainly the design of tailorable applications and tailoring as a work practice, and cooperation between users and tailors.

How Tailorable Software Should be Designed

When it comes to the design of tailorable systems, there is a broad range of different approaches. In (Mørch et al., 2004) the authors suggest new metaphors and techniques for choosing and bringing together components to facilitate end-user development. Stiemerling (2000) and Hummes and Merialdo (2000) also propose a component based architecture. Hummes and Merialdo also advocate dividing tailoring activities, as well as the application itself, into two parts: customization of new components and insertion of components into the application. The customization tool does not have to be a part of the application at all. This approach corresponds to Stiemerling's (2000) discussion of 'the gentle slope' where users can either just put together a few predefined components or, if more skilled, customize the components for more complex tasks.

Fischer and Girgensohn (1990) take up another side of tailorable systems. They state that even if the goal of tailorable systems is to make it possible for users to modify systems, it does not automatically mean that the users are responsible for the evolved design of the system. There will be a need for modifications of the users' design environment and Fisher and Girgensohn provide a rationale and techniques for handling this type of change.

An area that is also interesting is the mapping between the adaptable system and the users; which interfaces to provide. Mørch (1995) introduces three levels of tailoring, customization, integration and extension, which provide the users with increasing possibilities to tailor the system. Customization provides only opportunities to make small changes, whereas extension is when code is added, which means that more comprehensive changes can be made. Together with Mehandjiev, Mørch (2000) also presents how to support the three different types of tailoring by providing different graphical interfaces for each of the tailoring types.

Costabile et al. (2006) works with a methodology they call the software workshop approach. The software shaping workshop (SSW) makes it possible for users to develop software artefacts without using traditional programming languages. SSW means that the software is organized to fit various environments. The software is specific for different sub-communities. When a user (called domain-expert) wants to develop an artefact only the required tools are available. The users experience that they just manipulate objects as they do in the real world (Costabile et al., 2006).

Letondal (2006) is exploring how to "provide access to programming for non-professional programmers" (Letondal, 2006, p. 207). She makes it possible for users to do general programming at use time. Her approach also involves the possibility to modify the tool used.

The research approaches rarely address tailoring in the context of distributed systems. However in (Eriksson et al., 2003) a prototype that dynamically connects different physical devices (video cameras, monitors, tag readers, etc.) is presented. The tool can be regarded as tailoring the interaction between different intelligent devices. Stiemerling (2000) and his colleague (Stiemerling et al., 1998) show how to build a search tool by using customized Java Beans. The users customize search and visualization criteria. The tailorable search tool is used within a distributed environment provided by a groupware system. Neither of the distributed tailoring approaches is evaluated by users to explore beyond technical issues of how end-users can manage interaction between applications.

How the End Users Work with Tailoring

In (Mørch et al., 2004, p. 62) the authors state that an area for future research is "How to support cooperation among different users who have different qualifications, skills, interests, and resources to carry out tailoring activities." The area addressed is how the users work with tailoring. This area is well represented in the CSCW (computer supported cooperative work) community. In the following, some research in the category is presented.

MacLean et al. stated in 1990 (MacLean et al., 1990) that it is impossible to design systems that suit all users in all situations and they continue by expressing the need for tailorable systems. However, it is not enough to provide the users with a tailorable system. To be able to achieve flexibility there is a need for a tailoring culture, where it is possible for the users to have power and control over the changes. It also requires an environment where tailoring is the norm.

Wendy Mackay (1991) describes how she finds that although the users have tailorable software they do not customize the software, because it takes time from the ordinary work. There is a trade-off between how much time the tailoring takes to learn and how beneficial the change may be. To encourage users to customize the software, the customization has to allow users to work as before, and the customization must also increase productivity by just one single click of a button.

In another paper Mackay (1990) observes that customization of software is not mainly individuals changing the software for personal needs, but is a collaborative activity where users with similar or different skills share their files with each other. One group that has received attention is a group called 'translators'. Translators are users who are not as technically skilled as members of the highly technical group, but are people who are much more interested in making work easier for their colleagues. Mackay says that the translator role should be supported in organizations with tailorable systems. She also claims that not all sharing is good and that opportunities for sharing files have to be provided in the organization.

Gantt and Nardi (1992) find a role similar to the translator in a CAD (computer aided design) environment. They identify gardeners and gurus. Gardeners and gurus are domain experts, not professional developers, who have the role of local developers providing support for other users. Gardeners and gurus differ from other local developers in that they receive recognition for their task of helping fellow employees.

Costabile et al. classifies different user (domain-expert) activities. They group the activities into two classes. Class 1 means that the user chooses from predefined options. Class 1 contains the activities of parameterization and annotation. Parameterization means hat the user specifies some constraints in the data. Annotation is when users write comments next to the data to clarify what they mean. Class 2 contains several types of activities, All activities in Class 2 involves altering the artefact in some way (Costabile et al., 2006).

The research approaches discuss different kinds of collaborations, but not collaboration between different professions (e.g. end-users and developers) to provide for more extensive possibilities for end-users to do tailoring.

Carter and Henderson (1999) invented the expression *tailoring culture* to express the need for organizational support for tailoring. Kahler (2001) also points out that, in order to make tailoring successful, an organizational culture must evolve that supports the development and sharing of tailoring knowledge. Kahler also emphasizes three often coexisting levels of tailoring culture, identified and addressed by different researchers. First there is a level with equal users; people help each other to tailor the software (Gantt & Nardi, 1992) or there is a network of whom to ask when encountering trouble when tailoring the software (Trigg & Bødker, 1994). Second, there is a level with different competencies (Gantt & Nardi, 1992). The third level is a level of organizational embedment of tailoring efforts and official recognition of tailoring activities (MacLean *et al.*, 1990). We will return to this classification in the discussion of our results, as our findings propose the consideration of a fourth level of tailoring culture when implementing and deploying tailoring possibilities in an IT-infrastructure environment.

THE CASE STUDY

The research reported here is part of a long-term cooperation between the university and a major Swedish telecommunication provider, exploring the applicability of end-user tailoring in industrial contexts (Dittrich & Lindeberg, 2002; Eriksson, 2007). For the overall project we applied an approach we call cooperative method development (CMD) combining qualitative empirical research with improvements of processes, tools and techniques (Dittrich et al., 2007). The research reported here is part the deliberation whether to deploy new technologies in the design of applications and infrastructures. As the deliberation heavily depends on design, development and evaluation of our prototype, we complemented the CMD approach with design research for the deliberation phase (Nunamaker *et al.*, 1991). The question "What is necessary to allow end-users to tailor the interaction between flexible applications in an evolving IT-infrastructure?", addresses the design and deployment of a previously inexistent functionality. In design research, the design and development of a (prototypical) information system can be used both to answer technical questions and as a probe to explore requirements posed by the deployment of the technical possibilities. Hevner et al. (2004) especially emphasize the need for combining design research and behavioural science. The technical design of the prototype is discussed in (Eriksson, 2004).

The practical work was conducted during a period of slightly more than one and a half years. Prior research indicates that the collection of data to process the so-called extra payments was a bottleneck both for the users' work as well as for deploying the flexibility implemented in the existing systems. During the initial field studies focusing on the work practice of the business department, we visited our industrial partner once or twice a week to observe and interview both users and developers. These field studies informed the development of the overall research question and also the design of the prototype.

In the beginning of the design phase, workshops were arranged involving researchers, users and developers. When designing the prototype, one of the researchers was stationed at the company two or three days a week to ensure that the prototype conformed to existing company systems. Field notes were taken, and meetings and interviews were audio taped, during all phases of the case study.

The prototype was evaluated by all three employees involved in the collection of data and computation of the extra payments and by one developer involved in the maintenance of the payment system. These evaluations were video taped. The analysis in the section of this article entitled "Outcome of Evaluation" is mainly based on the latter tapes, but uses the other field material as a background. For secrecy reasons, videotaping is not allowed on the telecommunication provider's premises. We therefore installed the system on a stand-alone computer outside the actual work place. To allow the users to evaluate the prototype realistically, we reconstructed part of the IT-infrastructure in a local environment and populated it with business data, developing our prototype into a case-based prototype (Blomberg *et al.*, 1996). The users were given two tasks. One task was to construct the collection and assembly of data for an extra payment that they implemented regularly (in the manual fashion described above) as part of their normal work. For the second task they had to construct a totally new but realistic payment. The users were asked to *talk-aloud* while performing the task. This method is common when evaluating software in a use context (Ericsson & Simon, 1993; Robson, 2002). The researcher performing the evaluation observed and asked exploratory and open-ended questions to provoke reactions that differed from our expectations. The developer who worked with maintenance of the regular system evaluated the prototype in a workshop, and discussed advantages and drawbacks concerning use, tailoring, and expansion of the tailoring capabilities.

We analyzed the data in a manner that was inspired by grounded theory. A coding scheme was developed with its starting point in the transcripts of the evaluation sessions. The researchers coded the interviews independently from one other and then compared their results. The resulting categories were finally merged into three core categories, that is, design issues, user knowledge, and organizational and cooperative issues. The categorization can be found in the evaluation section.

History and Background

The subject of the study is a part of the telecommunication provider's back office support infrastructure for administering a set of contracts and computing payments according to these contracts. To compute payments, the system must be supplied with data from other parts of the IT-infrastructure. When creating new contract types based on different data, flexibility is constrained by the hard-coded interface to other systems. As a work-around, ASCII files can be created providing the necessary data sets – or events – to compute the payments. The data for these extra payments is handled and computed manually. To compute the data for an extra payment, members of the administrative department first run one or more SQL queries against the data warehouse. The result is stored in ASCII files. Next, the user copies the data from the ASCII files and pastes it into a prepared spreadsheet. When the user has thus accumulated the data, the user works through the spreadsheet in order to remove irregularities. The contents of the sheet are eventually converted again to an ASCII file that is imported into the payment management system. The manual procedure to compute the data for the extra payments has worked well until recently, although it is time consuming. The competitiveness of the telecom business is however continually forcing the company to come up with new services; more and more types of extra payments will be needed. This situation necessitates a tool to define and handle the new data sets or events. To make such event tool as flexible as possible, it must allow the collection and assembly of data from different kinds of systems. Experience suggests that it is impossible to anticipate the structure of future extra payments or which details will be needed. As a result, the tool must be able to communicate with any system in the IT-infrastructure. It is also essential that the tool allow for expansion of the tailoring capabilities, meaning that new data sources can be added. The addition of a new source should be as seamless as possible. Since different system owners and developers are responsible for these systems, it is their responsibility to make new data sources available. Such changes are part of the maintenance of the other systems, and here the limits of end-user tailoring are reached.

The Prototype

The prototype is divided into two parts, the Event Definer and the Event Handler (Figure 1). By using the Event Definer, the end-user can tailor communication and data interchange between systems, that is, the end-user defines the event types for the computation of the above-described extra payments. It allows the user to: define the assembly of data from different sources (Figure 1a), set up rules for aggregation and algorithms that will be performed on the data when aggregating the data (Figure 1b) and define how to map data sets to the format required by the receiving application. (Figure 1c). The Event Definer needs to be used only when defining new types of extra payments. The Event Handler handles the execution of extra payments or events and is to be used once a month to run the different extra payments.

Various solutions exist that provide the functionality needed to manage the connections between applications. These are found in tools for system integration that connect systems, in network management for monitoring the IT-infrastructure, in component management (if you choose to regard the different systems as components) and in report generation for assembling data. These tools are designed exclusively for system experts, not for end-users. A possible exception is report generation, which sometimes supports end-users but often needs support from developers to adapt it to fit new data sources. We found that none of these approaches was suitable for fulfilling the requirements for a tool for interapplication communication that can be adapted by users. Neither were the approaches suitable for the purpose of exploring what is necessary to allow end-users to tailor the interaction between flexible applications in an evolving IT-infrastructure. For our prototype we used an existing platform that supports integration between the telecommunication provider's back office applications. The integration platform makes it possible to publish events that other applications can subscribe to. We had a somewhat different intention when using the platform. We wanted to collect the informa-

Figure 1. The connection between the prototype and the surrounding systems

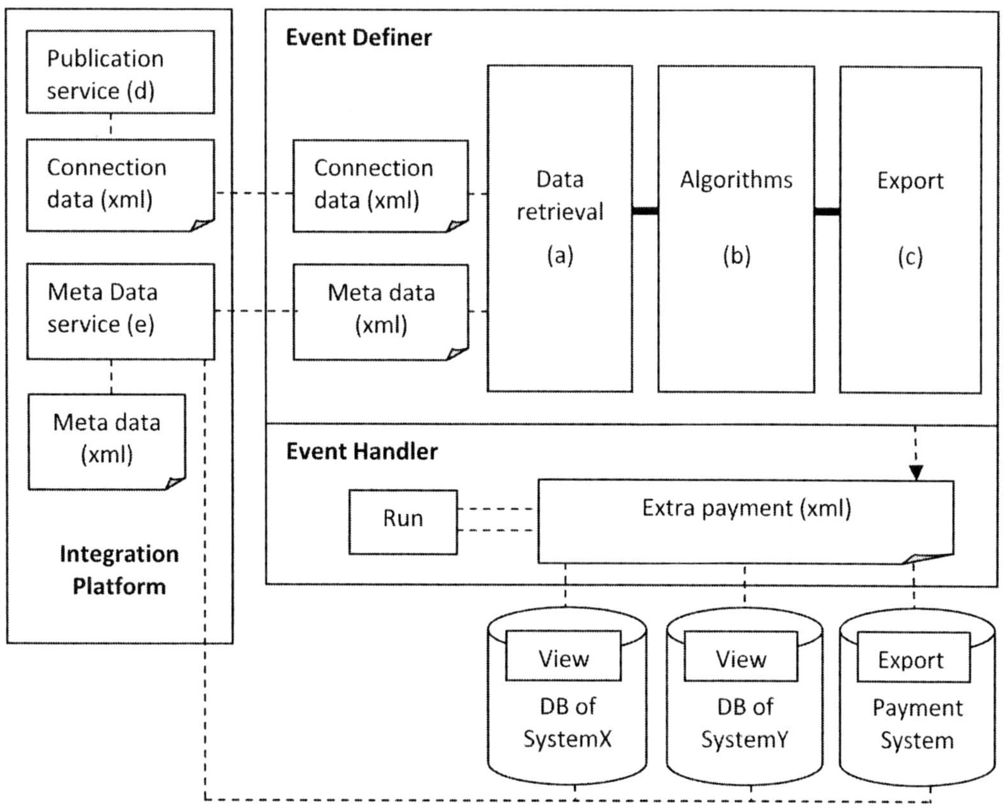

tion when needed, and we used the platform to provide the prototype with information about how to get in touch with desired resources and what data were accessible at these resources. We created a service (Figure 1d) on the integration platform that allowed the developers of the different systems to publish information about available data and showed how to connect to the respective database. To do so, the developers must set up a database *view* containing data that could be accessible to other systems (such as the prototype). The service produced an XML file containing connection data for all published data sources. When the Event Definer starts, the XML file is fetched from the integration platform. Yet another service (Figure1e) provided the prototype with metadata from the data sources, for example, which fields (attributes) could be accessed in a specific database, and the types of the fields.

Tailoring

The graphical tailoring interface of the Event Definer was constructed to consist of seven different steps. These steps are intended to guide the user through the process, but could also be used in an arbitrary order as the end-user chooses.

Step 1: Naming the extra payment.
Step 2: Choosing which databases to connect to.
Step 3: Choosing which fields to use from the selected databases.
Step 4: Setting up criteria for what data to collect from the different databases, that is, by drag and drop, the end-user chooses which field should be used and the end-user can also specify how the different views should be linked together, for example, fieldX in SystemX must be equal to fieldY in SystemY.
Step 5: Showing the specified criteria from Step 4 as SQL queries, that is, here the user can edit the SQL queries to set up more complicated (and unusual) conditions for data retrieval than can be accommodated by the graphical interface.
Step 6: Setting up algorithms for what to do with the collected data. (partially implemented.)
Step 7: Mapping the input table structure to the output table structure, that is, the end-user can map the assembled and computed data to a receiving system by dragging the fields from the assembled data table and dropping them in a table representing the receiving database.

All these choices, criteria, algorithms, mapping and so forth, were finally brought together and arranged into an XML file (extra payment in Figure 1).

Use

The XML files produced by the Event Definer are then used whenever the end-user decides to execute the extra payment. The Event Handler contacts the chosen systems one by one and collects the data specified in the XML file. When the data is collected and assembled in a single table it is displayed to the user to allow for checking and correcting the result where necessary. By clicking on a button, it is possible for the end-user to export the result to the system handling the payment data, in accordance with the mapping specification (Step 7).

Expansion of Tailoring Capabilities

There will inevitably be situations where end-users wish to define extra payments based on data that is currently unavailable. If the data and metadata are unavailable, the end-users are unable to perform new tasks. They have neither the authority nor the ability to alter or add views in surrounding systems. In this case the surrounding systems, as well as the tailorable system, have to evolve to meet the additional requirements from the end-users. The developer responsible for the respective system must then (a) alter the system by creating a new view or changing an existing view, so that it contains the required data, and (b) make the changes available through the integration platform. To support the latter, the publication of a new source was supported by a web interface where the developer (also system owner) could fill in the necessary data.

Outcome of Evaluation

The evaluation presented here focuses on issues beyond the technical design and the appearance of the graphical interface of this specific application. It addresses overall design issues for this kind of application, the end-user knowledge necessary to handle such complex tailoring tasks, and organizational issues to deploy such systems in a sustainable way. We have also evaluated the prototype against functional requirements, but the results are not reported here. Individual opinions held by only one or two of the subjects are disregarded in the following presentation.

Design Issues

In terms of technical support we focused on the different interfaces provided by the prototype: the tailoring interface, the deployment interface and the development interface.

The Tailoring Interface

Functionality for Controlling and Testing

All users appreciated the freedom to alternate between the seven steps. They found that the steps provided not only guidance and an overview but also the freedom to alter something performed in previous steps, without losing the overall view. To be able to overview all choices and trace them backwards was one way of providing control. But there was also a need for error control and limitation. The users, especially the beginners, wanted some kind of guidance in order to feel secure.

It became very obvious that the design must enable the end-user to test and control the correctness of the specification of extra payments. Control facilities must be provided to ensure security for the users in their work. Although control and test functionality was important for all users, the attitude towards test and control varied between the users. The better the knowledge of the task, the surrounding systems and possible errors, the less important explicit test and control seemed to be. Following statements exemplifies different attitudes towards control and test functionality:

When you make an extra payment for the first time you would probably like to make a test run to see that it really works correctly. (user comment, evaluation session, February 24, 2004)

and

there isn't the same protection as in SystemZ ... but to make a more flexible solution, then you can't expect it to be strictly user friendly (user comment, evaluation session, February 24, 2004).

Clear Division between Definition, Execution and the Tailoring Process

When tailoring, the user rises from one level of abstraction to another, higher level. From thinking only in terms of the execution of an extra payment the user had to think in more general terms of what characterizes this extra payment, what kind of data were fetched, what variables there were, and so forth. The users had to think in terms of levels, which is not an easy step to take. We found that a clear separation between execution and the tailoring process helped the users to make this step successfully.

The users also started to discuss the division of labour enabled by a system resembling the prototype. For example, one of the users said:

I think it is very good because then someone is very familiar with how to make a new extra payment and then all employees in the group can run the extra payment. (user comment, evaluation session, February 24, 2004)

Unanticipated Use Revealed to the Tailor

Systems that continuously evolve through tailoring aim to support unanticipated use. The possibilities for unexpected use are inevitably limited by the technical design. To support unanticipated ways of tailoring, the system has to provide additional information of what is possible to do and what the limitations are. In the prototype this was achieved by providing data for the user that is not directly applicable to the type of extra payments that exist today. As one of the users expressed it when seeing the opportunity for one of the export systems to also act as input source:

This is interesting! It opens up new opportunities. It might be like one extra payment uses another payment as a base (user comment, evaluation session, February 24, 2004).

Complexity

We found that the users preferred more information, rather than a less complex tailoring interface, resulting in more tailoring possibilities. Their opinion was that, as tailoring is not routine work, performed several times a day, it is allowed to take extra time. Then it is better to have a more complex interface providing more opportunities to tailor the system.

The Deployment Interface

Simplicity

One thing that was revealed and worth mentioning is that it seems that the deployment interface should be even simpler than an ordinary user interface. The users expressed the opinion that the tailoring interface and the tailoring process may be rather wide-ranging if that allows for a simpler deployment interface.

One Point of Interaction

The development interface in the prototype was a graphical Web interface where the developer could fill in the data that was to be published about the respective source system. During the evaluation of the development interface, the software engineer emphasized the importance of having one point where changes to the data sources are published. The developer should not be forced to make changes in several places in the application in order to extend the tailoring capabilities.

End-User Knowledge Required for Tailoring

Even previous to the evaluation session we had experienced the high expertise of the users not only regarding their tasks but also regarding the data available in the different databases that are part of the IT-infrastructure. The users acquired the knowledge in order to perform the assembly manually. The communication between different systems is normally hidden from the user in a data communication layer for the separate systems. Our prototype is designed to make exactly this communication tailorable. Its deployment depends on the respective expertise of the users.

Task Knowledge

Business knowledge about contracts and payments provides the base on which the users decided what data to collect. Extensive business knowledge was a prominent feature of the results of the evaluation. The users' reflections on which data to collect always concerned different aspects of the business tasks.

System Knowledge

To map requirements regarding the task at hand and the available data, demands expertise regarding the available data in the different systems. And the users knew where to find the data needed for defining a specific extra payment. The prototype just helped with the exact location of the data, for example it guided the user to which fields to use, by listing the fields with examples of the data they contained. However, the user had to understand the sometimes quite cryptic names and know where to look for specific data.

Error Knowledge

All users were extremely aware of which errors could occur, that is, errors concerning the use of the prototype, the IT-infrastructure and the task. Task-specific errors are particularly important for the end-user to overview since they may cause serious consequences for the company if the errors are not prevented. On several occasions during the user tests the users expressed concern about making errors. They made statements like:

when you work as we do you must know a little about database management, you have to understand how the tables are constructed and how to find the information. And also in some way understand the consequences of or the value of the payment. In other words how you can formulate conditions and what that leads to. (user comment, evaluation session, February 24, 2004)

Organizational and Cooperative Issues

The system for which the prototype was a test would depend on data published by many different surrounding programs. Each one of these systems is itself the subject of both tailoring and evolution. Both the users, and the software engineer who evaluated the prototype, addressed the necessary interaction with other system owners and the assignment of responsibilities regarding the publication and updating of the connection information and the kinds of data available.

Publication and Update Responsibilities

During the workshops it became apparent that there is already friction in the coordination between the payment system and the changes in the surrounding systems. When one system in the IT-infrastructure is changed, the changes are orally communicated to the owners of other systems that may or may not be affected by the change. For the prototype to function as designed, it was important that the systems that the prototype was expected to communicate with were visible and accessible. The design of the prototype solved this problem by requiring every change relevant to the prototype to be reflected in the published information. In other words, it was designed so that the respective system owners were responsible for keeping their system visible and showing its current status. As the prototype was dependent on accurate just-in-time information, the evaluation revealed a need for coordination concerning publication and updates of surrounding systems and tailoring activities in the prototype.

Collaboration between Developer and End Users

The fieldwork revealed, and the evaluation confirmed, that it is impossible to know what future contracts will look like. Therefore there will always come a time when the end-user wants to retrieve data that is not published in any available view. In this case the system that can provide the data has to be identified and the respective system owner or developer has to be persuaded to implement a new view of the system or update existing ones, and publish the relevant information.

Another issue related to communication and cooperation between users and developers concerned the decision of how much information to make available for the users to do a good job of tailoring. The users wanted to see as much information as possible, provided it was within reasonable limits. In order to have better control over the execution of the system and to decouple maintenance that would not necessarily impact the communication with the payment system, the developers would rather prefer to restrict the user's options. These two perspectives have to be negotiated.

In this company, cooperation between business units and the IT unit works very well. The users evaluating the system were quite aware of the limit of their own competences and knew when to consult the responsible developers. All users frequently referred to developers when they experienced that something was beyond them. None of them considered the necessary coordination and cooperation to be a serious problem.

Summary of Outcome of Evaluation

The evaluation revealed many issues to consider when making a system that continuously evolves through tailoring work in a rapidly changing business environment. The issues could be divided into three categories regarding design issues, user knowledge, and organizational and cooperative issues. Below, the issues are summarized and listed under the respective category.

Design Issues

1. Functionality for controlling and testing changes has to be integrated into the tailoring interface and there must be sufficient technical support for the end-user to estimate and check the correctness of the computation.
2. A tailorable system has to define a mental model that makes a clear division between definition, execution and tailoring. This mental model must be adopted in the tailoring interface and be shared by users, tailors and developers.
3. The tailoring interface also has to reveal potential for unanticipated use to the tailor. This means, that the information flow must, to a certain extent, exceed what is currently necessary.
4. The tailoring interface can be more complex, provided the tailoring process makes the deployment easier. The tailoring interface is not used as often as the deployment interface and additionally the tailoring itself often involves careful thought.
5. The deployment interface should be simpler than ordinary user interfaces.
6. The developer expanding the tailoring capability should only interact with one clearly defined point in the tailorable system, that is, changes are made at one point in the system.

End-User Knowledge

7. End-users must have sufficient knowledge of how the systems are structured and what the systems can contribute.
8. End-users must have solid knowledge of the nature of the task and what data is required to perform it.
9. End-users must have knowledge of which errors can occur and what the consequences of these may be.

Organizational and Cooperative Issues

10. System owners or developers must be responsible for making their systems publicly available within the company. System owners or developers must also be responsible for updating the systems according to external requirements.
11. The necessity to extend the possibilities for end-users to manage the interaction in an evolving IT-infrastructure requires effective

collaboration between the developer and end-users.

DISCUSSION

Our results indicate that tailoring in real world IT infrastructures needs to be complemented by and coordinated with software evolution, therefore requiring cooperation between software developers and users. That way the flexibility of tailorable software can support the changes in the business practices in a sustainable way. When tailoring is discussed in literature, the focus is mainly on how end users perform tailoring or how tailorable systems should be designed. The developer's role is only briefly touched upon. For example Stiemerling (Stiemerling, 2000) state that Human Computer Interaction efforts often focus on optimizing interfaces for non-programmers and that this effort often has "the nice side-effect of making life easier for programmers as well" (Stiemerling, 2000, p. 33). The professional developer is as essential as users and tailors for tailorable systems in a rapidly changing business environment, and to make the tailorable system work as intended, the activities of the three roles has to be coordinated. On the other hand, when discussing software evolution in the software engineering community, the end users are only mentioned briefly.

The users' and developers' perspectives are different. However, this is not necessarily a disadvantage: collaboration between different competences widens the boundaries for what is possible to do with a tailorable system. Nardi (1993) points out that end users with different skills cooperate when tailoring and she states that "...software design should incorporate the notion of communities of cooperative users..." which "...makes the range of things end users can do with computers much greater" (Nardi, 1993, p. 122). By extending the cooperation to involve professional developers too, 'things the end user can do with computers' may even increase.

Regarding the design of tailoring functionality, our results both confirm and extend existing research. Users ask for additional functionality to guide the tailoring and test the outcome (Burnett *et al.*, 2003). We found that users wished to incorporate control of the tailoring process in the form of an outline, preferably in a step-by-step fashion. They also asked for visualization and test facilities in order to check the impact of the separate steps on the end results. The evaluation of the interface allowing software engineers to expand the tailoring possibilities confirms and expands previous research results addressing the developer responsible for the evolution of tailorable systems as an additional stakeholder whose requirements also have to be considered (Eriksson et al., 2003; Lindeberg et al., 2002).

Our results further indicate that tailoring in an IT-infrastructure of networked applications provides additional challenges for the design of the software, the competence of the users and tailors, and the cooperation between users and developers. Changes – independently of whether they are implemented by tailoring or by evolving the software – can depend on and affect changes in other applications of the IT-infrastructure and the interaction between applications. This requires coordination between tailoring and development, and cooperation between the persons responsible for tailoring and developing the different applications. And this, in turn, requires a different set of competences from users and developers. The use of an application such as the prototype discussed here, for example, required knowledge of the surrounding systems and their data structures. Developers as well as users have to understand not only the system they are responsible for but also the dependencies between different systems and tasks. Several researchers have discussed collaboration between users and tailors, but not between users, tailors, and professional developers.

In order to make tailoring sustainable, it must be made possible for the tailorable system to evolve beyond the initial intention when building the tailorable system. Kahler's three levels (Kahler, 2001) of tailoring culture – cooperation between tailoring end-users, cooperation between tailors and users, and the organizational recognition and coordination of tailoring efforts - have to be extended with a fourth level, of organizational support for coordinating tailoring and development activities *involving the cooperation not only between users and tailors but also between end-users, tailors and software developers.*

CONCLUSION

Allowing end-users to tailor the interaction between flexible applications in an evolving IT-infrastructure requires that the tailoring activities are supported by the design of the system, for example by providing a clear division between execution and tailoring, by revealing potential for unanticipated use, and by supporting single interfaces for changes to the software. It is also essential that the competence of the end-users is sufficient in terms of knowledge of how the systems are structured and what the systems can contribute. End-users must also have substantial knowledge of the task and which errors can occur and what the consequences of these may be. To allow end-users to tailor the interaction between applications in an evolving IT-infrastructure, the organization has to allow for cooperation between users and developers.

The evaluation clearly showed the dependencies between tailoring and the further development of the tailoring capabilities. The evaluation also made it apparent how the different actors were aware of their colleagues' skills and of what each individual could contribute. To ensure a sustainable tailorable system when deploying a system intended to evolve continuously through tailoring, it is necessary to take into account resources concerning various skills and collaboration between users and developers. Without smooth collaboration between the parties an extended fourth level of tailoring culture will not be provided for, and therefore the system will soon become partially obsolete and the competitive advantages provided by the system will decrease dramatically. The results challenge the clear division between software use and evolution on one side and software development on the other side, when developing and maintaining an IT-infrastructure. Collaboration between the end-user and the developer must work satisfactorily in order to achieve tailorable, sustainable software. In other words, in a rapidly changing business environment with continuously changing requirements, such as the one presented in this paper, the tailoring activities have to be coordinated with the software evolution activities.

ACKNOWLEDGMENT

This work was partly funded by The Knowledge Foundation in Sweden under a research grant for the project "Blekinge - Engineering Software Qualities (BESQ)" (http://www.bth.se/besq).

REFERENCES

Bennett, K. H., & Rajlich, V. T. (2000). Software maintenance and evolution: A roadmap. *The conference on the future of software engineering.* Limerick, Ireland: ACM Press.

Bleek, W.G. (2004). *Software Infrastruktur. Von analystischer Perspective zu konstruktiver Orientierung.* Hamburg: Hamburg University Press.

Blomberg, J., Suchman, L., & Trigg, R.H. (1996). Reflections on a work-oriented design project. *Human-Computer Interaction, 11*(3), 237-265.

Burnett, M., Rothermel, G., & Cook, C. (2003). Software engineering for end-user programmers.

Proceedings of the Conference on Human Factors in Computing Systems (CHI'03), 12-15.

Carter, K., & Henderson, A. (1999). Tailoring culture. *Proceedings of the 13th Information Systems Research Seminar (IRIS'13)*, 103-116.

Costabile, M. F., Fogli, D., Mussion, P., & Piccinno, A. (2006). End-user development: The software shaping workshop approach. In H. Lieberman, F. Paternò & V. Wulf (Eds.), *End user development* (1st ed., pp. 183-205). Netherlands: Springer.

Dittrich, Y., & Lindeberg, O. (2002). Designing for changing work and business practices. In N. Patel (Eds.), *Evolutionary and Adaptive Information Systems* (pp. 152-171). USA: IDEA Group Publishing.

Dittrich, Y., Rönkkö, K., Eriksson, J., Hansson, C., & Lindeberg, O. (2007, December). Co-operative method development – combining qualitative empirical research with process improvement. *Empirical Software Engineering Journal.*

Ericsson, K.A., & Simon, H.A. (1993). *Protocol Analysis: Verbal Reports as Data.* Cambridge, MA: MIT Press.

Eriksson, J., Warren, P., & Lindeberg, O. (2003). An adaptable rchitecture for continuous development - User perspectives reflected in the architecture. *Proceedings of the 26th Information Systems Research Seminar (IRIS'26)*, Finland.

Eriksson, J. (2004). Can end-users manage system infrastructure? - User-adaptable inter-application communication in a changing business environment. *WSEAS Transactions on Computers*, *6*(3), 2021-2026.

Eriksson, J. (2007). Usability patterns in design of end-user tailorable software, *The Seventh Conference on Software Engineering Research and Practice in Sweden, SERPS 07.* Gothenburg.

Fischer, G., & Girgensohn, A. (1990). End-user modifiability in design environments. *The conference on Human Factors in Computing Systems, CHI'90.* Washington, USA.

Gantt, M., & Nardi, B.A. (1992). Gardeners and gurus: Patterns of cooperation among CAD Users. *Proceedings of the SIGCHI Conference on Human Factors in Computing Systems*, (pp. 107-117).

Henderson, A., & Kyng, M. (1991). There's no place like home: Continuing design in use. In J. Greenbaum, & M. Kyng (Eds.), *Design at Work* (pp. 219-240). Hillsdale, NJ: Lawrence Erlbaum.

Hevner, A.R., March, S.T., Park, J., & Ram, S. (2004). Design science in information systems research. *MIS Quarterly*, *28*(1), 75-105.

Hummes, J., & Merialdo, B. (2000). Design of extensible component-based groupware. *Computer Supported Cooperative Work (CSCW)*, *9*(1), 53-74.

Kahler, H. (2001). *Supporting collaborative tailoring.* Roskilde: Roskilde University.

Lehman, M.M. (1980). Programs, life cycles, and laws of software evolution. *Proceedings of the IEEE*, *68*(9), 1060-1076.

Letondal, C. (2006). Participatory programming: Developing programmable bioformatics tools for end-users. In H. Lieberman, F. Paternò & V. Wulf (Eds.), *End user development* (1st ed., pp. 207-242). The Netherlands: Springer.

Lindeberg, O., Eriksson, J., & Dittrich, Y. (2002). Using metaobject protocol to implement tailoring; possibilities and problems. *Proceedings of the 6th World Conference on Integrated Design & Process Technology (IDPT '02)*, Pasadena, USA.

Mackay, W. E. (1990). Patterns of sharing customizable software. *The Conference on Computer Supported Cooperative Work (CSCW'90).* Los Angeles, California, USA: ACM Press.

Mackay, W. E. (1991). Triggers and barriers to customizing software. *The conference on Human factors in Computing Systems (CHI'94)*. Boston, Massachusetts, USA.

MacLean, A., Carter, K., Lövstrand, L., & Morgan, T. (1990). User-tailorable systems: Pressing the issues with buttons. *Proceedings of the Conference on Human Factors in Computing Systems (CHI'90)*, 175-182.

Mørch, A. (1995). Three levels of end-user tailoring: Customization, integration, and extension, *the 3rd Decennial Aarhus Conference*. Aarhus, Denmark.

Mørch, A., & Mehandjiev, N. (2000). Tailoring as collaboration: The mediating role of multiple representations and application units. *Computer Supported Cooperative Work (CSCW), 9*(1), 75-100.

Mørch, A. I., Stevens, G., Won, M., Klann, M., Dittrich, Y., & Wulf, V. (2004). Component-based technologies for end-user development. *Communications of the ACM, 47*(9), 59-62.

Nardi, B.A., & Miller, J.R. (1991). Twinkling lights and nested loops: distributed problem solving and spreadsheet development. *International Journal of Man-Machine Studies, 34*(1), 161-184.

Nunamaker, J., Chen, M., & Purdin, T. (1991). System development in information systems research. *Journal of Management Information Systems, 7*(3), 89-106.

Robson, C. (2002). *Real world research.* Oxford, UK: Blackwell Publishers Ltd.

Stiemerling, O. (2000). *Component-based tailorability.* Bonn: Bonn University.

Stiemerling, O., & Cremers, A.B. (1998). Tailorable component architectures for CSCW-systems. *Proceedings of the 6th Euromicro Workshop on Parallel and Distributed Programming*, 302-308.

Trigg, R., & Bødker, S. (1994). From implementation to design: Tailoring and the emergence of systematization in CSCW. *Proceedings of the Conference of Computer Supported Cooperative Work (CSCW 94)*, 45-54.

Chapter III
Usability, Testing, and Ethical Issues in Captive End-User Systems

Marvin D. Troutt
Graduate School of Management, Kent State University, USA

Douglas A. Druckenmiller
Western Illinois University – Quad Cities, USA

William Acar
Graduate School of Management, Kent State University, USA

ABSTRACT

This chapter uses some special usability and ethical issues that arise from experience with what can be called captive end-user systems (CEUS). These are systems required to gain access to or participate in a private or privileged organization, or for an employee or member of another organization wishing to gain such access and participation. We focus on a few systems we list, but our discussion is relevant to many others, and not necessarily Web-based ones. The specific usability aimed at in this chapter is usability testing (UT), which we use in its usually accepted definition.

INTRODUCTION

This chapter reviews, extends and updates earlier work (Troutt, 2007) that introduced the term *captive end-user systems* (CEUS) and the basic ideas. The earlier paper focused on some special usability, usability testing (UT), and ethical issues that arise from experience with such systems. These are systems that are required to gain access to, and participate in, a private or privileged organization. They do also apply to situations where an employee or member of another organization who may wish to gain similar access and participation. The examples that come to mind are generally web-based and required for submitting such material as:

- Articles to academic journals (editorial systems)
- Job applications
- Student applications to universities and academic programs
- Faculty curriculum vitae (CV) material into a database

Our focus is on systems like those listed, but many others, not necessarily web-based ones, could also be listed. Automated phone-answering systems come to mind. Government forms such as tax filing forms also qualify, although in that case, commercial tax preparation software is available in the form of several competing products.

As an attempt at a more precise definition, we suggest the following. A **captive end-user system** (CEUS) is a computer-based system whose intended end users will not have had input into the systems analysis and user testing of the system and or did not have an opportunity to shop among similar systems.

BACKGROUND: A GROWING PROBLEM

A First Example

A case familiar to one of the authors involves a change to the e-mail system at a Midwestern US university. Some people, like his spouse, hate the new system which is called Zimbra. This has now become a new curse word in the office (as in "I've been Zimbra-ed" or just "Oh Zimbra!"). While sympathetic to his spouse's viewpoint, the coauthor in question doesn't have the same problems using it. This is in part due to many hidden features that are part of the interface, hence undocumented and not obvious to the general user. For instance, this mail client is a Web 2.0 application. People were used to using MS Outlook and all of the features it had, as well as its user-interface components. Since this interface doesn't work the same way, it feels unfriendly and not as functional, until one figures out how to do the same things in it.

This usability deficiency was described to an expert in this sort of situation, most prevalent in the conversion of an old system or the adoption of a new system. The new system may be actually a better design, but because it doesn't allow them to do things in the same old way, it first appears unfriendly and not a good fit with established procedures and habits. As a forerunner of our discussion, let us reveal at this stage that the expert's recommended solution is to first find out what outcome or objective the end users are seeking, and only then show them how this can be accomplished in the new system. At the 2007 HICSS conference, a presentation specifically looked at that issue as a shared mental model problem in virtual team support (Thomas & Bostrom, 2007) The research showed that training and the development of shared mental models among users has a direct impact on successful appropriation of newly introduced tools. Such training and development activity is many times replaced by coercive actions that mandate change without justification or explanation. Such coercive approaches are rarely successful.

The General Issue

CEUS and their problems are likely to expand as technology is brought to bear on more and more business matters and everyday life. In these systems, usually the end user has little opportunity to influence the choice of the product or its design and testing. At the same time, such systems impact very large numbers of users so that their initial usability becomes very important. Also, these systems often impose substantial data-entry burdens on end users. Thus, they raise concerns about economizing end-user time involvement and preventing loss of productivity. For organizations, problems arise from the impact on the individual end user, who may be frustrated, have his or her morale unfavorably impacted, and perhaps underutilize or ineffectively use the system. These *technical* problems quickly add up to *managerial* ones.

Because the end user has little control over such systems, they also raise ethical concerns. Since large numbers of people are often affected, they may have broad social and economic productivity impacts in case of poor design. This chapter calls attention to some problems and potential needs with respect to this class of systems.

Our Central Example

As a central illustrative example for this paper, a business college familiar to the other two authors recently began to require all faculty and staff to enter professional data into a commercial database system. The adopted system was one of very few available. An impending re-accreditation review was fast approaching so that urgency in selection was required. The choice of one particular commercial software was made without benefit of faculty review or shopping. Given those circumstances, there was no opportunity to shop for a best-of-breed alternative product. In addition, it was awkward to voice complaints other than through the system vendor's support personnel. Most importantly, UT information was not made available.

It was soon discovered that the selected system had a number of problems that suggested inadequate UT had been carried out to economize time and effort of the end users, and therefore loss of productivity to the college. This caused considerable loss of time for a large number of faculty members and other staff. In contrast, systems developed within an organization for use by its own employees, or for E-commerce website applications, are generally subjected to careful UT (Bruegge & Dutoit, 2004; Dennis & Wixon, 2000; Hoffer, George & Valacich, 2002; Lazar, Adams & Greenridge, 2005; Rubin, 1994; Schach, 2005; Whitten, Bentley & Dittman, 2001).

The system vendor advertised numerous universities as prior adopters, giving the impression that the product should be far along the UT and maturity cycle. However, a surprising number of problems were found that seem reasonably identifiable by subject area expert (SAE) end users. A few of these are described next.

A Sample Slate of Problems Encountered

Expert judgment, in the subject area expert (SAE) sense of expertise, was often required from the users of that system to figure out how best to describe certain academic curriculum vitae (CV) features such as presentations only, as opposed to presentations with proceedings publication, letters to an editor, proceedings reprinted in refereed journals, and so on. A basic problem was that an inadequate number of categories relevant to faculty publication activity were provided by the software. Also, it was found that one of the more important categories of publication (articles accepted but not yet published) did not appear on CV reports generated by the system. Yet such reports were essential checks on the system's accuracy and completeness. If these were not handled properly, there was little confidence that accreditation review reports would avoid omission of essentials. While later a "work-around" was found by some users by declaring these as actually published instead of their true status, it could not be easily communicated to the entire affected faculty, thus resulting in lack of standardization in a system precisely aiming at the systematic collection of academic data.

The literature on UT has long noted that the user profile should be broken down into different classes of users (Caulton, 2001; Rubin, 1994). This is also known as *audience analysis* in the interface design and human computer interaction (HCI) literatures (Drommi, 2001). The faculty database example uncovered an interesting dimension along similar lines. There were various categories of academic entries, especially publications. It was reasonably easy to add one publication. However, as the number of entries increased several problems were noticed.

First, only one view of the already entered publications was available during data entry activity. It was in alphabetical order on the apparent key field of the publication title. This was at odds with the typical user CV practice of ordering entries by publication category and dates in reverse publication date order. Thus, it was difficult to return to specific items to check whether they had been entered and completed. Complaints to system support led to a change to a different view that was in reverse date order but mixed all publication types together. Later, a further change also grouped the titles within year of publication. Interestingly, this kind of problem might not have been detected in usability tests that involved just one or a few publication entries. UT would have benefited by emphasis on what may be called specific category heavy users. That is, the user profile or audience should be divided into sub-classes including those who have heavy (high-volume) publication activity, those who have heavy service activity, etc.

Also another glitch was that, once a publication was edited, page return went to the top of list on the page so that searching from the top had to be repeated to go past the previously just edited item. Thus even locating and finding publication items continue to be problematical and unnecessarily time consuming.

Other problems were also present. Two entry boxes were provided for the end user's own name. One contained college faculty in a drop-down list and the default version of the user's name. This would have been useful, particularly for sole or first authored publications of the user if the default name was in the form the faculty member has chosen as routine for publication authorship purposes. It appeared that the system analysis effort did not become aware of the usual practice of publication name standardization followed by most academics. However, it was difficult to delete the default name for publications for which the user was not first author.

In addition, when an article's volume number was entered a drop-down list appeared showing past volume numbers that had been used – a case in which past choices do not have any relevance whatever. The same was true of issue numbers for articles. To make matters worse, these drop-down lists covered the entry boxes for data below them, thereby further slowing the entry process. This latter problem was later found by some users (but not fast enough by others) to have a solution by working upward from the bottom of the input screen.

Inconvenience and Inefficiency

Problems like the foregoing not only waste time directly for end users but also cost extra time in developing work-around solutions, and describing and communicating them to colleagues and vendor support staff. In such systems, end users are essentially captives of the situation and cannot participate in competitive shopping. Without representation, the end user can become an unwilling part of a "design by fixing complaints" mode of system analysis and design. Developers of such systems could be easily tempted to rely mostly on user feedback after roll-out in order to accomplish UT and iterative improvements as a by-product of actual sale and use.

It should also be noted that the kind of time loss during system use addressed here is not the same as user-learning time or *Learnability* (Bennett, 1984). In fact, the faculty data system in question seemed fairly easy to learn. The time concern here is more closely related to Constantine and Lockwood's (1999) *Efficacy* and *Context* factors, as well as Bennett's (1984) *Throughput* dimension. Efficacy requires that a system should not interfere with or impede use by a skilled user who has substantial experience with the system. The CEUS time concern is similar except that it should hold without substantial experience with the system. Context factors deal mainly with whether the system is suited to the actual context of use; it applies directly to CEUS with an emphasis on the subject area expert's (SAE's) understanding of the context. As to efficacy, it refers to the speed of task execution, and thus also directly applies to the CEUS setting.

Since this occurs at the expense of the end users and/or their employer, it raises clear *ethical* concerns in addition to loss-of-productivity concerns. This type of ethical concern fits well in several of the general ethical systems mentioned in Hoffman (2004). Hoffman stresses professional ethics in particular and the obligations of Information technology professionals in development of systems. Specifically IT professionals who develop and deploy information systems have an obligation to the intended users of the systems they deploy. This obligation is three-fold. It must first satisfy the requirements of the intended user as seen by that user, not the view of the designer. Second the system must meet requirements of all stakeholders or users. This includes accountability and auditability for outside interests or users of the information generated by the system. Lastly the system under development must use technology that balances the needs of the user and the needs of the larger enterprise the user is part of and not the preferences of the developer. Clearly there are competing interests in the development of information systems and it is the professional obligation of the developer to balance those interests in the largest possible context. In the case of CEUS, this means that developers must consider all intended users of a system. Individual and organizational user interests must be balanced and fit to the appropriate technical platform.

Hoffman (2004) also points out that these ethical concerns apply to not just the professional developers of information systems but to all individuals involved in the development of organizational information systems. Because of the ubiquity of end-user development and easy to use web development tools, many non-professional individuals are now engaged in the development of information systems for use within organizational environments. These same ethical concerns apply to all developers of information systems. Organizational leadership has an obligation to see that standardized engineering processes are used at all levels of the organization, not just in the information systems department.

SOLUTION SUGGESTIONS AND APPROACHES

Diagnosing the Problem

It seems that the problem of CEUS is one of the fit between the actual end-user requirements and what the software provides when it meets the requirements. Then another level of problems is in the documentation and training. In this sphere of problems the software actually fits the user requirements but the features are not properly documented, or the user is not trained. This is usually shortcut both in the development and user testing. UT should be performed on documentation and training as well when following a socio-technical systems approach. And, as can be gathered from the previous example, another level is represented by the user innovations commonly called work-arounds. These are innovations that users have developed for working around the system to make it fit their unique ways of doing work. These work-arounds are interesting because they represent the users' own ways of developing compensatory mechanisms for the lack of fit both in the software and in the training for how to use the system.

What suggestions can be brought to bear on CEUS-related problems? Hopefully, increasing awareness and further discussion of the issue will help in itself. In addition, increased stress on ethics in the business curriculum should help, especially with the aid of cases built along these lines. International standards possibilities also exist. In fact, the main international standard affecting the product development process is *ISO 13407:* human-centered design processes for interactive systems (UsabilityNet, 2008). This standard outlines four design activities including carrying out user-based assessment. The same website lists several related ISO guidelines. However, such ISO type guidelines, while clearly helpful, tend to stress process rules and benchmarks rather than measurable outcomes. Moreover, in the present case without competitive influences, they are on a voluntary basis.

Efforts such as the Software Engineering Institute (SEI) at Carnegie Mellon University (SEI/CMU, 2008) might be adapted to address CEUS needs. Among many other activities, SEI works through the global community of software engineers to amplify the impact of new and improved practices by encouraging and supporting their widespread adoption. SEI offers continuing education courses based on matured, validated and documented solutions, and licenses packaging and delivery of new and improved technologies. Especially relevant is the CMMI project for process improvement. This process is also linked to ISO standards initiatives. CMMI or Capability Maturity Model Integration is a product that supports process improvement in organizational systems development. The TRUMP (TRial, Usability, Maturity, Process) (http://www.usabilitynet.org/trump/resources/standards.htm) project has similar aims.

ATTEMPTS AT IMPROVEMENTS

From several Systems Analysis & Design and Software Engineering books reviewed so far in this research, it appears that UT efforts in those fields have concentrated on goals such as user friendliness, ease of understanding, minimizing time to learn, and similar criteria. However, most of the above problems center on problems of efficiency and end-user productivity in interface design (Chen & Sharma, 2002). CEUS considerations suggest adding emphasis on specific task or scenario times with the final product and time spent on error reporting during beta tests. Once a product is released it is those kinds of times that most impact potential loss of end-user productivity. Thus, while the areas of human factors and human-computer interaction are being extensively explored, more research is needed from the point of view of society and employers as stakeholders. Recent research has acknowledged the limitations of UT efforts that are focused on interface design alone and the need to apply testing to the social context and process that surrounds the use of such systems (Druckenmiller, Acar & Troutt, 2007). The newly emerging field of collaboration engineering looks beyond the specific configuration of software tools used in group decision-making to the facilitation process that utilizes such tools (Briggs, de Vreed & Nunamaker, 2003; Briggs, de Vreed & Kolfschoten, 2007). UT needs to be applied to these processes as well for the development of effective socio-technical systems.

Various general UT process theories and techniques have been proposed (Dillon, 2001; Lewis et al., 1990; Molich & Nielson, 1990; Nielson, 1994). Also, Ju & Gluck (2005) argue that *UT with real users* is the most fundamental method and is essentially irreplaceable. The points above underscore this point even more for CEUS. In addition, the points surfaced above suggest that, while general UT techniques are essential, the specific context of a system plays a critical role.

In short, the right UT is needed for the right end users.

Projects such as SEI are voluntary but participating firms can gain a kind of "Good Housekeeping Seal of Approval" distinction, while end users and/or their employers can gain a measure of confidence that sound usability practices have been promoted. These kinds of safeguards work best under competitive pressure, however, so that their application cannot be expected to fix most of the problems with CEUS. Research into the prevalence and impact of such systems within firms, as well as across the whole economy, could be influential. If the lost productivity is as substantial as we fear, special ISO guidelines might be promulgated, and possibly tied-in with general ISO quality assurance guidelines. As research in the area becomes more available, the possibility of legislative action at state or federal levels might also be deliberated. Such efforts would need to be weighed against or perhaps combined with other approaches like professional licensure of software engineers (Ficarrotta, 2004).

Further research might also be directed towards establishing categories based on estimated numbers of end users affected by various kinds of such systems. For example, a cutoff level of 100 or more expected end users, say, might be set as a flag for vendors required to have stricter CEUS-related UT standards.

Organizations and individuals also need to be protective of their interests and perhaps resist the urge to quietly accept such imposed systems. In the case of a college seeking accreditation, there is a natural tendency to acquiesce to a strongly recommended or mandated system so as not to be viewed unfavorably. However, the college's evaluation itself depends on good employee compliance in use of the system. A poorly designed system can lead to underutilization or non-utilization and hurt the college's standing in a different and perhaps more substantial but less direct way. Namely, a favorable review by an accreditation agency may depend critically on thorough and complete reporting of all of the college's academic assets.

LIMITATIONS

The captive condition in which the users are held often silences them and precludes the search for solutions such as those proposed above. An example that we have experienced is that of a well-known journal in the Operations Management and Information Systems fields. In its article submission system, the user is required to select up to three items from a list of subject areas. After three are in the selected list, one cannot compare a possible new choice for the list. That is, to bring another into the list required dropping one or more of the current selection of three, although any such drop might be reconsidered later for the final overall selection. This is awkward and not very intuitive. Perhaps at the least, the user deserves a warning on such restrictive navigation limitations.

Luckily, though, not all cases that seem to be *prima facie* instances of CEUS turn out to belong to that vexing category. For instance, government forms such as tax filing forms appear to qualify. But in that case, commercial tax preparation software is readily available in the form of several competing products. Here, competitive pressures require intensive consideration of customer wants and needs, and the self-selected buyers cannot be considered as captive users in the above sense. Market forces eventually kick in. However, this isn't a major limitation of the concerns expressed above, because of the time lag between the captive conditions and one's eventual release from their shackles.

An example of the damage caused by such a time lag is provided by the manner in which WordPerfect was initially preferred by many to Microsoft Word because of its efficient Reveal Codes and macro/scripting capabilities. Although still favored by some special academic, governmental and legal users, a span of a couple of years between the introduction of MS Office and WordPerfect's own introduction of a "utilities library" allowed in the early 1990s a large segment of its captive users to escape, and reinvest time and effort in acquainting themselves with the seemingly better related Microsoft products. We note that the dates of publication for WordPerfect utilities ranged from 1988-1996 (http://www.bookcase.com/library/software/msdos.apps.wordperfect.html, accessed 15 February 08).

CONCLUSION

In a classic article in *Management Science,* Drucker (1959) criticizes quantitative methods for their reliance on data estimation, if not naïve forecasting, and their lack of probing into the future. This chapter has discussed some special usability and ethical issues arising from experience with what we called *captive end-user systems* (CEUS). Here we look to the future in the present context as well. The specific usability aimed at in this chapter is **usability testing** (UT). Looking at the long-term future as exhorted by Drucker, we discussed a growing problem and outlined several potential solutions to it. As a limitation of our discussion, we acknowledge that sometimes market forces, acting serendipitously, manage to dampen or altogether remedy situations without user involvement. Our words of caution are still justified because the number of such lucky happenstances is likely to be dwarfed, at least in IS/IT applications, by the gushing speed of proprietary innovations to which fear of falling behind shackles competing institutions.

ACKNOWLEDGMENT

This paper benefited greatly from stimulating conversations with Gerald DeHondt II and Kholekile Gwebu.

REFERENCES

Bennett, J. (1984). Managing to meet usability requirements. In Bennett, J., Case, D., Sandelin, J. & Smith, M., (eds.), *Visual Display Terminals: Usability issues and Health concerns* Englewood Cliffs, NJ: Prentice-Hall.

Briggs, R. O., de Vreed, G. J., & Kolfschoten, G. L. (2007). *Report of the HICSS-40 Workshop on Collaboration Engineering.*

Briggs, R. O., de Vreed, G. J., & Nunamaker, J. F. (2003). Collaboration engineering with thnkLets to pursue sustained success with group support systems. *Journal of Management Information Systems, 19,* 31-64.

Bruegge, B., & Dutoit, A. H. (2004). *Object-oriented software engineering using UML, Patterns, and Java,* 2nd Ed. Upper Saddle River: Pearson Prentice Hall.

Caulton, D. A. (2001). Relaxing the homogeneity assumption in usability testing. *Behavior and Information Technology,* 20, 1-7.

Chen, Q., & Sharma, V. (2002). Human factors in interface design: An analytical survey and perspective. In Snodgrass, C. R., & Szewczak, E. J. (eds.) *Human Factors in Information Systems.* Hershey, PA: IRM Press.

Constantine, L. L., & Lockwood, L. A. D. (1999). *Software for use: A practical guide to the models and methods of usage-centered design.* New York: ACM Press.

Dennis, A., & Wixon, B. H. (2000). *Systems analysis and design—An applied approach.* New York: John Wiley & Sons, Inc.

Dillon, A. (2001). Usability evaluation, in W. Karwowski (ed.), *Encyclopedia of Human Factors and Ergonomics.* London: Taylor and Francis.

Drommi, A. (2001). Interface design: An embedded process for human computer interactivity.

In Chen, Q. (ed.), *Human Computer Interaction: Issues and Challenges*. Hershey, PA: Idea Group Reference.

Druckenmiller, D. A., Acar, W., and Troutt, M. D. 2007. Usability testing of an agent-based modeling tool for comprehensive situation mapping. *International Journal of Technology Intelligence and Planning, 3*(2), 193-212.

Drucker, P. (1959). Long-range planning challenge to management science, *Management Science, 5*(3), 238-249.

Ficarrotta, J. C., (2004). Software engineering as a profession: A moral case for licensure. In Brennan, L. L., & Johnson, V. E. (eds.), *Social, Ethical, and Policy Implications of Information Technology*. Hershey PA: Information Science Publishing.

Hoffer, J., George, J., & Valacich, J. (2002). *Modern systems analysis and design* (3rd ed.). Reading, MA: Addison-Wesley.

Hoffman, G. M. (2004). Ethical challenges for information system professionals. In Brennan, L. L., & Johnson, V. E. (eds.), *Social, Ethical and Policy Implications of Information Technology*. Hershey PA: Information Science publishing.

Ju, B., & Gluck, M., (2005). User-process model approach to improve user interface usability. *Journal of the American Society for Information Science and Technology, 56*(10), 1098-1112.

Lazar, J., Adams, J., & Greenridge, K-D. (2005). Web-STAR: Development of survey tools for use with requirements gathering in Web site development. In *Issues of Human Computer Interaction* (Sarmento, A., ed.). Hershey, PA: IRM Press.

Lewis, C., Polson, P.G., Wharton, C., & Rieman, J. (1990). Testing a walkthrough methodology for theory-based design of walk-up-and-use interfaces. *Proceedings of the ACM CHI '90*, (April), 234-242.

Molich, R., & Nielsen, J. (1990). Improving a human-computer dialogue. *Communications of the ACM, 33*(3), 338-348.

Nielsen, J. (1994). Heuristic evaluation. In J. Nielsen and R. Mack (eds.) *Usability inspection methods*. (pp. 25-62). New York: Wiley.

Rubin, J. (1994). *Handbook of usability testing: How to plan, design, and conduct effective tests.* New York: Wiley.

Schach, S. R. (2005). *Object-oriented & classical software engineering*, 6th Ed. Boston: McGraw-Hill.

SEI/CMU (2008). http://www.sei.cmu.edu/. (Accessed 7 February 08.) SEI/CMU is the Systems Engineering Institute at Carnegie Mellon University.

Thomas, D. M., & Bostrom, R. P. (2007). The role of a shared mental model of collaboration technology in facilitating knowledge work in virtual teams. *Proceeding of the 40th Hawaii International Conference on Systems Sciences*. http://csdl2.computer.org/comp/proceedings/hicss/2007/2755/00/27550037a.pdf

Troutt, M. D. (2007). Some usability and ethical issues for captive end-user systems. *Journal of Organizational and End User Computing, 19*(3), i-vii.

UsabilityNet (2008). http://www.usabilitynet.org/tools/r_international.htm. (Accessed 7 February 08.) UsabilityNet is a project funded by the European Union (EU) to provide resources and networking for usability practitioners, managers and EU projects.

Whitten, J. L., Bentley, L. D., & Dittman, K. C. (2001). *Systems analysis and design methods*. 5th Ed. Boston: McGraw-Hill Irwin.

Chapter IV
Do Spreadsheet Errors Lead to Bad Decisions?
Perspectives of Executives and Senior Managers

Jonathan P. Caulkins
Carnegie Mellon University, USA

Erica Layne Morrison
IBM Global Services, USA

Timothy Weidemann
Fairweather Consulting, USA

ABSTRACT

Spreadsheets are commonly used and commonly flawed, but it is not clear how often spreadsheet errors lead to bad decisions. We interviewed 45 executives and senior managers/analysts in the private, public, and non-profit sectors about their experiences with spreadsheet quality control and with errors affecting decision making. Almost all of them said spreadsheet errors are common. Quality control was usually informal and applied to the analysis and/or decision, not just the spreadsheet per se. Most respondents could cite instances of errors directly leading to bad decisions, but opinions differ as to whether the consequences of spreadsheet errors are severe. Some thought any big errors would be so obvious as to be caught by even informal review. Others suggest that spreadsheets inform but do not make decisions, so errors do not necessarily lead one for one to bad decisions. Still, many respondents believed spreadsheet errors were a significant problem and that more formal spreadsheet quality control could be beneficial.

INTRODUCTION

Spreadsheets are used in diverse domains by decision makers at all levels (Gerson, Chien, & Raval, 1992; Chan and Storey, 1996; Seal, Przasnyski, & Leon, 2000; Croll, 2005); the entire July-August, 2008 issue of the journal *Interfaces* is devoted to spreadsheet modeling success stories. However, laboratory studies and field audits consistently find that a large proportion of spreadsheets contain errors. In a dozen studies reviewed by Kruck, Maher, & Barkhi (2003), the average proportion of spreadsheets with errors was 46%. Panko's (2000a, 2005) synthesis of spreadsheet audits published since 1995 suggested a rate of 94%. Powell, Baker, & Lawson (2007a, 2008b) critique past work and greatly advance methods of defining and measuring spreadsheet errors but at the end of the day reach the same overall error rate of 94%.

Hence, one might expect that (1) spreadsheet errors frequently lead to poor decisions and (2) organizations would invest heavily in quality control procedures governing spreadsheet creation and use. We investigated both hypotheses through 45 semi-structured interviews with executives and senior managers / analysts in the public, nonprofit, and private sectors. Field interviews raise fewer concerns about external validity than do laboratory studies, and they focus on the overall decision making process, not just the spreadsheet artifact as in audit studies. However, our approach has two important limitations. First, the respondents are a convenience sample. Second, self-report can be flawed, whether through imperfect memories, self-serving bias, conscious deception, and/or limited self-awareness. Given these limitations, we focus on broad qualitative conclusions. In brief, we found that most respondents could describe instances in which spreadsheet errors contributed to poor decisions, some with substantial consequences, yet few reported that their organization employs quality control procedures specific to spreadsheet analysis.

The literature on spreadsheet errors in general is large (see Panko, 2000b and Powell, Baker, & Lawson, 2008a for reviews), but much less has been written on these specific questions. Regarding the frequency with which spreadsheet errors lead to bad decisions, the European Spreadsheet Research Interest Group (EUSPRIG) maintains a webpage of news stories reporting the consequences of spreadsheet errors (http://www.eusprig.org/stories.htm). However, spreadsheets are used by so many organizations that even if only a small proportion were hurt badly by spreadsheet errors, there could still be scores of examples. We started with a population of individuals and organizations for which we had no a priori reason to think spreadsheet errors were a particular problem. This approach has been taken by others (e.g., Cragg and King, 1993) to explore the prevalence of defective spreadsheets, but like Powell, Baker, & Lawson (2007b), we shift the focus to assessing the impact of those spreadsheet errors.

A considerable corpus on controlling spreadsheet errors concerns what organizations *should* do. Classic recommendations lean toward application of good software engineering principles (Mather, 1999; Janvrin and Morrison, 2000; Rajalingham, Chadwick, & Knight, 2000; Grossman and Özlük, 2004) or formal theories (Isakowitz, Schocken, & Lucas, 1995). Kruck and Sheetz (2001) combed practitioner literature for practical axioms validated by empirical results, supporting aspects of the spreadsheet lifecycle theory (e.g., include planning / design and testing/debugging stages) and recommendations to decrease formula complexity.

There is also some literature describing what organizations actually do. Notably, Finlay and Wilson (2000) surveyed 10 academics and 10 practitioners on the factors influencing spreadsheet validation. Those most commonly mentioned were (a) aspects of the decision and (b) aspects of the spreadsheet underlying the decision context. However, Grossman (2002) argues that it would be valuable to have greater knowledge of what error control methods are currently used.

We seek to fill that gap. The next section describes data and methods. The third discusses results pertaining to types of errors, error control procedures and policies, and reported effects on decisions. The paper closes by discussing the decision processes within which spreadsheets were embedded and implications for practice and future research.

DATA AND METHODS

Sampling

Data collection methods were similar to Nardi and Miller (1991). Interview subjects were identified primarily by referral through personal and institutional contacts. Only one person approached through these contacts declined to be interviewed. In contrast, earlier attempts at cold-calling often led to refusals or perfunctory interviews. Even when given every assurance of anonymity, respondents seemed more wary of admitting bad decisions to a complete stranger than to someone referred by a mutual acquaintance.

Fifty-five people were interviewed, but ten were excluded from the analysis below. Seven were excluded because they were duplicate interviews within the same work group; we retained only the respondent with the most sophisticated perspectives concerning spreadsheet use within that work group. Two were eliminated because they had worked for more than one organization, and it became ambiguous which of their comments concerned which organizations. One was excluded because s/he used spreadsheets only for list-tracking and database functions.

Sample Characteristics

All but one interview was done in person, so most interviewees (73%) represented organizations located in one region. The one phone interview stemmed from meeting in person with the CEO of a manufacturing firm who suggested that his CFO, in another state, had more insights into the topic.

We interviewed subjects from three sectors (for-profit, non-profit, and government) and two organizational levels (executives vs. senior managers/analysts). Since we found few pronounced differences across groups, results are primarily reported in aggregate. Overall, 96% of the executives were male, as were 45% of the manager/analysts. All executives and 90% of manager/analysts were non-Hispanic White. The executives came from small to medium size organizations (ranging from several dozen to several thousand employees). Private sector managers' organizations, in contrast, spanned the full range of sizes, up to organizations with tens of thousands of employees. Some respondents in the non-profit sector worked for large health care or educational institutions, but most came from smaller organizations. Educational attainment ranged from several PhDs to a local government manager with a high school degree. No respondent remembered any recent formal training in Excel. Most reported learning spreadsheet techniques informally, in the office environment or with the help of practical manuals.

We classified interviewees by their highest reported spreadsheet use as basic calculation (n=6), simple modeling (n=20), or advanced modeling (n=19). This distinction was based on the highest mathematical complexity of formulas and models, as well as spreadsheet size (in number of cells and/or file size), links, functionality and features, such as optimization with Solver. Respondents reported using an average of 2.7 other types of software to support decision making, with database and accounting software mentioned most frequently.

Table 1. Description of sample

		Gender		Location	
Group	**N**	**Women**	**Men**	**Local**	**Other**
Public Executives	6	0%	100%	100%	0%
NonProfit Executives	13	8%	92%	85%	15%
Private Executives	6	0%	100%	50%	50%
Public Managers	7	57%	43%	57%	43%
NonProfit Managers	6	83%	17%	100%	0%
Private Managers	7	29%	71%	43%	57%
Total	45	27%	73%	73%	27%

Interview Protocol

The interview protocol was finalized after conducting ten exploratory interviews with subjects who are not part of this analysis. Those interviews revealed that open-ended questions elicited richer responses than did tightly-scripted or multiple choice questions. Indeed, senior executives resisted highly structured questions. Furthermore, the most effective sequence in which to cover topics varied from interview to interview. Hence, we adopted a semi-structured interview protocol to ensure that a specific set of topics with associated probes were covered in every interview, while allowing the conversation to move nonlinearly through the topics.

Interviewees were sent a description of the research project and the interview protocol. The protocol addressed individual and organizational experience with spreadsheets, spreadsheet errors, error control, and effects on decision making. (See Appendix A.) Variables were coded by the primary interviewer based on audio recordings and detailed notes into categorical variables representing groupings of common responses. To test inter-rater reliability, a subset of the interviews were coded by both primary interviewers. Seventy-six percent of those items were coded consistently, and most discrepancies were instances in which one interviewer drew a conclusion and the other thought there was insufficient information to decide, not from the interviewers drawing contradictory conclusions. The most discordant variables pertained to advantages and disadvantages of using spreadsheets for decision making and were excluded from the analysis.

RESULTS

Types of Errors Reported

All but one respondent reported encountering errors in spreadsheets. The most commonly mentioned types were *inaccurate data* (76%), *errors inherited from reuse* of spreadsheets (49%), *model errors* (33%), and *errors in the use of functions* (also 33%). We recorded the presence or absence of a type of error, but have no information on the frequency of occurrence.

These error types emerged from respondents' statements, not a literature based classification as in Rajalingham et al. (2000), Purser and Chadwick (2006), Panko (2007), or Powell et al. (2007a) because we merely asked respondents for their opinion of the "cause" or "source" of the error, without more specific prompts. Powell et al. (2007b) found that many errors in operational spreadsheets have little or no quantitative impact. For instance, an erroneous formula could generate the correct answer if the error pertained to a condition that did not occur in the given data.

Table 2. Commonly mentioned errors, by sophistication of highest spreadsheet use and by sector

Error Type	All	By Sophistication of SS Use			By Sector		
		Advanced	Simple	Basic	Public	Nonprofit	Private
Inaccurate data	76%	74%	80%	67%	77%	95%	46%
Errors inherited from reusing spreadsheets	49%	63%	40%	33%	31%	63%	46%
Model error	33%	42%	25%	33%	23%	16%	69%
Error in use of functions	33%	21%	45%	33%	46%	26%	31%
Misinterpretation of output/report	27%	47%	15%	0%	15%	32%	31%
Link broken/failed to update	22%	37%	15%	0%	8%	26%	31%
Copy/Paste	22%	21%	30%	0%	23%	26%	15%
Other	11%	11%	15%	0%	8%	5%	23%
Lost file/saved over file	7%	5%	10%	0%	0%	5%	15%
No errors	2%	5%	0%	0%	8%	0%	0%
N	45	19	20	6	13	19	13

Although we did not ask respondents explicitly to distinguish between errors that did and did not have an impact, since the context of the interviews was effects on decision making, we believe most respondents discussed only errors that did have an impact or would have had an impact if they had not been detected and corrected.

Inaccurate Data

We expected that inaccurate data would come primarily from miskeyed data and other "mechanical errors" to use Panko's (1999, 2007) term. Such "typos" were frequently mentioned, but there were other sources of data value errors.

One was bad data piped into a spreadsheet automatically, for example, from a database query or web-based reporting service. A health insurance firm reported that a change in prescription pill units triggered automatic reimbursements that were too high because of a units inconsistency between the revised database and the spreadsheet. Such systems integration problems illustrate an issue raised by multiple respondents and the literature (Panko, 2005). Figures in spreadsheets can sometimes be attributed with an aura of inerrancy, lulling users into not reviewing them as critically as they might have in another context.

Respondents mentioned human bias as another source of inaccurate data, akin to Panko's (2007) "blameful acts", varying in culpability from "wishful thinking" to outright fraud. Wishful thinking included choosing "base case" parameter values that "made sense" to the analyst because they gave the (perhaps unconsciously) preferred conclusion. The more extreme version was willful and self-serving manipulation of input values. Bias can contaminate any analysis, but bias buried in a spreadsheet can be hard to detect.

Different types of quality control are required for these different sources of inaccurate data. Asking analysts to check parameter values a second time might help catch typos, but it would do little to address a fraudulent self-serving bias.

Errors Inherited from Reusing Spreadsheets

Almost all respondents said they reused their spreadsheets, and almost half described errors from reuse of their own or colleagues' spreadsheets. One respondent inherited a model containing a vestigial 'assumptions page' that did not link to the model. Several described spreadsheet errors that endured for an extended period of time and noted that small errors replicated many times led to substantial losses over time. One nonprofit had relied on a faulty ROI worksheet for several years, affecting contracts worth ~$10 million. Reports of reuse errors increased with the complexity of spreadsheet applications.

Opinions differed on the value of reuse. For many respondents, the ability to reuse templates was a key advantage of spreadsheets. Some echoed the observation by Nardi and Miller (1991) that templates enable sharing of domain expertise as well as spreadsheet skills. However, a sizable minority felt reused spreadsheets were difficult to control, since updating worksheets can introduce more errors.

Errors in the Use of Functions

Thirty-three percent of respondents described errors in functions, ranging from inappropriate use of built-in functions to mistaken operators and cell addressing problems. Careful inspection of formulas was reported to be rare unless motivated by an observed discrepancy. In the words of one respondent, "Formulas are only examined in depth if there's a reason." The most frequently mentioned explanation was the difficulty of review. One senior analyst observed: "Spreadsheets are not easy to debug or audit…It's a very tedious process to check someone else's cells, especially two or three levels [of cell references] down."

Model Error

One-third of respondents reported model errors, including errors in assumptions, overall structure, errors of omission, and other major distortions of the modeled situation (as identified by the respondent). Model errors are not programming mistakes, but rather erroneous judgments about how to model real world situations, so they cannot be detected solely by reviewing the spreadsheet (Grossman, 2003). The same can be said for the fifth most commonly cited type of error, namely misinterpretation of spreadsheet output, since a correct spreadsheet may be misinterpreted by a person who has flawed understanding of the spreadsheet model's assumptions or limitations.

Spreadsheet Quality Control Methods

Most respondents thought about quality control in terms of detecting errors ("inspecting in quality") rather than in terms of preventing them ("building in quality") from the outset. The twelve quality control methods mentioned repeatedly by respondents can usefully be divided into three categories: (1) informal quality control, (2) organizational methods such as peer review, and (3) technical tools that are specific to spreadsheets. (See Table 3.)

The academic literature focuses on extending formal software quality control practices to the end user environment (e.g., Rajalingham, Chadwick, Knight, and Edwards, 2000), but the respondents did not report following common design and development recommendations (Teo and Tan, 1999, Clermont, Hanin, & Mittermier, 2002), a software lifecycle approach, or formal software tools (Morrison, Morrison, Melrose, & Wilson, 2002). Instead, the methods described might be characterized as stemming primarily from the application of general managerial acumen. They do not differ in obvious ways from quality control steps that would be employed to

Table 3. Proportions reporting use of various quality control methods, of those for whom sufficient information was gathered to ascertain whether the method was used (Method Type 1 = Informal methods; Type 2 = Organizational methods; Type 3 = Technical tools)

				Spreadsheet Sophistication (Highest Level)		
Quality Control Method	Method Type	N	Avg of All Groups	Advanced Modeling	Simple Modeling	Basic Calculation
Gut check against the bottom line	1	45	96%	100%	90%	100%
Review by developer	2	44	86%	84%	85%	100%
Review by someone other than developer	2	44	73%	67%	70%	100%
Crossfooting	3	45	60%	63%	55%	67%
Review by multiple reviewers other than developer	2	44	45%	56%	45%	17%
Documentation	2	45	42%	53%	40%	17%
Keep it simple	1	43	33%	33%	30%	40%
Input controls	3	45	22%	37%	15%	0%
Prevent deletion of calculation cells (protection)	3	45	20%	21%	25%	0%
Other		45	16%	11%	15%	33%
Test cases	3	45	13%	26%	5%	0%
Separate page for audit/change tracking	3	45	13%	21%	10%	0%
Audit tools	3	45	7%	16%	0%	0%

review any other form of analysis. What seems to distinguish the conscientious from the lackadaisical was not necessarily technical sophistication. Rather, it was the formality with which general purpose quality control procedures such as peer review were employed.

Informal Quality Control: The 'Sniff Test' and 'Keeping It Simple'

The most frequently cited quality control procedure was the 'gut check' or 'sniff test,' a cursory examination of bottom line figures for reasonableness. Many respondents acknowledged the limitations of 'sniff tests'. One analyst reported finding a major modeling error worth several million dollars by "dumb, blind luck" the night of the presentation to the board. Pointing to a blank cell accidentally excised a significant portion of the analysis, but even that obvious error passed many sniff tests. However, a minority held the contrary view that elementary review would detect all errors of consequence before they misinformed a decision.

The other informal method mentioned was to 'keep it simple.' This included simplifying the analysis as well as the spreadsheet itself. Advanced developers reported using advanced quality control methods in addition to, not instead of, these informal methods, although it is possible that what advanced developers call a gut check or keeping it simple might be richer than what less experienced mean when they use the same term.

Formal Quality Control Methods that Are Organizational in Nature

Respondents mentioned two methods that were organizational in nature: review and documentation. Most (73%) respondents reported that spreadsheets were reviewed by someone besides the developer. For the forty-five percent using multiple outside reviewers, this ranged from two colleagues in the office to review by several executives in teleconference. One organization mentioned inadvertent outside review. A public manager confessed that their annual budget, which invariably contained errors, would be scrutinized by unions and other groups for their own interests, helping to correct the document. Although review seems common, one-quarter of respondents did not mention any kind of review by others. Furthermore, almost no respondents reported spending even the minimum time on validation that is suggested by Olphert and Wilson (2004). Documentation concerns focused on structural assumptions and parameter values, rather than formulas and was reported more frequently than in Baker, Powell, Lawson, & Foster-Johns (2006a).

Technical Tools

We define technical tools as quality control methods that are specific to spreadsheets, such as protecting portions of the spreadsheet as a form of input control. Advanced modelers were the most likely to report using these technical tools, particularly those beyond crossfooting (redundant calculations). Test cases and separate pages for audit/change tracking were rarely cited. As in Baker et al. (2006b), least frequently mentioned were automatic audit tools, such as add-ins or Excel's own (limited) built-in audit feature, used by just three advanced modelers.

Other Methods

Several other error control methods were mentioned by only one respondent, but are still interesting since they represent "outside the box" thinking relative to methods typically discussed in the literature. Two were personnel-related: firing people who produced flawed spreadsheets and, on the positive side, hiring talented people in the first place, where talent referred to general analytical competence not spreadsheet skills. These actions might be seen as consistent with Panko's (2007) point that spreadsheets are not the cause of spreadsheet errors, and are usefully seen as a special case of human error more generally. Another method cited was to avoid using spreadsheets altogether, for example, by converting the organization to some enterprise resource planning (ERP) system or by encouraging use of other analytical software.

Spreadsheet Quality Control Policies

Consistent with Cragg and King (1993) and Baker et al. (2006a, 2006b), few respondents reported that their organizations had formal policies intended to ensure spreadsheet quality. Financial services sector organizations were the exception. They often reported using standardized spreadsheet models, created by IT personnel at corporate offices and distributed to branches.

Even without formal policies, several workgroups described high quality environments where review and documentation were routine. While there has been little success identifying correlates to an individual's error rate (Howe and Simkin, 2006), we felt it was likely that some backgrounds facilitate more awareness within organizations. Of our respondents who stressed quality control, one was a computer scientist familiar with the risks of programming errors and two others were trained in accounting controls and taught spreadsheet modeling to graduate students.

For most of our respondents, the absences of policies were not the result of any thoughtful decision balancing the benefits of improved quality against the overhead associated with rigorous quality control (cf., Grossman, 2002). Indeed, many responded by saying, in essence, "Never thought about it, but it sounds like a great idea."

Others could cite reasons why a more structured approach might not work, including lack of time. One government analyst reported that she often had to create or modify spreadsheets in as little as half an hour. Sometimes errors were later discovered in those spreadsheets, but timeliness was paramount.

Several managers perceived that checking spreadsheets more carefully would require hiring another employee, and one public manager in particular had resisted an auditor's recommendations to hire additional staff to check spreadsheet accuracy, citing budget constraints. These views may help explain the absence of formal policies even if they are inconsistent with the conventional wisdom in software development that time invested in quality control is more than recouped in reduced rework.

The literature suggests that formal policies often encounter resistance because a principal advantages of spreadsheets is empowering end users to complete analysis independently (Cragg and King, 1993 and Kruck and Sheetz, 2001). Some respondents expressed a related yet distinct concern. They worked in small, non-corporate environments and suggested that guidance must be informal and/or implicit in these close-knit workplaces. Their arguments centered around unintended consequences of formality on workplace culture, such as signaling lack of trust in staff competence, rather than effects on spreadsheet productivity per se.

Most organizations likewise had no formal policies governing how to respond when errors were detected, and respondents described a range of responses. A common response was to fix the specific error but do nothing else. Forty percent claimed to go further by investigating the associated processes to detect other, related errors. Other responses were less obviously effective. One individual threw out the computer on which a perpetually buggy spreadsheet was being run, in the belief that the hardware was somehow at fault.

Thirty percent of respondents mentioned sometimes rebuilding a spreadsheet from scratch when errors were detected. This could be eminently sensible if the original design was not well conceived. Often a cumbersome exploratory spreadsheet can be replaced by one whose design is better engineered and less error-prone. On the other hand, this may also reflect undue confidence that rebuilding the spreadsheet will not introduce new, more serious errors. Some respondents seemed to view errors as aberrations that can be eliminated if the work is redone carefully, rather than a predictable outcome, as in the software engineering perspective anticipating a certain number of errors per thousand lines of code (cf., Panko, 1999).

Other Factors Mediating the Application of Spreadsheet Quality Control

Distinct from the question of what methods are used is how *consistently* they are are applied. Our protocol did not address this directly, so we do not have data from all 45 respondents, but it came up spontaneously in many interviews.

One ideal espoused in the literature is *not* to apply all methods to all spreadsheets. Rather, the risk analysis philosophy suggests investing more heavily in quality control for certain high-risk and/or high-stakes spreadsheets (Whittaker, 1999; Finlay and Wilson, 2000; Grossman, 2002; Madahar, Cleary, & Ball, 2007). However, at least eight respondents mentioned situations in which the level of review for important spreadsheets was less, not more, rigorous:

- Highly independent executives often completed one-off, ad hoc and first-time analysis for an important decision without the benefit of review.
- When the spreadsheet was highly confidential, few people in the organization had

access to it, making effective review difficult.
- Important decisions were often associated with time pressures that precluded formal review.

Several respondents noted that details of the decision context matter, not just the stakes, including whether the decision was internal as opposed to being part of a public or adversarial proceeding such as labor negotiations or law suits. One example stemmed from a highly partisan political budgeting battle. The respondent noted that the opposition's budget literally did not add up. The sum of itemized subcategories did not quite match the category total. It was a minor discrepancy both in absolute dollars and as a percentage of the total budget. However, the respondent was able to exploit that small but incontrovertible error to cast doubt on the credibility of all of the other party's analysis, leading fairly directly to a dramatic political victory before the legislative body.

Spreadsheets Role in Decision Making

This project began with a vision of spreadsheets and decision making that was shaped by experience teaching decision modeling courses. It might be caricatured as follows. "Leaders sometimes analyze decisions with spreadsheets that estimate bottom-line consequences of different courses of action. The spreadsheet includes cells whose values can be selected to achieve managerial goals. Decision makers combine that analysis with expert judgment and factors outside the model to select a course of action."

This view suggests an almost one-to-one connection between spreadsheet errors and decision errors. Instead, we observed a continuum in terms of how spreadsheet output interacted with human judgment that is much broader than our original "academic" view of organizational decision making. We discretize this continuum into a five-part typology based on how tightly coupled the spreadsheet analysis is to the decisions:

Automatic data processing: When spreadsheets are used for automatic data processing, errors do directly translate into adverse consequences. A typical administrative horror story was a mis-sort, such as a mailing list matching incorrect names to addresses. As a result, many organizations avoided using spreadsheets for processing data automatically. As one respondent put it, "It's not like we're cutting a check off a spreadsheet." Yet others described instances in which output from spreadsheets was formalized without any review, including a spreadsheet-based program to automatically generate invoices and this could be problematic.

Making recommendations that are subject to human review: Some applications matched our academic view in the sense of computing a bottom-line performance metric that pointed directly to a recommended course of action, but the results were still subject to human review. A typical example would be a spreadsheet used to project the ROI from a contemplated investment. In principle, the decision recommendation is simple; if the ROI is favorable, then take the contemplated action. However, borderline cases triggered more intense scrutiny. In effect, the spreadsheet plus minimal human oversight made the easy calls, but for tough decisions spreadsheet analysis was just the first step. Furthermore, the user was ultimately responsible, and when bad decisions were made, our respondents did not scapegoat the spreadsheets. As one executive in the financial sector said, quantitative analysis is only "the first 75% of a decision. ... this [spreadsheet] is a decision making tool, not an arbiter." Another senior financial manager emphasized that it was "his job" to interpret the spreadsheet analysis in light of other critical qualitative factors. Because human decision making based on experience and domain knowledge was critical, these interviewees were less concerned about the impact of spreadsheet errors.

Projecting decision consequences: Many spreadsheets supported "what if" analysis but not of bottom-line performance metrics. Their output becomes just one of several inputs to a higher-level decision process. A typical example was a senior financial manager at a school district modeling the budget impacts of alternative property tax rate scenarios. The spreadsheet output was directly relevant to a specific decision before the school board, yet the board members could have been at least as interested in effects on educational attainment, relations with the teachers' union, and/or voter anger and their reelection prospects, considerations that were entirely outside of the spreadsheet model.

Understanding a system's relationships: Managers sometimes used spreadsheets to understand interrelationships among a system's variables even if they did not model a decision's consequences directly. Respondents in this category might "use the spreadsheet to think with," as one respondent put it. Another respondent noted, "The spreadsheet exists as part of the analytical framework in any strategic or operational decision, but it's not the spreadsheet alone." In these situations, the spreadsheet is used more for insight than for computing a specific number. If spreadsheet errors distorted the key relationships, those errors could harm understanding and, hence, decision making. However, the effect would be mediated through something else, namely a person's understanding. Furthermore, the understanding that was the proximate source of any decision errors was subject to independent quality control not tied to the spreadsheet itself.

Providing background, descriptive information: Respondents described many instances where the spreadsheet was used as an information management system to synthesize, organize, and process facts that were relevant to the decision maker, but the spreadsheet did not project the consequences of any specific decision per se. Decision makers might use the spreadsheet to compare current year-to-date costs to those of the previous year, to check progress relative to plan, or to estimate budgets for the coming year. Such basic information could inform myriad decisions, but there is no sense in which the spreadsheet is modeling the consequences of a particular decision. Rather, such spreadsheets estimated parameters that fed into some other decision making system, typically human judgment.

Severity of the Impact of Spreadsheet Errors on Decision Making

A slim majority of subjects whose responses could be coded (25 of 44) expressed strong concern about the consequences of spreadsheet errors in their organization. Nine of the nineteen who were not strongly concerned said they simply did not use spreadsheets in ways that were integral to high-stakes decision making. Almost by definition, spreadsheet errors could not cause those organizations grave harm. Looking solely at organizations that reported using spreadsheets to inform high-stakes decisions, the slim majority of concerned subjects (25 of 44 or 57%) becomes a substantial majority (25 of 35 or 71%). Still, ten organizations reported both using spreadsheets to inform important decisions and experiencing spreadsheet errors, yet they still had no major concern about adverse impact on the decisions made.

Opinions about whether spreadsheet errors led to bad decisions could also be categorized into three groups in a slightly different way: (1) a small minority who thought spreadsheet errors were always caught before decisions were made, (2) a larger group who acknowledged that not all errors are detected but who thought any errors of consequence would be detected before they misinformed a decision, and (3) the plurality who thought spreadsheet errors could have a significant adverse impact on decisions.

These responses support two conclusions. First, spreadsheet errors sometimes lead to major losses and/or bad decisions in practice. Indeed, we heard about managers losing their jobs because of inadequate spreadsheet quality control. Second, many decision makers whose organizations produce erroneous spreadsheets do not report serious losses or bad decisions stemming from those flawed spreadsheets. Those reassuring self-reports could simply be false; our respondents may have been over-confident about their organizations ability to withstand spreadsheet errors even if all but one admitted they had flawed spreadsheets. However, the interviews suggested another possibility, namely that organizational factors can help prevent errors in spreadsheets that inform decisions from automatically or inevitably leading to bad decisions. Investigating that possibility was not part of our interview protocol and we do not have systematic data concerning this hypothesis, so we elaborate it in the next section, as part of the discussion.

DISCUSSION

We found that spreadsheet errors do lead directly to bad decisions, but the more common scenario may be spreadsheet errors contributing indirectly to suboptimal decisions. That is, our respondents could offer their share of classic spreadsheet errors (e.g., incorrect formulae) causing someone to make the wrong decision. Often, however, it may be more useful to think of spreadsheet errors as misinforming deliberations rather than as recommending the wrong course of action. Spreadsheets are used for monitoring, reporting, and a host of other managerial activities besides deliberative decision making, and even when a well-defined decision was being made, spreadsheet analysis often provided merely descriptive or contextual information. Spreadsheets were often more of a management information system, than a decision support system. Furthermore, when spreadsheets are used in decision support mode, there is still, by definition, a human in the loop.

Couched in terms of Simon's (1960) intelligence, design, choice decision making framework, spreadsheet errors may lead to errors in the intelligence phase, but their ramifications are buffered by human judgment at the choice phase. Furthermore, it has long been observed that managers' activities do not fit well into the classic vision of planning, organizing, and coordinating (Mintzberg, 1975) and that in practice decision making in organizations departs fundamentally from the classical decision making paradigm (March and Simon, 1958; Cyert and March, 1963; Lipshitz, Klein, Orasanu, & Salas, 2001).

The information context underpinning an organization's decision making process process is almost always murky, ill structured, and incomplete. Since poor information is the norm, decision making processes in successful organizations have evolved to cope with that reality, and a piece of bad information is not like a monkey wrench lodged in the gears of a finely tuned machine. Respondents' reports were consistent with Willemain, Wallace, Fleischmann, Waisel, & Ganaway's (2003) argument that there is a "robust human ability to overcome flawed decision support." Indeed, a flawed spreadsheet might even help dispel some of the fog of uncertainty surrounding a decision, just less effectively than it would have if it had not contained an error (cf., Hodges, 1991; Grossman, 2003). This metaphor of spreadsheet errors clouding an already hazy decision situation is relevant whether the poor information deriving from the erroneous spreadsheet pertains to historical facts, future projections/random quantities, relationships among variables, and/or basic structural assumptions about how to think about the problem.

These observations do not minimize the problem of spreadsheet errors or question the premise that reducing spreadsheet errors can improve organizational performance. However, they may help explain why, if spreadsheets are as riddled with errors as Panko (2005) and Powell et al. (2007a, 2008b) suggest, organizations continue to use them to support key decisions and why a substantial minority of respondents seemed relatively unconcerned about the ramifications of spreadsheet errors.

Implications for Research

To the extent that results from this sample generalize, they have implications for research on spreadsheet errors and decision making. An obvious point is simply that field interviews contribute interesting insights on this important topic. There is a small but important organizational literature that studies spreadsheet use in situ (Nardi and Miller, 1991; Nardi, 1993; Hendry and Green, 1994; Chan and Storey, 1996). Extending its focus slightly from spreadsheet errors to spreadsheet errors' effect on decision making appears useful.

A second point is that the spreadsheet itself is not the only productive unit of analysis. This research took the organization as the unit of analysis. Further research should also consider the decision and/or decision process as the unit of analysis. The impact of spreadsheet errors depends on various factors beyond the spreadsheet itself or even the decision. Our research suggests also paying attention to (1) how the spreadsheet informs the decision and (2) the larger organizational decision making context, such as whether the decision in question is made by an individual or group, whether it is a final decision or a recommendation, and whether the spreadsheet will remain private or be subject to (potentially hostile) external review.

In particular, if our five-part typology of spreadsheet's roles in decision making generalizes to other samples, it would be interesting to assess how concern about errors' effects on decision making varies across those five types. Likewise, it would be useful to move beyond self-report to obtain objective measures of the consequences of flawed spreadsheets misinforming decisions in each of the ways outlined in the typology.

Implications for Practice

Discussion of implications for practice must be prefaced by a large caveat. We did not ask about interventions to improve spreadsheet quality, let alone collect evaluation data demonstrating that an intervention improved organizational performance. All we can do is offer some opinions.

Our first suggestion is that organizations such as those in this sample ought to consider investing more in spreadsheet quality control. There was a yawning chasm between what the research literature suggests and what respondents' described as typical practice. There was also a great gulf between the most and least quality conscious organizations encountered. The gap between literature and practice was most apparent in the types of quality control tools used (formal and technical vs. informal and organizational); the gap among respondents was apparent in the varying intensity with which informal and organizational methods were pursued. Hence, most organizations can ramp up quality control efforts even if they lack the technical sophistication to use high-end methods described in the literature.

A related suggestion is that an organization should not eschew appointing a "chief spreadsheet quality control officer" just because it does not employ software engineers. Certain other educational backgrounds appear to prepare people to appreciate readily the concepts of spreadsheet quality control, notably industrial engineering, systems analysis, and accounting.

Third, the range of methods relevant for preventing bad decisions is broader than is the range of methods relevant for preventing spreadsheet errors. There are two ways an organization can prevent spreadsheet errors from leading to bad decisions: (1) preventing spreadsheet errors and (2) preventing any spreadsheet errors that do occur from translating into bad decisions. For our respondents, the latter was as important as the former. Even when it comes to preventing spreadsheet errors, some actions are organizational not technical, such as insisting that spreadsheet training include error control methods not just functionality, explicitly budgeting time for spreadsheet quality assurance testing, and having reviewers publicly sign-off on a spreadsheet before it is used, the way a professional engineer must certify the quality of building plans before construction begins.

The final suggestion is for executives simply to raise awareness in their organizations about the idea of establishing spreadsheet quality control standards and procedures. Many managers seemed not to have thought about the possibility of being proactive in spreadsheet quality management.

SUMMARY

Our interviewees affirmed two common findings: (1) Spreadsheets are frequently used to inform decisions and (2) spreadsheets frequently have errors. Given this, one might expect these respondents to be able to recount many instances of spreadsheet errors leading to bad decisions. Indeed, the majority could cite such instances and viewed them as a serious problem. However, a significant minority did not view them as a serious problem and even among those who did, the sky was not falling. No respondent suggested that the proportion of flawed decisions in any way approached the proportion of spreadsheets the literature finds to be flawed.

Disaster was not being avoided because of systematic application of formal, spreadsheet-specific quality control policies and procedures. Indeed, few organizations outside the financial sector had such policies, and the actual practices seem to reflect common concern for the quality of analysis generally more than they did technical or spreadsheet-specific tools or procedures.

Three alternative but not mutually exclusive explanations emerged as to why spreadsheet errors lead to some, perhaps even many, but still not an overwhelming number of flawed decisions. The first view, espoused by a significant minority of respondents, was that informal quality control methods work for precisely those errors that could be most problematic. When the spreadsheet analysis is wildly off, experienced decision makers can sniff that out. Small errors might not be noticed, but small errors were believed to have minor consequences.

The second explanation is that for some organizations (nine of forty-five in our sample), spreadsheets are not used in ways that are tied to specific high-stakes decisions. The spreadsheets might be used for various types of information processing, ranging from database-like functions to synthesizing and graphing data drawn from another system, but the spreadsheets are not being used in ways that guide specific decisions.

The third explanation is that even if large errors might go undetected in spreadsheets that inform specific, strategic decisions, the spreadsheet analysis is merely informing not driving the decisions. The image one should have is not that an analyst enters all relevant considerations into a spreadsheet, analyzes that spreadsheet, and the organization implements whatever course of action the spreadsheet suggests. Instead, there is some organizational decision process often involving multiple people. Those people bring to the table a great deal of judgment and wisdom, as well as a range of data, mental models, forecasts, and so forth. Spreadsheets may have been used to inform or even generate some of those data, mental models, and forecasts, but other sources of information are also drawn upon. At the end of the day, it is humans exercising human judgment that make the decision.

Usually that judgment is exercised in the face of terribly incomplete and imperfect information. A good spreadsheet analysis might fill in some but not all of that incomplete information. A bad spreadsheet analysis might increase the amount of imperfect information. The murkier the information, the greater the risk of bad decisions, so spreadsheet errors contribute to bad decisions. Ultimately, however, organizational decision processes do not necessarily break down in the face of some bad information, whether it comes from a spreadsheet error or some other source.

REFERENCES

Baker, K.R., Powell, S.G., Lawson, B., & Foster-Johns, L. (2006a). *A survey of spreadsheet users* (in submission).

Baker, K.R., Powell, S.G., Lawson, B., & Foster-Johns, L. (2006b). Comparison of characteristics and practices among spreadsheet users with different levels of experience. Presented at the *European Spreadsheet Risks Interest Group 6th Annual Symposium*, Cambridge.

Chan, Y.E., & Storey, V.C. (1996). The use of spreadsheets in organizations: Determinants and consequences. *Information and Management, 31*(3), 119-134.

Cragg, P.B., & King, M. (1993). Spreadsheet modeling abuse: *An opportunity for O.R J., 44*(8), 743-752.

Clermont, M., Hanin, C., & Mittermier, R. (2002). A spreadsheet auditing tool evaluated in an industrial context. https://143.205.180.128/Publications/pubfiles/pdffiles/2002-0125-MCCH.pdf.

Croll, G.J. (2005). The importance and criticality of spreadsheets in the city of London. Paper presented to the *EuSpRIG Conference*. Available at http://www.eusprig.org/tiacositcol4.pdf.

Cyert, R., & March, J. (1963). *A behavioral theory of the firm*. Englewood Cliffs, NJ: Prentice Hall.

Finlay, P.N., & Wilson, J.M. (2000). A survey of contingency factors affecting the validation of end-use spreadsheet-based decision support systems. *JORS, 51*, 949-958.

Gerson, M., Chien, I.S., & Raval, V. (1992). Computer assisted decision support systems: Their use in strategic decision making. *Proceedings of the 1992 ACM SIGCPR Conference on Computer Personnel Research* held in Cincinnati, Ohio. New York: ACM Press (pp. 152-160).

Grossman, T.A. (2002). Spreadsheet engineering: A research framework. *European Spreadsheet Risks Interest Group 3rd Annual Symposium*, Cardiff . Available at: http://www.usfca.edu/sobam/faculty/grossman_t.html

Grossman, T.A. (2003). Accuracy in spreadsheet modeling systems. *European Spreadsheet Risks Interest Group 4th Annual Symposium*, Dublin.

Grossman, T.A., & Özlük, O. (2004). A paradigm for spreadsheet engineering methodologies. *European Spreadsheet Risks Interest Group 5th Annual Symposium*, Klagenfurt, Austria. Available at: http://www.usfca.edu/sobam/publications/AParadigmforSpreadsheetEngineeringMethodologies2004.pdf.

Hendry, D.G., & Green, T.R.G. (1994). Creating, comprehending and explaining spreadsheets: A cognitive interpretation of what discretionary users think of the spreadsheet model. *International Journal of Human-Computer Studies, 40*, 1033-1065.

Hodges, J. (1991). Six (or so) things you can do with a bad model. *Operations Research, 39*, 355-365.

Howe, H., & Simkin, M.G. (2006). Factors affecting the ability to detect spreadsheet errors. *Decision Sciences Journal of Innovative Education, 4*(1), 101-122.

Isakowitz, T., Schocken, S., & Lucas, Jr., H.C. (1995). Toward a logical/physical theory of spreadsheet modeling. *ACM Transactions on Information Systems, 13*(1), 1-37.

Janvrin, D., & Morrison, J. (2000). Using a structured design approach to reduce risks in end user spreadsheet development. *Information and Management, 37,* 1-12.

Kruck, S.E., Maher, J.J., & Barkhi, R. (2003). Framework for cognitive skill acquisition and spreadsheet training. *Journal of End User Computing, 15*(1), 20-37.

Kruck, S.E., & Sheetz, S.D. (2001). Spreadsheet accuracy theory. *Journal of Information Systems Education, 12*(2), 93-107.

Lipshitz, R., Klein, G., Orasanu, J., & Salas, E. (2001). Taking stock of naturalistic decision making. *Journal of Behavioral Decision Making, 14,* 331-352.

Madahar, M., Cleary, P., & Ball, D. (2007). Categorisation of spreadsheet use within organisations incorporating risk: A progress report. In *Proceedings of the European Spreadsheet Risks Interest Group 8th Annual Conference,* University of Greenwich, London, (pp.37-45).

March, J.G., & Simon, H.A. (1958). *Organizations.* New York: Wiley.

Mather, D. (1999). A framework for building spreadsheet based decision models. *Journals of the Operational Rsearch Society, 50,* 70-74.

Mintzberg, H. (1975, June to August) The manager's job: Folklore and fact. *Harvard Business Review,* 49-61.

Morrison, M., Morrison, J., Melrose, J., & Wilson E.V. (2002). A visual code inspection approach to reduce spreadsheet linking errors. *Journal of End User Computing, 14*(3), 51.

Nardi, B.A. (1993). *A Small matter of programming: Perspectives on end user computing.* Cambridge, MA: MIT Press.

Nardi, BA., &. Miller, J.R. (1991). Twinkling lights and nested loops: Distributed problem solving and spreadsheet development. *International Journal of Man-Machine Studies, 34,* 161-184.

Olphert, C.W., & Wilson, J.M. (2004). Validation of decision-aiding spreadsheets: The influence of contingency factors. *Journal of the Operational Research Society, 55,* 12-22.

Panko, R.R. (1999). Applying code inspection to spreadsheet testing. *Journal of Information Management Systems, 16*(2), 159-176.

Panko, R. (2000a). Two corpuses of spreadsheet error. IEEE. In *Proceedings of the 33rd Hawaii International Conference on Systems Sciences,* Retreived from http://panko.cba.hawaii.edu/ssr/Mypapers/HICSS33-Panko-Corpuses.pdf

Panko, R.R. (2000b). Spreadsheet errors: What we know. What we think we can do. In *Proceedings of the European Spreadsheet Risks Interest Group Conference* , University of Greenwich, London, (pp. 7-17). Retrieved from www.arxiv.org

Panko, R.R. (2005). What we know about spreadsheet errors. Available at: http://panko.cba.hawaii.edu/ssr/Mypapers/whatknow.htm, updated 2005. Previously published in Journal of End User Computing 10(2).

Panko, R.R. (2007) Thinking is bad: Implications of human error research for spreadsheet research and practice. *Proceedings of the European Spreadsheet Risks Interest Group 8th Annual Conference,* pp. 69-80, available at www.arxiv.org.

Powell, S.G., Lawson, B., & Baker, K.R. (2007a). Errors in operational spreadsheets. Under review at *Journal of Organizational and End User Computing.*

Powell, S.G., Lawson, B., & Baker, K.R. (2007b). Impact of errors in operational spreadsheets. In *Proceedings at the European Spreadsheet Risks Int. Grp.,* (pp. 57-68).

Powell, S.G., Lawson, B., & Baker, K.R. (2008a). A critical review of the literature on spreadsheet errors. *Forthcoming in Decision Support Systems.*

Powell, S.G., Lawson, B., & Baker, K.R. (2008b). An auditing protocol for spreadsheet models. Forthcoming in *Information & Management.*

Purser, M., & Chadwick, D. (2006). Does an awareness of differing types of spreadsheet errors aid end-users in identifying spreadsheet errors. *Proceedings of the European Spreadsheet Risk Interest Group Annual Conference,* Cambridge, UK, (pp. 185-204).

Rajalingham, K., Chadwick, D., & Knight, B. (2000). Classification of spreadsheet errors. *British Computer Society (BCS) Computer Audit Specialist Group (CASG) Journal, 10*(4), 5-10.

Rajalingham, K., Chadwick, D., Knight, B., & Edwards, D. (2000, January). Quality control in spreadsheets: A software engineering-based approach to spreadsheet development. *Proceedings of the Thirty-Third Hawaii International Conference on System Sciences,* Maui, Hawaii.

Seal, K., Przasnyski, Z., & Leon, L. (2000, October). A literature survey of spreadsheet based MR/OR applications:1985-1999. *OR Insight, 13*(4), 21-31.

Simon, H.A. (1960). *The new science of management decision.* New York: Harper & Row.

Teo, T.S. H., & Tan, M. (1999). Spreadsheet development and what if analysis: Quantitative versus qualitative errors. *Accounting, Management and Information Technology, 9,* 141-160.

Whittaker, D. (1999). Spreadsheet errors and techniques for finding them. *Management Accounting, 77*(9), 50-51.

Willemain, T.R., Wallace, W.A., Fleischmann, K.R., Waisel, L.B., & Ganaway, S.N. (2003). Bad numbers: Coping with flawed decision support. *Journal of the Operational Research Society, 54,* 949-957.

APPENDIX A: INTERVIEW PROTOCOL (SS = SPREADSHEET)

1. Introduction of researchers and project topic
   - Managers frequently use SS to analyze and inform decisions; research has shown that many of these SS contain errors.
   - This project will investigate how these errors affect the quality of decision making, and propose recommendations on the best ways of reducing these errors.

2. How often do you build SS decision making tools?
   - Do you personally create SS to support decisions?
   - How complex are these SS (in terms of the calculations performed in them, or the amount of data contained in them)?

3. How often do you use SS for making decisions?
   - Does your staff present you with SS or SS-based analysis on a regular basis?
   - How complicated are the SS you encounter (in terms of the calculations performed in them, or the amount of data contained in them)?
   - What decisions are these SS being used to support?
   - What makes SS useful for this decision making?
   - Do you use other quantitative tools for decision making?

4. What is your level of expertise with SS modeling and/or other development environments?
   - Have you had formal training in Excel or software programming/development?
   - Have you ever held a position that required daily creation and manipulation of SS?
   - Which features of Excel are familiar to you?

5. Please describe your experiences with SS that were known to contain errors
   - Do SS errors have the potential to cause significant impact?
   - Was the source of the error(s) ever determined?
   - Were the errors caught before damage was done? If not, what was the extent of damage?
   - Describe the errors and what you think caused them.
   - How were the errors fixed?

6. What are the advantages and disadvantages of using SS for decision making?
   - Are there particular features or tools that you have found most useful in your SS?
   - What are the limitations of SS?
   - Is the quality and reliability of your SS a concern for you?
   - Is there anything that might reduce your concerns?

7. Please describe any processes or tools you have used to ensure the integrity of SS
   - Have you or your staff used Excel's built in tools for error-checking?
   - Have you or your staff used Add-ins provided by another vendor?
   - Does your organization follow a particular development process for creating SS models?
   - What other methods are used to detect errors?

continued on following page

APPENDIX A CONTINUED

8. Are there any other issues related to the topic that you would like to talk about?
    - Do you have advice for other decision-makers?
    - Any stories/anecdotes about particularly helpful solutions to SS problems, or horror-stories about the impact of errors?
    - Recommended readings, web sites, other resources?

Chapter V
A Comparison of the Inhibitors of Hacking vs. Shoplifting

Lixuan Zhang
Augusta State University, USA

Randall Young
The University of Texas-Pan American, USA

Victor Prybutok
University of North Texas, USA

ABSTRACT

The means by which the U.S. justice system attempts to control illegal hacking are practiced under the assumption that hacking is like any other illegal crime. This chapter evaluates this assumption by comparing illegal hacking to shoplifting. Three inhibitors of two illegal behaviors are examined: informal sanction, punishment severity, and punishment certainty. A survey of 136 undergraduate students attending a university and 54 illegal hackers attending the DefCon conference in 2003 was conducted. The results show that both groups perceive a higher level of punishment severity but a lower level of informal sanction for hacking than for shoplifting. Our findings show that hackers perceive a lower level of punishment certainty for hacking than for shoplifting, but students perceive a higher level of punishment certainty for hacking than for shoplifting. The results add to the stream of information security research and provide significant implications for law makers and educators aiming to combat hacking.

INTRODUCTION

Interest in hacking has increased in popularity due to high-profile media coverage of system breaches. In June 2005, the information belonging to 40 million credit card holders was hacked through a credit card processor (Bradner, 2005). In 2006, about 18,000 personal records in the U.S. Department of Veterans' Affairs had been compromised. A recent analysis of compromised electronic data records shows that about 1.9 billion records were reported compromised between 1980 and 2006. This means that for every U.S. adult, nine records have been compromised in aggregate. About 32% of the 1.9 billion comprised records were related to hackers (Erickson and Howard, 2007). Companies are reluctant to publicize that they have experienced information security breaches because of the negative impact such incidents have on their public image leading to loss of market value. Cavusoglu, Mishra and Raghunathan (2004) estimate the loss in market value for organizations to be 2.1% within two days of reporting an Internet security breach which represents an average loss of 1.65 billion.

The rise of computer and Internet use has coincided with an increase in ability of users to commit computer abuses (Parker, 2007) along with an increase in the number of unethical, yet attractive situations faced by computer users (Gattiker and Kelley, 1999). Recently, Freestone and Mitchell (2004) examined the Internet ethics of Generation Y. They found that hacking is considered less wrong than other illegal Internet activities such as "selling counterfeit goods over the Internet." We recognize that illegal hacking activities encompass a wide array of violations of varying degrees of seriousness. For this study, we are not interested in any specific type of illegal hacking but rather illegal hacking activities in general.

Hacking is one of the technologically-enabled crimes (Parker, 2007). Originally the term hacker was a complimentary term that referred to the innovative programmers at MIT who wanted to explore mainframe computing and were motivated by intellectual curiosity and challenges (Chandler, 1996). However, the term became derogatory as computer intruders pursued purposefully destructive actions that caused serious damage for both corporations and individuals. American Heritage Dictionary (2000) defines a hacker as "*one who uses programming skills to gain illegal access to a computer network or file*".

Hacking is a relatively new crime and, as such, is potentially perceived differently from other crimes. Most recently, there has been demand for research which will aid in developing an understanding of how computer crimes differ from more traditional crimes (Rogers, 2001). Due to cost-effectiveness concerns, the chief avenue utilized by the United States government to deter illegal behavior is to increase the severity of punishment (Kahan, 1997). This approach is also used to deter illegal hacking behavior. However, this approach to control illegal hacking is practiced with the assumption that the factors affecting illegal hacking are similar to the factors that influence other types of crime. We set out to evaluate this assumption by comparing illegal hacking activities to shoplifting. The decision to use shoplifting for comparison to illegal hacking was motivated by three reasons:

First, the act of shoplifting is in some ways similar to hacking in that both are acts of illegally obtaining something (i.e. illegal hacking is an act of acquiring access and/or information). Hackers, especially those that are motivated by greed and profit, commit a crime that is analogous to trespassing and taking others' property with the intention of keeping it or selling it. Both of these crimes increase an organization's security costs and overburden the courts. Secondly, the social stigma associated with shoplifting is not as extreme as for crimes like auto theft, burglary of a residence, and money laundering. And as such we believe that there is a higher probability that our target population has heard discussion of shoplifting or knows someone that has engaged in the activity. Thirdly, many people who commit these two crimes are juveniles. According to the statistics of the National Association for Shoplifting Prevention, 25% of the shoplifters are young juveniles and 55% of shoplifters started shoplifting in their teens. Research on hackers also shows that most hackers are between 12 and 28 years old (Rogers, 2001). Therefore, it is relevant to examine the differences in perception between these two crimes.

College students have been identified as a high risk population for supporting hacking activities due to their computer literacy and the general openness of university systems (Hoffer and Straub, 1989). For example, two students at Oxford University hacked into the school computer system to access students' email passwords and other personal information (McCue, 2004). In 2002 and 2003, a former student at the University of Texas named Christopher Andrew Phillips stole more than 37,000 social security numbers that resulted in more than $100,000 worth of damage (Kreytak, 2005).

This accentuates the importance of the computer user in the computer and information security domain. However, several researchers have pointed out the lack of research on antecedents of illegal behavior in the information security domain (James, 1996; Stanton et al., 2003). Therefore, this study attempts to examine some of the factors affecting the act of illegal hacking by answering the following question: How do hackers and students perceive the inhibitors of hacking compared to shoplifting?

The chapter is organized as follows: First, the theoretical foundation of the chapter is discussed and the relevant alternative hypotheses are presented. Next, the research instrument and data collection activities are outlined and results are analyzed and reported. Finally, we summarize the key findings, highlight the implications, discuss the study's limitations and propose future research directions.

THEORETICAL FOUNDATIONS

General deterrence theory is often used to examine crimes, including computer crimes (Workman and Gathegl, 2007; Young et al., 2007). The principal components of the theory are certainty of punishment, severity of punishment and a set of average socioeconomic forces. Among these three components, the first two – certainty and severity of punishment – are the core components of the deterrence theory (Becker, 1968). The theory assumes that criminals are rational individuals. They will not commit crimes if the expected cost is greater than the expected gain. Therefore, any increase in severity of punishment or certainty of punishment will increase the expected cost of committing a certain crime. Since punishment may deprive the criminals of their freedom and social status, individuals may not want to take the chance of being caught breaking the law. The theory posits that severity and certainty of punishment are factors that can serve the purpose of reducing and preventing crimes.

Besides the two components of the general deterrence theory, researchers find informal sanctions to be a significant factor influencing criminal decision-making. In fact, it is claimed that adding the informal sanction construct to the theory is perhaps the most important contribution to deterrence theory (Jacobs et al., 2000). Some researchers even suggest that informal sanctions may be even more salient than formal sanctions (Kahan, 1997; Katyal, 1997; Paternoster and Iovanni, 1986). Informal sanction is defined as social actions of others in response to crime. Some examples of informal sanctions are loss of respect from family and friends (Liu, 2003), social stigma (Grasmick and Scott, 1982) and shame and embarrassment (Blackwell, 2000).

Building on the previous theories, the study examines the three inhibitors of crime: informal sanction, punishment severity and punishment certainty. We assume that any potential offender will consider the three inhibitors before he/she commits deviant behavior. Few studies have examined the perception of these three inhibitors in regards to hacking. In addition, few studies have compared the perception of these three inhibitors between two or more crimes. This study intends to fill the gap by examining these inhibitors with respect to hacking and then compare them to inhibitors associated with shoplifting.

HYPOTHESES DEVELOPMENT

Informal Sanction

Informal sanctions are the actual or perceived responses of others to deviant acts (Liu, 2003). These sanctions can serve to reinforce or weaken people's deviant behavior. Sanctions from close friends or family members are more potent than those from distant relationships (Kitts and Macy, 1999). Informal sanctions are not sufficient to deter crimes, but it puts social pressures on the individuals who intend to engage in deviant acts.

Social views suggest that hacking behavior is less frowned upon than other crimes (Coldwell, 1995). In fact, hackers are viewed by some as talented individuals, and there are incidences of gifted hackers being treated like celebrities. For example, Mark Abene received a one year prison term for his hacking activities, but after his release, a large party was thrown for him at an elite Manhattan club. Also, he was voted one of the top one hundred smartest people in New York by New York magazine. In addition, famous hackers are invited to conferences, and granted interviews along with writers, scientists and film stars (Skorodumova, 2004). Researchers also find that there is a strong sense of peer group support in hacker chat rooms and there is no fear of social disapproval (Freestone and Mitchell, 2004). However, when asked about the perception of shoplifting, both shoplifters and non-shoplifters were negative towards shoplifting believing that it is a somewhat serious offense and is not acceptable behavior (El-Dirghami, 1974). Therefore we propose the following hypotheses:

H1a: Hackers perceive less informal sanction for hacking than for shoplifting.

H1b: Students perceive less informal sanction for hacking than for shoplifting.

Punishment Severity

Severity of punishment refers to the magnitude of the penalty if convicted. It is a core component of deterrence theory (Becker, 1968). Based on the assumption that people engage in criminal and deviant activities if they are not afraid of punishment, deterrence theory focuses on implementing laws and enforcement to make sure that these activities will receive punishment. Deterrence theory proposes that an individual makes the decision to exhibit or not exhibit deviant behavior based on an internal perception of the benefits and costs of the respective behavior. The idea that the cost of committing a crime must exceed the benefit to reduce crime is a staple in the United States (U.S.) criminal justice system (Kahan, 1997).

The U.S. government regards computer crimes as both traditional crimes using new methods and new crimes requiring new legal framework. The Computer Fraud and Abuse Act (CFAA) is the main statutory framework for many computer crimes (Sinrod and Reilly, 2000). According to CFAA, punishment depends upon the seriousness of the criminal activity and the extent of damage. Researchers have argued that the CFAA has been overly punitive (Skilbell, 2003). However, others state that the penalties for computer crimes need to be stiffened (Worthen, 2008).

In the United States, shoplifting is classified as a misdemeanor crime committed against a retail establishment. Punishment for shoplifting varies from state to state. In Georgia, the law calls for misdemeanor punishment for shoplifting goods worth $300 or less. Shoplifting goods worth more than $300 is a felony which can result in punishment of up to ten years in jail. In California, the punishment for shoplifting is more severe. A person entering the store with the intent to shoplift constitutes burglary regardless of the value of the goods that are shoplifted. Any accused shoplifter who has a prior theft conviction will be charged with a felony.

Nevertheless, in most states, shoplifting is not prosecuted heavily. For example, in Texas, the amount in controversy determines the severity of the offense and, therefore, the range of punishment. However, if a person intends to obtain a benefit by breaching computer security, he or she would commit a felony. In addition, researchers find that a large proportion of apprehended shoplifters are never formally changed. Shoplifters with no prior arrests or only one prior arrest are more likely to be dismissed (Adams and Cutshall, 1984). Therefore, we propose the following hypotheses:

H2a: Hackers perceive a higher level of punishment severity for hacking than for shoplifting.

H2b: Students perceive a higher level of punishment severity for hacking than for shoplifting.

Punishment Certainty

Certainty of punishment measures the probability of an individual receiving a legally-imposed penalty. Severity of punishment has little or no effect when the likelihood of punishment is low (Von Hirsch et al., 1999). However, when individuals perceive that there is a greater likelihood that they will get caught in committing a crime, they are less likely to engage in the associated crime.

Because hacking is committed anonymously and from any place in the world, it is not easy to catch an offender. For example, one security officer estimates that the chances of convicting a hacker are at best one in 7,000 and could be as low as one in 600,000 (Worthen, 2008). Besides, many hacking cases are not even revealed to the public or reported to the police. Corporations and government agencies being attacked by computer hackers are reluctant to report the breaches. An FBI survey finds that 90 percent of corporations and agencies detected computer security breaches in 2001 but only 34 percent reported those attacks to authorities (Anonymous, 2002). The reason for this may be the fear of losing consumers' confidence.

Shoplifters also have a low likelihood of getting punished. Researchers find that salesclerks and store security are not effective in deterring shoplifting (El-Dirghami, 1974). According to the statistics from National Association for Shoplifting Prevention, shoplifters are caught an average of only once in every 48 times they steal and they are turned over to police only 50% of the time when they are caught. Therefore, it is unlikely for a shoplifter to get caught and then reported to police. In a study using observation data, researchers estimated that in 2001 around 2,214,000 incidents of shoplifting occurred in a single pharmacy in Atlanta. However, only 25,721 of the shoplifting cases were officially reported to the police (Dabney et al., 2004).

Although some hackers have limited computer hacking ability (script kiddies), the majority of hackers either have a sound level or a high level of computer knowledge (Chantler, 1996). The popular press has described the attitudes of the hackers attending the DefCon conference as: "There is a core of arrogance, of genuine belief that hackers are somehow above not only laws, but people around them, by sheer virtue of intellect" (Ellis and Walsh, 2004). Because they possess unique hacking skills and techniques, hackers may feel safer in conducting hacking than shoplifting. Therefore, we propose the following hypotheses:

H3a: Hackers perceive a lower level of punishment certainty for hacking than for shoplifting.

Prior work supports the contention that students do not believe that store security is an effective deterrent to shoplifting because of the low probability of punishment (El-Dirghami, 1974). However, there is a paucity of studies that examine how students perceive the likelihood of punishment for hacking. Some researchers indicated that students perceived a low level of punishment certainty for other computer crimes, such as software piracy (Peace et al., 2003; Higgins et al., 2005). Since hacking can be conducted anywhere behind a computer screen, we propose the hypothesis below:

H3b: Students perceive a lower level of punishment certainty for hacking than for shoplifting.

METHOD

Instrument Development

Most measures were developed by the authors through literature review. Grasmick and Bryjak (1980) discussed at length the various means of measuring punishment severity and certainty of punishment along with strengths and weaknesses. In accordance with their suggestions we chose survey items that ask for the respondent's estimate of punishment severity in general terms and avoid asking about specific penalties (i.e. prison time, fines, etc.) . Specific penalties such as fines may be viewed as severe to a financially-insecure individual but viewed inconsequential to a wealthy individual. The items measuring informal sanction are adapted from the measures in Liu's (2003) manuscript which includes disapproval from families and friends. The items were measured using a 5 point Likert scale with anchors ranging from 5 which represents "Strongly agree" to 1 which represents "Strongly disagree". Appendix 1 shows the items in the survey. The hackers and students first answered the questions about shoplifting and then about hacking.

Content Validity

Content validity is based on the extent to which a measurement reflects the specific intended domain of content (Carmine and Zeller, 1979). The constructs that we used are relatively newer constructs although references supported their development. To ensure that we are measuring what we intend to measure, a literature review was conducted to identify, select and phrase the items to measure these constructs. This activity was followed with a panel of subject-matter-experts that were asked to indicate whether or not an item in a set of measurement items is "essential" to the operationalization of a theoretical construct.

Specifically, two scholars who are familiar with research in criminology were interviewed and asked for input on the constructs, their measurement domains, and the appropriateness of the measures we selected from the prior studies. As a result of this input, some items were modified to make them easier to understand.

Sample

Data was collected from 136 undergraduate students attending a university located in the southern United States. All students were enrolled in a lower level MIS course. The majority of the respondents have a high degree of computer literacy. Correspondingly, it was presumed that they are reasonably apt to understand hacking and their responses confirmed this presumption. Table 1 shows the profiles of the student sample.

Data was also collected through handout surveys distributed to participants of the 2003 DefCon hacker convention in Las Vegas, the largest annual computer hacker convention. The majority of the attendees are hackers or people who have an interest in hacking activities. Participation in the study was strictly voluntary. The handout survey was considered, by the researchers, an ideal approach to the study because of the adequacy of the respondents and the higher response rates

Table 1. Profile of the students

Gender	
Male	66
Female	60
Missing	10
Marriage status	
Single	90
Married	34
Divorced	1
Widowed	1
Missing	10
Current university classification	
Freshman	1
Sophomore	2
Junior	57
Senior	65
Missing	11
Number of students who shoplifted last year	8
Number of students who hacked illegally last year	1

and satisfaction scores associated with handout surveys compared to mail surveys (Gribble and Haupt, 2005). Additionally, all responses were anonymous and for purposes of this study all data are reported in aggregate. A total of 155 surveys were collected during the three day conference. Twenty eight surveys are deemed unusable as the respondents either answered every question the same or failed to answer the majority of the questions. Therefore, the usable responses from DefCon were 127.

Since not all attendees of the DefCon conference are hackers, to ensure that we only examine people that have committed illegal hacking acts, the respondents were asked if they had participated in a hacking activity that would be considered outside the bounds of that allowed by the court system within the past year. The answer for the question is worded in a yes or no format. 54 respondents answered yes for the question. Although it is a conservative estimate since we only inquired about illegal hacking activity within the past year, we can be assured that the 54 respondents are truly active illegal hackers. Therefore, data from the 54 hackers were used for the following analysis. Table 2 shows the profile of the 54 hackers. All hackers are single males.

DATA ANALYSIS AND RESULTS

Principal components factor analysis with an oblique rotation was used to assess construct validity for items measuring perceptions of inhibitors of shoplifting and of hacking. SPSS was used to perform factor analysis on our instrument based on data from 54 hackers and from DefCon and 136 students. Factor analysis shows six factors. Factor loadings above 0.5 on one construct and no cross-loadings over 0.4 provide evidence

Table 2. Profile of the illegal hackers

Gender	
Male	54
Female	0
Marriage status	
Single	54
Married	0
Divorced	0
Widowed	0
Number of hackers who shoplifted last year	6

Table 3. Factor Loadings of hackers' and students' perception on shoplifting and hacking

	HKinformal	**SLcertainty**	**SLinformal**	**HKseverity**	**SLseverity**	**HKcertainty**
HKinformal2	**0.855**	-0.004	0.278	0.285	0.264	-0.121
HKinformal1	**0.823**	0.106	0.320	0.175	0.286	-0.300
SLcertainty2	0.077	**0.883**	0.149	0.121	0.179	-0.401
SLcertainty3	-0.049	**0.799**	-0.218	0.222	-0.029	-0.323
SLinformal1	0.370	0.048	**0.891**	0.139	0.174	-0.264
SLinformal2	0.340	-0.052	**0.849**	0.203	0.132	0.028
HKseverity1	0.047	0.100	0.181	**0.879**	0.054	-0.145
HKseverity2	0.517	0.239	0.110	**0.798**	0.333	0.001
HKseverity3	0.506	0.355	0.172	**0.750**	0.381	-0.081
SLseverity3	0.185	0.091	0.133	0.171	**0.926**	-0.135
SLseverity2	0.286	0.173	0.198	0.200	**0.909**	-0.090
SLseverity1	-0.106	0.104	0.271	0.301	**0.632**	-0.507
HKcertainty3	0.027	0.323	0.024	0.071	0.102	**-0.879**
HKcertainty2	0.216	0.305	0.203	0.108	0.172	**-0.757**
Variance explained	31.51%	15.65%	12.04%	10.15%	6.36%	5.29%
Cronbach's alpha	0.870	0.600	0.700	0.748	0.827	0.686

Note:

- *HKinformal: informal sanction for hacking*
- *HKseverity: punishment severity for hacking*
- *HKcertainty: punishment certainty for hacking*
- *SLinformal: informal sanction for shoplifting*
- *SLseverity: punishment severity for shoplifting*
- *SLcertainty: punishment certainty for shoplifting*

See Appendix for survey items

Table 4. Comparison of perception between shoplifting and hacking

		Informal Sanction	Punishment Severity	Punishment Certainty
Hackers	Hacking	1.86	4.84	2.23
	Shoplifting	2.99	3.67	3.92
Students	Hacking	3.84	3.79	3.64
	Shoplifting	4.28	3.31	3.22

Note: 1- strongly disagree and 5- strongly agree

of convergent and discriminant validity (Campbell and Fiske, 1959). Table 3 provides the factor loadings for shoplifting and for hacking. Internal reliability was assessed by using Cronbach's alpha. As shown in Table 3, it varied from 0.600 to 0.870. According to Hair et al (1998), 0.60 is satisfactory for exploratory studies. Other IS researchers have similar reliability estimates in their exploratory studies (Ma et al., 2005). Table 4 shows the means of these constructs for hackers and students.

Repeated measures analysis of variance was used to test hypotheses H1a, H2a, and H3a using student data and the same statistical technique was used to test H1b, H2b and H3b using hacker data. Repeated analysis is appropriate when measurements are taken on the same unit of analysis. In our case, each individual answered questions about their perception of the consequences of shoplifting, and then again answered those about perception related to those of hacking. Therefore, multiple measurements are used on the same unit of analysis.

Hypothesis 1a proposed that hackers perceive less informal sanction from hacking than from shoplifting. Repeated measures ANOVA supported this proposition (Within subject $F=142.03$, $p<0.01$), showing that there is a significant difference regarding hackers' perception of informal sanctions associated with hacking and shoplifting. Since Table 4 shows that informal sanction for shoplifting is higher than that of hacking, H1a is supported.

Hypothesis 1b posited that students perceive less informal sanctions associated with being caught hacking than from shoplifting. Tests supported this proposition, showing that students did have a significantly different perception of informal sanctions regarding hacking and shoplifting (Within subjects $F=19.489$, $p<0.01$). Table 4 shows that the informal sanction for shoplifting is higher than hacking. Therefore, our data supports H1b.

Hypothesis 2a stated that hackers perceive a higher level of punishment severity from hacking than from shoplifting. Table 4 shows that hackers perceive a higher level of punishment severity from hacking than from shoplifting. The repeated measures ANOVA test shows that the difference is significant (Within subject $F=78.51$, $p<0.01$) and confirms this hypothesis.

Hypothesis 2b proposed that students perceive a higher level of punishment severity for hacking than for shoplifting. Table 4 shows that the students indeed perceive a higher level of punishment severity for hacking when compared to shoplifting. The repeated measures ANOVA test shows that the difference is significant (Within subjects $F=23.446$, $p<0.01$). Therefore, our data supports H2b.

Hypothesis 3a proposed that hackers perceive punishment certainty from hacking will be lower than shoplifting. Table 4 shows that their estimated possibility of getting caught from hacking is indeed lower than shoplifting. The repeated measures ANOVA test shows that the difference

Table 5. Results of hypotheses testing

Hypotheses	p-value	Results
H1a: Hackers perceive less informal sanction for hacking than for shoplifting.	p<0.01	Supported
H1b: Students perceive less informal sanction for hacking than for shoplifting.	p<0.01	Supported
H2a: Hackers perceive a higher level of punishment severity for hacking than for shoplifting.	p<0.01	Supported
H2b: Students perceive a higher level of punishment severity for hacking than for shoplifting.	p<0.01	Supported
H3a: Hackers perceive a lower level of punishment certainty for hacking than for shoplifting.	p<0.01	Supported
H3b: Students perceive a lower level of punishment certainty for hacking than for shoplifting.	p>0.1	Rejected

is significant (within subject $F=190.06$, $p<0.01$) and confirms this hypothesis.

Hypothesis 3b stated that students perceive a lower punishment certainty for hacking than for shoplifting. From Table 4, we can see that students perceive a higher possibility of punishment for hacking than for shoplifting. The repeated measures ANOVA test shows that the difference is significant (Within subjects $F=18.95$, $p<0.01$). Therefore, hypothesis H3b is rejected. The results of the hypothesis testing are shown in Table 5.

DISCUSSION AND IMPLICATION

In summary, this study intends to address one important question: whether or not students and hackers have different perceptions of inhibitors against hacking compared to inhibitors against shoplifting. The results of the study show that both hackers and students have a lower informal sanction for hacking than for shoplifting. In other words, they believe that a greater social pressure is associated with shoplifting than hacking. Among the student group, although the respondents regard hacking as fairly unacceptable, it is still more acceptable than shoplifting.

Ultimately the success of an organization's information security program is dependent on the behavior of the users (Stanton et al., 2003). As this population (student and hacker) accepts corporate and government job responsibilities, it is imperative for the host organization to recognize the lack of social pressure working to discourage hacking activities. Research within the information systems domain, oftentimes, assumes that a single culture (organizational culture) impacts individual behavior effectively ignoring ethical and social factors (Straub et al., 2002). Research must evaluate the impact of these other factors on the organization's information security risk.

Many researchers in the criminology discipline suggest that the key deterrent effect lies in the threat to expose individuals to social disapproval (Kahan, 1997; Katyal, 1997). The academic community must do more to change the perception that illegal hacking is more socially acceptable than other crimes like shoplifting. Illegal hacking is not only illegal but the penalties are far more severe. One conviction can have a severe impact on an individual's future career goals. There are indications that issues of ethics, trust, integrity, and responsibility will become more crucial to organizations striving to protect themselves from information security risks (Dhillon and Backhouse, 2000) and students with a circumspect past that include illegal hacking will likely face a difficult prospect when it comes to landing a job. The academic community, largely, has failed to address ethical issues associated with computer and information use (Couger et al., 1995). The academic community should be proactive toward shaping an ethical student population that will one day accept corporate and government job responsibilities. As the ethical standards increase we may see social pressure to conform increase as well.

Both hackers and students perceive that hacking receives more severe punishment than shoplifting. Because of the fear and unknown doubt of high-tech crimes such as hacking, the U.S. government has aggressively prosecuted criminal hackers. According to the National Association of Criminal Defense Lawyers, computer crimes are punished more harshly compared to other crimes. These findings suggest that the U.S. government has been successful in communicating the serious negative consequence of hacking to the public.

Our findings show that hackers and students have different perceptions about punishment certainty. Hackers perceive a lower punishment certainty for hacking, while students perceive a lower punishment certainty for shoplifting. One possible explanation for these differing perceptions is the students may have little understanding about how hacking activities are committed and how little monitoring is done on the Internet. Students are possibly biased due to media publications of potential government monitoring of network activities (i.e. Carnivore) while hackers understand the difficulty of monitoring the extreme volume of network activities that exist on the Internet or even a corporate LAN. Another possible reason is that hackers may be very confident of their hacking skills while students have much less computer skills.

The lower probability of punishment may work to diminish the effect of severe punishment. Researchers have found that the punishment certainty has far greater deterrent effect on crimes than punishment severity (Von Hirsch et al., 1999). A criminal is more likely to engage in criminal activity despite the threat of severe punishment when he or she believes that there is only a small or no chance of being caught. This suggests that security enforcers should make investments that will improve their ability to detect illegal hacking activities and assist in criminal prosecution. For example, the government may need to allocate more of the computer security budget to hiring competent security and law enforcement personnel, as well as to increase employee training in computer security monitoring and investigation.

LIMITATION AND FUTURE DIRECTION

One limitation of the study is the restriction imposed by the sample. While students enrolled in a lower-level MIS course represent those intending to major in any of the business disciplines, care must be taken in generalizing to students in other majors and young adults not attending college. Future research needs to investigate the generalizabilty of these measures to other majors, as well as non-students in a similar age group. Of particular interest is to examine the perception of students majoring in computer science. These students have a comparatively higher level of computer knowledge than other students and may be more likely to be hackers than other students. Another limitation of the research is that more valid measures need to be developed. Since this is an exploratory study, most of the items were developed by the authors.

Researchers have called for the need for more end-user oriented research examining IT security issues (Troutt, 2002). The present study offers important insights about hackers' and students' awareness of ethics and moral issues associated with computer technology in comparison with the traditional crime of shoplifting. This study advances our knowledge of hackers' and students' moral judgment of inhibitors relevant to hacking and shoplifting. These preliminary results suggest that law enforcement officials and educators should consider the means that will strengthen informal sanctions associated with illegal hacking.

REFERENCES

Adams, S., & Cutshall, C.R. (1987). Refusing to prosecute minor offenses: The relative influence of legal and extralegal factors. *Justice Quarterly, 4,* 595-609.

Anonymous. (2002). Many hack attacks go unreported. Retrieved March, 3, 2008,from http://www.cbsnews.com/stories/2002/04/07/tech/main505563.shtml

Becker, G.S. (1968). Crime and punishment: An economic approach. *Journal of Political Economy, 78*(2), 169-127.

Blackwell, B.S. (2000). Perceived sanction threats, sex, and crime: A test and elaboration of power-control theory. *Criminology, 38*(2), 439-488.

Bradner, S. (2005). The winner so Far: Cardsystems solutions. *Network World, 22*(25), 30.

Campbell, D. T., & Fiske, D. W. (1959). Convergent and discriminant validation by the multitrait-multimethod matrix. *Psychological Bulletin, 56*(1), 81-105.

Carmine, E. G., & Zeller, R.A. (1979). *Reliability and Validity Assessment.* Newbury Park, CA: Sage Publications.

Cavusoglu, H., Mishra, B., & Raghunathan, S. (2004). The effect of Internet security breach announcements on market value: Capital market reactions for breached firms and Internet security developers. *International Journal of Electronic Commerce, 9*(1), 69-104.

Chandler, A. (1996). The changing definition and image of hackers in popular discourse. *International Journal of the Sociology of Law, 24,* 229-251.

Chantler, N.(1996). Risk: The profile of the computer hacker. Ph.D. dissertation, Curtin University of Technology, Perth, Western Australia.

Coldwell, R. (1995). Australian attitudes toward legal intervention into hacking. *Communications of the ACM, 38*(11),115-117.

Couger, J, D., Davis, G.B., Dologite, D G., & Feinstein, D. L. (1995). IS' 95: Guidelines for undergraduate IS curriculum. *MIS Quarterly, 19*(3), 341-359.

Dabney, D.A., Hollinger, R.C., & Dugan, L. (2004). Who actually steals? A study of covertly observed shoplifters. *Justice Quarterly, 21*(4), 694-728.

Dhillon, G., & Backhouse, J. (2000). Information system security management in the new millennium. *Communications of the ACM, 43*(7), 125-128.

El-Dirghami, A. (1974). Shoplifting among students. *Journal of Retailing, 50*(3), 33-42.

Ellis, P., & Walsh, P. (2004). Hacking the planet: DefCon harbors sex, geeks and rock 'n' roll in one sleepless package, Retrieved March 3, 2008, from http://www.defcon.org/html/links/dc_press/archives/12/citylife_hackingtheplanet.htm.

Erickson, K., & Howard, P.N. (2007). A case of mistaken identity? News accounts of hacker, consumer, and organizational responsibility for compromised digital records. *Journal of Computer-Mediated Communication, 12,* 1229-1247.

Freestone, O., & Mitchell, V-W. (2004). Generation Y attitudes towards E-Ethics and Internet-related misbehavior. *Journal of Business Ethics, 54,* 121-128.

Gattiker, U.E., & Kelley, H. (1999). Morality and computers: Attitudes and differences in judgments. *Information Systems Research, 10*(3), 233-254.

Grasmick, H., & Bryjak, G. (1980). The deterrent effect of perceived severity of punishment. *Social Forces,* 59(2), 471-491.

Gribble, R.K., & Haupt, C. (2005). Quantitative and qualitative differences between handout and mailed patient satisfaction surveys. *Medical Care, 43*, 276-281.

Grasmick, H.G., & Scott, W.J. (1982). Tax evasion and mechanisms of social control: A comparison with grand and petty theft. *Journal of Economic Psychology, 2*, 213-230.

Hair, J.E., Anderson, R.E., Tatham, R.L., & Black,W.C. (1998). *Multivariate data analysis with readings*. Englewood Cliffs: Prentice Hall.

Higgins, G.E., Wilson, A.L., & Fell, B. D. (2005). An application of deterrence theory to software piracy. *Journal of Criminal Justice and Popular Culture, 12*(3), 166-184.

Hoffer, J.A., & Straub, D.W. (1989). The 9 to 5 underground: Are you policing computer crimes? *Sloan Management Review, 30*(4), 35-43.

Jacobs, B.A., Topalli, V., & Wright, R. (2000). Managing retaliation: Drug robbery and informal sanction threats. *Criminology, 38*(1), 171-197.

James, H.L. (1996). Managing information systems security: A soft approach. *Proceedings of the Information Systems Conference of New Zealand*, 10-20.

Kahan, D. (1997). Between economics and sociology: The new path of deterrence. *Michigan Law Review, 95*(8), 2477-2498.

Katyal, N. (1997). Deterrence's difficulty, *Michigan Law Review, 95*(8), 2385-2476.

Kitts, J.A., & Macy, M.W. (1999). Structural learning: Attraction and conformity in task-oriented groups. *Computational & Mathematical Organization Theory, 5*(2), 129-145.

Kreytak, S. (2005, June 7). Hacker at UT was harmless, his lawyer says as trial starts. *Austin American-Statesman*..

Liu, R.X. (2003). The moderating effects of internal and perceived external sanction threats on the relationship between deviant peer associations and criminal offending. *Western Criminology Review, 4*(3), 101-202.

Ma, Q., Pearson, M.J., & Tadisina, S. (2005). An exploratory study into factors of service quality for application service provider. *Information and Management, 43*, 1067-1080

McCue, A. (2004). *Oxford student "hackers" suspended.* Retrieved March 3, 2008 from http://software.silicon.com/security/0,39024655,39125462,00.htm

Parker, D. B. (2007). The dark side of computing: SRI International and the study of computer crime. *IEEE Annals of the History of Computing, 29*(1), 3-15

Paternoster, R., & Iovanni, L. (1986). The deterrent effort of perceived severity: a reexamination. *Social Forces, 64*, 751-777.

Peace, A.G., Galletta, D., & Thong, J.Y.L (2003). Software piracy in the workplace: A model and empirical test. *Journal of Management Information Systems,* 20(1), 153-177

Rogers, M. (2001). *A social learning theory and moral disengagement analysis of criminal behavior: An exploratory study.* Ph.D. Thesis, Dept. of Psychology, University of Manitoba, Winnipeg.

Skibell, R. (2003). Cybercrimes & Misdemeanors: A reevaluation of the computer fraud and abuse act. *Berkeley Technology Law Journal,* 18(3), 909-944

Skorodumova, O. (2004). Hackers as information space phenomenon. *Social Sciences, 35*(4), 105-113.

Sinrod, E.J., & Reilly, W.P. (2000). Hacking your way to hard time: application of computer crime laws to specific types of hacking attacks. *Journal of Internet Laws, 4*(3), 1-14.

Stanton, J.M, Guzman, I., Stam, K.R., & Caldera, C. (2003). Examining the linkage between organizational commitment and information security. *International Conference Systems, Man, and Cybernetics, 3*, 2501-2506.

Straub, D, Loch, K, Evaristo, R, Karahanna, E., & Strite, M. (2002). Towards a theory-based measurement of culture. *Journal of Global Information Management, 10*(1), 13-23.

Troutt, M. D. (2002). IT security issues: The need for end user oriented research. *Journal of Organizational and End User Computing, 14*(2), 48-49.

Von Hirsch, A., Bottoms, A.E., Burney, E., & Wickstrom, P.O. (1999). *Criminal deterrence and sentence severity*. Hart, Oxford.

Young, R., Zhang, L., & Prybutok, V.R. (2007). Hacking into the minds of hackers. *Information Systems Management, 24*(4), 281-287.

Workman, M., & Gathegl, J. (2007). Punishment and ethics deterrents: A study of insider security contravention. *Journal of the American Society and Technology, 58*(2), 212-222.

Worthen, B. (2008) Laws go soft on hackers. *Wall Street Journal,* Feb 22nd, Retrieved February 25, 2008, from http://blogs.wsj.com/biztech/2008/02/22/laws-go-soft-on-hackers/

APPENDIX

Shoplifting

Informal Sanction

SLinformal1: My friends would think less of me if I were caught shoplifting.
SLinformal2: My family would think less of me if I were caught shoplifting.

Punishment severity

SLseverity1: The punishment for shoplifting is severe.
SLseverity2: If you were caught shoplifting, your life would be severely disrupted.
SLseverity3: If you were caught shoplifting, it would have a detrimental effect on your life.

Punishment certainty

SLcertainty1: People who shoplift are caught eventually.
SLcertainty2: If you were to shoplifting, the chances of you being caught are small *
SLcertainty3: The chance that an average person being caught shoplifting is small. *

Hacking

Informal Sanction

HKinformal1: My friends would thinks less of me if I were caught hacking illegally.
HKinformal2: My family would think less of me if I were hacking illegally.

Punishment severity

HKseverity1: The punishment for being caught illegally hacking is severe.
HKseverity2: If you were caught hacking illegally, your life would be severely disrupted.
HKseverity3: If you were caught hacking illegally, it would have a detrimental effect on your future.

Punishment certainty

HKcertainty1: People who hack illegally would be caught eventually.
HKcertainty2: If you were to hack illegally, the chances you would be caught are small. *
HKcertainty3: If other people were to hack illegally, the chances they would be caught are small. *

* *Items are reverse coded.*

Chapter VI
Developing Success Measure for Staff Portal Implementation

Dewi Rooslani Tojib
Monash University, Australia

Ly Fie Sugianto
Monash University, Australia

ABSTRACT

The last decade has seen the proliferation of business-to-employee (B2E) portals as integrated, efficient, and user-friendly technology platform to assist employees to increase their productivity, as well as for organizations to reduce their operating costs. To date, very few studies have focused on determining the extent to which the portal implementations have been successful. Such a study is crucial, considering that organizations have committed large investments to implementing the portals and they would certainly like to see the return on their investments. Our study aims to develop a scale for measuring user satisfaction with B2E portals. The four steps of scale development: conceptual model development, item generation, content validation, and an exploratory study, are reported in this chapter. Evidence about reliability, content validity, criterion-related validity, convergent validity, and discriminant validity is presented.

MOTIVATION FOR DEVELOPING MEASUREMENT

Business-to-employee (B2E) portals have been widely implemented across industries in the past decade (Cedar Crestone Survey, 2007). This system, more commonly known as *staff portal*, integrates a number of applications such as e-mail, reporting tools, employee self service (ESS) and manager self service (MSS), and presents them in a customized and personalized interface to the users. More importantly, users can access the portal at any time and anywhere from their desktops, notebooks, or even personal digital assistants (PDAs) through the Internet connection. The flexibility offered by the portals allows them to be the desktop medium through which employees can perform their work-related and personal-related tasks. B2E portals have also met with the approval of a number of organizations such as IBM, Toshiba, HP, General Motors, and Ford, to name a few. This is attributed to their perceived potential benefits, namely, improved corporate communication, increased competitive advantage, increased employee productivity, and reduced organizational costs (Sugianto & Tojib, 2006). In particular, the self-service model introduced by the portal is very attractive, providing convenience for the employees, and cost savings for the organizations.

Despite these substantial benefits, there are still some drawbacks associated with portal implementation, particularly when B2E portals promote a new way of working and communication among employees, which can directly influence employees' work practices. For instance, the automation of business processes may discourage employees who are used to a paper-based culture to use the portal, as they will have less time for meeting and socially interacting with their colleagues in person. Moreover, the ability to access the portal at any time anywhere may imply that a stronger work commitment is demanded of the employees. All these concerns may inhibit the portal usage within the workplace, which in turn will influence the success of portal implementation. Thus, an accurate success measure is required to assist organizations in assessing the value of such portal implementation.

There have been a number of approaches for measuring the success of Information Systems (IS) such as system quality (Srinivasan, 1985), information quality (Mahmood & Medewitz, 1985), system use (Igbaria, Pavri & Huff, 1989), individual impact (Bergeron, 1986), organizational impact (Jenster, 1987), and user satisfaction (Baroudi & Orlikowsky, 1988). Having reviewed these different methods of measuring IS success, the two most frequently used are user satisfaction and system use (Galletta & Lederer, 1989). However, since the use of B2E portal is usually not mandatory, measuring B2E portal success through system use may be of limited value. Hence, in our study, we adopted the user satisfaction concept when developing a perceptual measure of B2E portal success.

Extensive research has been done in the past on the measurement of user satisfaction. Since the 1980s, considerable conceptual and empirical studies have been devoted to establishing a standard user satisfaction scale. The scales of Bailey and Pearson (1983) and Doll and Torkzadeh (1988) are those most frequently adopted or adapted when measuring user satisfaction with IS applications (Wang, Tang & Tang, 2001). The former was initially developed to measure general user information satisfaction for the traditional data processing (TDP) environment, while the latter was developed to measure user satisfaction with specific application for the end user computing (EUC) environment. Apart from these scales, previous researchers have developed user satisfaction scales for different types of applications such as enterprise resource planning (ERP) systems (Calisir & Calisir, 2004), knowledge management systems (Ong & Lai, 2004), and asynchronous electronic learning systems (Wang, 2003).

Our review of the literature that addresses user satisfaction measurement showed that studies which specifically examine user satisfaction with the B2E portal are virtually non-existent. The existing user satisfaction scales in the IS field are not entirely appropriate for measuring user satisfaction with the B2E portal. Firstly, both the Bailey and Pearson (1983) and Ives et al. (1983) scales were designed to measure general user satisfaction in the TDP environment. These scales tend to overlook environments in which end users have less direct interaction with the IS staff (Jones & Beatty, 2001). Moreover, Doll and Torkzadeh (1988) argued that, if users have limited interaction with the IS staff, those scales are not appropriate. The B2E portal users directly interact with the portal without the intervention of IS staff, thus, neither the Bailey and Pearson (1983) nor the Ives et al. (1983) scale should be adopted for the B2E portal environment. Secondly, another standard measure of user satisfaction, the Doll and Torkzadeh (1988) scale, was designed to measure user satisfaction with specific IT applications in the EUC environment. Previous researchers who adopted this scale to measure user satisfaction with web-based IS found that the scale is still valid; however, certain modifications are needed to adapt to the web-based environment. Xiao & Dasgupta (2002) suggested adding some factors such as privacy and security that could be relevant to the web-based environment, while Abdinnour-Helm, Chaparro & Farmer (2005) suggested refining the Timeliness factor for future use. Therefore, adopting Doll and Torkzadeh's (1988) scale for the B2E portal is also inappropriate. The third reason that other user satisfaction scales (for example, Huang, Jin, Yang & Chiu, 2004; Muylle, Moenaert, Despontin, 2004) do not suit the B2E portal environment has to do with the B2E portal technology. Embedded within such portals are technologies with functionalities that are distinct from those employed within the EUC or TDP environment such as search and retrieval processes, work flow systems, online self service applications, and collaboration tools (Tojib, 2003). These reasons led us to conclude that user satisfaction with the B2E portal should be examined using a different scale from any proposed in the literature. Hence, there is a need to develop a multidimensional scale measuring user satisfaction with the B2E portal.

This chapter aims to explore the development of a measure of user satisfaction with B2E portals. To ensure the reliability and validity of the scale, this study follows four sequential studies, namely: conceptual model development, item generation, content validation, and exploratory study. The preliminary test of the scale psychometric properties was also conducted and reported.

THE B2EPUS CONCEPTUAL MODEL DEVELOPMENT

Domain Identification

Prior to establishing the conceptual model, we first defined user satisfaction with the B2E portal. In our study, we adapted the Doll and Torkzadeh (1988) definition, considering that the way in which the users interact with the portal is similar to the way users interact with a specific application in the EUC environment. Thus, user satisfaction with the B2E portal is then defined as 'an affective attitude towards the B2E portal by employee who interacts with the portal directly'.

We then conducted a review of the existing user satisfaction scales to identify the domains of the B2EPUS construct. Two widely accepted scales for measuring user satisfaction in IS research, the Bailey and Pearson (1983) and Doll and Torkzadeh (1988) scales, were carefully examined. In addition, other scales measuring user satisfaction with specific IT applications (Chen, Soliman, Mao, & Frolick., 2000; Huang et al., 2004; Wang, 2003; Cho and Park, 2001; Ong and Lai, 2004; Muylle et al., 2004) and websites (Loiacono, Watson, Goodhue, 2002; Yoo and Donthu, 2001; Yang, Cai, Zhou & Zhou, 2005), were also considered.

As shown in Table 1, a range of dimensions contributing to user satisfaction with general IS, certain types of IT applications, as well as end user perception of the quality of websites and web portals, were identified and grouped into three main categories: information quality (IQ), system quality (SQ), and system design quality (SDQ). This categorisation process was based on the current literature, which emphasises these three categories when measuring user satisfaction construct (Bailey & Pearson, 1983; Doll &

Torkzadeh, 1988; Yoo & Donthu, 2001; Muylle et al., 2004). Other dimensions, which do not fall into these three categories, were excluded from further investigation.

Since there were still far too many possible dimensions of the B2EPUS construct, we introduced two additional criteria to identify dimensions exclusive to the investigated domain. Firstly, the dimensions must have been used in most measures of user satisfaction with various types of IS and IT applications. We believe that dimensions which have conceptual and empirical relevance to most general user satisfaction scales can be appropriately included in the B2E portal user satisfaction domain. Secondly, the dimensions must be theoretically associated with the B2E portal environment. We examined the features and contents of B2E portals offered by both large and small portal vendors and found that they have identical primary characteristics, namely, the portal must: be accessible at any time whenever there is an Internet connection, incorporate a single log-on procedure, provide role-based content to each employee, enable employees to do more tasks electronically with the integration of self-service applications, and act as a medium of communication between the organization and its employees as well as employees and their colleagues. Thus, these major characteristics were considered when investigating dimensions of the B2E portal user satisfaction.

Careful examination of the above criteria and characteristics resulted in the identification of nine dimensions of the B2E portal user satisfaction: *information content, ease of use, convenience of access, timeliness, efficiency, security, confidentiality, communication, and layout.* It should be noted that these dimensions (as depicted in Figure 1) are an initial identification of the B2EPUS construct and they will be subjected to exploratory study.

Figure 1. An evolution of the B2EPUS dimensions

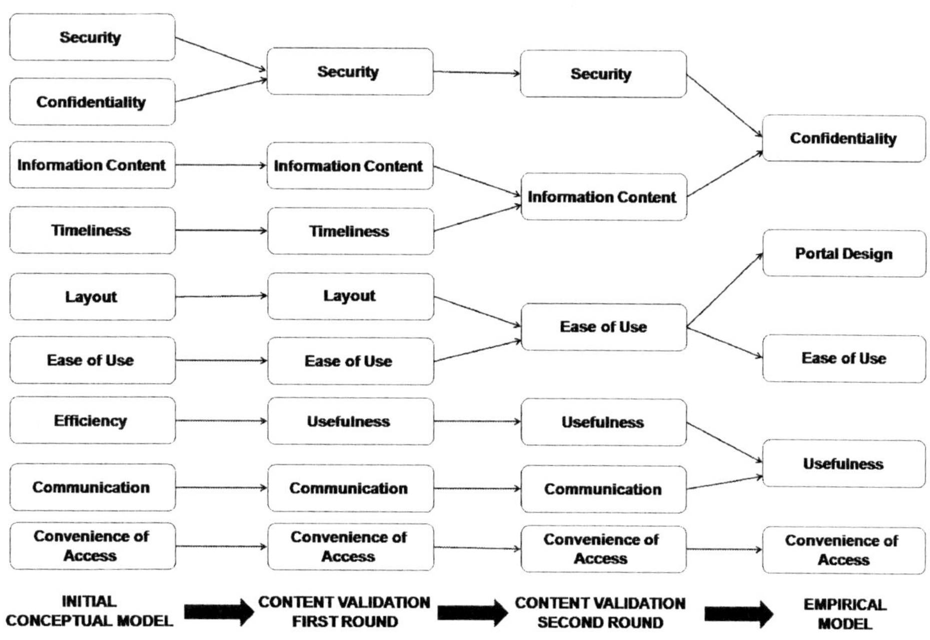

Table 1. Dimensions derived from existing user satisfaction scales

No	Group	Factors
INFORMATION QUALITY		
1	Information Content	Content (Doll & Torkzadeh, 1988), Knowledge content (Ong & Lai, 2004), Usefulness of content & Adequacy of information (Yang et al, 2005), System output (Nath, 1989), Accuracy, format, preciseness (Chen et al, 2000), Information (Huang et al, 2004 ; Bailey & Pearson, 1983), Reliability, Currency, Completeness & Relevancy (Bailey & Pearson, 1983), Information comprehensiveness, Information relevancy & Information comprehensibility (Muylle et al, 2004)
2	Format of output	Format of output (Bailey & Pearson, 1983), Format (Doll & Torkzadeh, 1988)
3	Volume of output	Volume of output (Bailey & Pearson, 1983)
SYSTEM QUALITY		
4	System accuracy	Accuracy (Huang et.al, 2004; Bailey & Pearson, 1983; Doll & Torkzadeh, 1988)
5	Timeliness	Timeliness (Doll & Torkzadeh,1988; Bailey & Pearson, 1983), Response/turnaround time (Bailey & Pearson, 1983), Website speed (Muylle et al, 2004), Processing speed (Yoo & Donthu, 2001), Response time (Loiacono et al, 2002)
6	Ease of use	Ease of use (Doll & Torkzadeh, 1988; Cho & Park, 2001; Muylle et al,2004; Yoo & Donthu, 2001), Learner interface (Wang, 2003), Ease of understanding (Loiacono et al,2002), Intuitive operation (Loiacono et al,2002), Usability (Chin, Diehl, & Norman, 1988; Yang et al, 2005), Knowledge manipulation (Ong & Lai, 2004)
7	Flexibility	Flexibility (Bailey & Pearson, 1983)
8	Usefulness	Information fit to task (Loiacono et al,2002), Perceived utility (Bailey & Pearson, 1983), Fulfillment of end users needs (Chen et al, 2000)
9	Convenience	Convenience (Huang et al, 2004), Convenience of access (Bailey & Pearson, 1983), Accessibility (Yang et al, 2005)
10	Security	Security (Huang et al, 2004), Security of data (Bailey & Pearson, 1983), Security (Yoo & Donthu, 2001), Trust (Loiacono, Watson & Goodhue, 2002)
11	System efficiency	System efficiency (Nath, 1989)
12	System capabilities	System capabilities (Chin, Diehl & Norman, 1988), System characteristics (Nath, 1989)
13	Personalisation	Personalisation (Wang, 2003; Ong & Lai, 2004)
14	Knowledge map	Knowledge map (Ong & Lai, 2004)
15	Communication	Learner community (Wang, 2003), Knowledge community (Ong & Lai, 2004), Interaction (Yang et al, 2005)
16	Hyperlink connotation	Hyperlink connotation (Muylle et al, 2004)
17	Language customisation	Language customization (Muylle et al, 2004)
18	Tailored communication	Tailored communications (Loiacono et al,2002)
SYSTEM DESIGN QUALITY		
19	Layout	Interface (Huang et al, 2004), Screen (Chin et al,1988), Site design (Cho & Park, 2001), Entry guidance, Website structure & Layout (Muylle et al,2004), Aesthetic design (Yoo & Donthu, 2001), Visual appeal (Loiacono et al 2002)
20	Innovativeness	Innovativeness (Loiacono et al,2002)

B2E Portal User Satisfaction (B2EPUS) Dimensions

Confidentiality. This dimension was adopted from the *Security* dimension introduced by Bailey and Pearson (1983), Huang et al., (2004), Loiacono et al., (2002), and Yoo and Donthu (2001). In the case of the B2E portal, the integration of Employee Self Service (ESS) applications allows employees to submit or retrieve their personal information electronically. Hence, privacy or confidentiality issues have been a serious concern within the online B2E portal environment (Yang et al., 2005). The ability of the portal to maintain confidentiality is likely to be associated with user satisfaction. In this study, the term *Confidentiality* rather than *Security* is examined from a different angle. The dimension *Confidentiality* is defined as the ability of the portal to provide a sense of assurance that any personal information retrieved or submitted from and through the portal will not be misused by authorised people.

Security. While past studies on Internet security refer to the misappropriation or unauthorised alteration or loss of data (Bailey & Pearson, 1983), the security of personal and financial information (Yoo & Donthu, 2001), and information privacy (Loiacono et al., 2002), the *Security* dimension in this study is intended to cover a more encompassing aspect: that is, the ability of the portal to provide secure access to all the available applications and facilities. Hence, this dimension includes issues relating to security breaches, such as data theft, which increases in proportion to the number of organizations storing their personnel files electronically (Zviran & Erlich, 2003).

Information content. This dimension was derived from the *Information Content* dimension, which has been typically measured in previous studies in terms of its accuracy, relevancy, currency, reliability (Doll & Torkzadeh, 1988; Bailey & Pearson, 1983; Ong & Lai, 2004; Yang et al., 2005; Chen et al., 2000; Muylle et al., 2004). In this study, *Information Content* is referred to as the relevancy, accuracy, currency, and reliability of information presented to each employee based on his/her role in the organisation.

Timeliness. This dimension was derived from Bailey and Pearson (1983), Muylle et al. (2004), and Yoo and Donthu (2001). They all affirmed response time or processing speed as an indicator of user satisfaction with IT applications. Following their research, dimension *Timeliness* in this study attempts to measure the ability of the portal to deliver requested information within a reasonable response time.

Layout. This dimension was derived from Chin, Diehl and Norman (1988), Cho and Park (2001), Huang et al. (2004), and Loiacono et al. (2002). They confirmed the importance of system design quality in measuring user satisfaction with IT applications. In measuring this dimension, they have focused on the look of the interface design in terms of colour schemes, fonts, images, and background. In this study, the *Layout* dimension measures the aforementioned aspects as well as design clarity, attractiveness, and appealing designs.

Ease of use. This dimension measures the user friendliness of the system. It has been considered an important aspect of user satisfaction with various types of systems and applications in the previous research (Cho & Park, 2001; Doll & Torkzadeh, 1988; Harrison & Rainer, 1996; Muylle et al., 2004; Wang, 2003; Yang et al., 2005). Similar to previous research, the dimension *Ease of Use* in the current study also measures user friendliness of the system, which includes ease of navigation, the training issue, feeling of being in control, and learnability.

Efficiency. Previous research found that user satisfaction with a system is likely to increase when users believe that using a system will increase their performance and productivity (Mawhinney & Lederer, 1990; Vlahos & Ferratt, 1995). This view is also supported by Bhattacherjee (2001) who reported that user satisfaction is influenced by the system's perceived usefulness. In this study, the term *Efficiency* was used as opposed to Usefulness as the emphasis of this dimension goes beyond the extent to which the portal can assist employees in performing their work. This dimension aims to assess the extent to which the portal is useful to them in terms of bringing efficiency/productivity to their work.

Communication. Previous research conducted by Wang (2003), Ong and Lai (2004), and Yang et al. (2005) found that the ability of the system to enable system users to interact with others may influence user satisfaction. Similar to their research, the dimension of *Communication* in this study attempts to measure the extent to which the portal can mediate interaction, information sharing, or collaborating, between employees and the organizations as well as employees and their colleagues.

Convenience of access. This dimension was derived from Huang et al. (2004) and Yang et al. (2005) who asserted that the ability to access a portal at all times is an important dimension of user satisfaction. The B2E portal clearly provides convenience for employees, particularly those who spend more time out of the office during working hours or those who work remotely from home. They could remain updated with the organization's news and at the same time perform their work-related and personal tasks by accessing the portal. Thus, in this study, *Convenience of Access* refers not only to the ability of the portal to be accessed at any time and anywhere, but also the accessibility of the portal through different types of media.

Item Generation

The nine dimensions of the B2EPUS construct were subsequently used as a basis to generate items. At least three items were generated for each dimension since a multiple-item scale tends to be more reliable (Allen & Yen, 1979) and the repetition of similar items in different ways decreases the chance of a random answer (Galletta & Lederer, 1989). A pool of 47 items (available from the authors) was generated, refined and then arranged in a suitable sequence for the purpose of content validation.

CONTENT VALIDATION PROCESS

The main objective of examining the content validity of the scale is to ensure that only the best items are retained. The content validity process generally involves asking a specific number of experts to evaluate the validity of individual items and the instrument as a whole. In our study, we utilised the panel selection criteria of Grant and Kinney (1992) and Grant and Davis (1997).

First Round Content Experts' Judgement Process

We sent invitations to 34 potential candidates (for expert panel) who have been actively conducting research on portals as well as members of EDUCAUSE Web Portals Constituent Group. Although 16 experts agreed to participate, only 11 experts returned their feedback in time. Furthermore, only 6 responses of those returned were fit for further analysis.

The expert panel expresses their judgement on the content validity using the 5-point Likert scale (1 represents 'Extremely Unimportant', 2 represents 'Somewhat Unimportant', 3 represents 'Neutral', 4 represents 'Somewhat Important', and 5 represents 'Extremely Important'). The expert panel rates all 9 dimensions as being important. The mean values indicating the degree of importance of all 9 dimensions range from 3.67 to 3.84.

Following Fehring's (1987) method as part of the content validation, the expert panel also categorised each item into a dimension and ranked the relevancy of the item to the assigned dimension using a 5-point rating scale (1 represents 'not at all', 2 represents ' very little', 3 represents 'somewhat', 4 represents 'well' and 5 represents 'very well'). As per Lynn (1986), it was concluded that 15 out of 47 items were incorrectly assigned to the nominated dimension. Thus, these items were excluded from the scale. Fehring's (1987) weighted ratios were also calculated to confirm the deletion of these 15 items.

When reviewing the remaining items, we noticed that there was only one item left to assess the dimension *Timeliness* and three items remained to assess the dimension *Information Content*. Hence, five more items were added to the scale to avoid unreliability issues (Hinkin & Schriesheim, 1989). Other suggestions from the experts were also accommodated, such as combining *Confidentiality* to *Security*, and renaming *Efficiency* to *Usefulness*. Since the scale had undergone significant changes during the content validity phase, we conducted another round of content validity.

Second Round Content Experts' Jjudgement Process

In the second round, personalized email invitations were sent out to 9 potential candidates to sit as an expert panel. Although 6 experts agreed to participate, only 3 responses were fit for further analysis.

Following the feedback from the first round content experts, we simplified the task in the second round process. As per Lynn's (1986) method, the content experts were presented with the eight dimensions along with their individual definitions and corresponding items. They were required only to evaluate the extent to which each item is relevant to the assigned factor using a 4-point rating scale (1 represents 'irrelevant', 2 represents 'somewhat relevant if phrasing is profoundly adjusted', 3 represents 'relevant with some adjustment as to phrasing', and 4 represents 'very relevant'). We then calculated the Content Validity Index (CVI) value for each item, which was determined by the proportion of experts who rated it as content valid (a rating of 3 or 4). As there were only three responses, all three had to give a rating of 3 or 4 in order to retain the item. It was found that three items were considered irrelevant (a rating of 1) by the content experts and thus were removed from the scale. We also calculated the CVI value for the entire scale, which is determined by the proportion of total items judged content valid (a rating of 3 or 4). There were 34 items rated 3 or 4 by the content experts. The CVI value for the entire scale was 91.89 %, which clearly exceeded the expected minimum CVI of 0.80 for a new scale (Davis, 1992).

Further item analysis was performed to ensure that unnecessary or redundant items were excluded from the scale. During this process, six more items were eliminated. Other suggestions from the experts were also accommodated, such as combining *Layout* to *Ease of Use*, and combining *Timeliness* to *Information Content*. As a result, twenty-eight items measuring dimension *Usefulness*, *Communication*, *Convenience of Access*, *Ease of Use*, *Information Content*, and *Security* were retained (see Figure 1) and used for the subsequent exploratory study.

EXPLORATORY STUDY

Data Collection: Sample and Procedure

The exploratory study was conducted in the Australian higher education sector considering that portal is an important and increasingly used tool in academic institutions (Deans & Allmen, 2002) and that there has been increasing scholarly attention in higher education literature on the subject of portal use (Katz, 2002). Ten out of thirty-eight Australian universities indicated that they have built and implemented B2E portals for more than one year. Five of these ten universities agreed to participate in the study.

Since it was not possible to have direct access to the employees' contact details, a non-probability convenience sampling method was adopted. To ensure the anonymity and confidentiality of the respondents, an email invitation was broadcasted by the University contact person informing staff about our study and the location of the online survey. Respondents were mainly asked to rate the extent to which they agree with the items prepared to measure their attitudes toward certain aspects of the staff portal they were using at the time of the study. A seven-point Likert-type scale ranging from *Strongly disagree* (1) to *Strongly agree* (7) was chosen to increase the response variability and to minimise a ceiling effect (Zimet, Dahlem, Zimet, Farley, 1988) as well as to reach the upper limit of reliability (Nunnally, 1978). Furthermore, the items were not assigned to their associated dimensions or factors, but were placed at random to improve reliability and validity (Bagozzi and Baumgartner, 1994).

Three hundred and two responses were collected at this stage. These responses were then randomly split into two parts: 145 cases were used for the exploratory study and the remaining 157 cases were used for the confirmatory study (reported in Sugianto, Tojib, and Burstein, 2007). This split-sample method is a common approach utilised by IS researchers as can be seen from the work of Moore and Benbasat (1991), Basselilier et al., (2003), and Yang et al., (2005). Sample size for both studies is considered adequate as it meets the parameter estimate ratio of 5 as suggested in Bentler (1995).

Identifying the Factor Structure of the B2EPUS Construct

To be able to identify the underlying dimensions of the B2EPUS construct, we performed Exploratory Factor Analysis. The Bartlett's test of sphericity (significant at $p<0.01$) and the Kaiser-Meyer-Olkin test (KMO= 0.867) indicated that there was sufficient minimum sample size and that the data were appropriate for factor analysis (Pallant, 2005). The 145 cases were then subjected to a principal component factor analysis. Promax was chosen as the rotation method since the data demonstrated high correlations among the extracted factors. Items were deleted if they did not load on any factors and they were cross-loaded on more than one factor. Two commonly employed decision rules were applied to identify the factors underlying the construct, that is, an Eigenvalue of one as the cut-off value for extraction (Tabachnick & Fidell, 2001), and factor loadings of less than 0.45 on all factors (Comrey & Lee, 1992). In addition, a scree plot was carefully examined.

The iterative sequence of factor analysis and item deletion was repeated, resulting in a final scale of 22 items belonging to five distinct factors associated with the B2EPUS construct. These factors accounted for 73.63 % of the variance. Table 2 summarises the factor loadings for the condensed 22-item scale. The significant loading of all the items on the single factor indicates unidimensionality. The fact that no item had multiple cross loading was found to support the preliminary discriminant validity of the scale.

The results of the factor analysis revealed a different pattern from the expected conceptual model previously discussed. However, nearly all factors hypothesized in the conceptual model appeared in the EFA (see Figure 1). Thorough explanations of the extracted five factors are given below.

The results showed that items initially generated to measure the dimension *Communication* were grouped together with items associated with

Table 2. Factor loading of the 22 final items

	Factor					Corrected Item-to-Total Correlation
	1	2	3	4	5	
Usefulness						
Sharing or exchanging information with team member colleagues	**0.937**					0.61
Facilitating collaboration with all colleagues	**0.916**					0.62
Discussing work issues with work colleagues	**0.908**					0.58
Performing more work electronically	**0.792**					0.52
Streamlining work processes	**0.767**					0.62
Sharing information within the whole organization	**0.683**					0.58
Confidentiality						
Certainty of appropriate use for submitted information		**0.871**				0.55
Confidentiality of submitted information		**0.839**				0.59
Trustable retrieved information		**0.793**				0.61
Dependable retrieved information		**0.777**				0.65
Security of portal		**0.745**				0.41
Provided reliable information		**0.643**				0.58
Ease of Use						
Self-explanatory use			**0.934**			0.42
Navigatibility			**0.788**			0.62
Learnability			**0.770**			0.63
Feeling of being in control			**0.629**			0.64
Convenience of Access						
Accessible from home				**0.892**		0.39
Gaining access easily				**0.723**		0.64
Accessible 24/7				**0.693**		0.56
Portal Design						
Aesthetic design					**0.954**	0.68
Attractive design					**0.949**	0.65
Availability of help functions and useful buttons and links					**0.519**	0.75
Corrected Cronbach's Alpha	0.917	0.893	0.838	0.809	0.886	
Eigenvalue	8.918	3.071	2.014	1.163	1.032	
Cumulative variance explained (%)	40.536	54.494	63.650	68.937	73.628	

the dimension *Usefulness* in Factor One. Careful investigation of these six items found that they measured a similar concept from two slightly different points of view: the ability of the portal to assist users in performing their tasks at work, and the provision of communication facilities. It was then decided to rename Factor One *Usefulness.*

The EFA results also indicated that items belonging to the *Security* factor were grouped together with three items associated with the dimension *Information Content* factor as Factor Two. An interpretation of these items revealed that they tapped a similar concern (i.e., the ability of the portal to provide a sense of assurance that any information submitted would remain confidential and any information retrieved would be trustworthy). Hence, Factor Two was then renamed *Confidentiality.*

Next, the EFA results also indicated that items conceptually associated with the dimension *Ease of Use* were divided into two factors: Factor Three and Factor Five. All of the items placed in Factor Three were clearly measuring the user friendliness of the staff portal. Thus, Factor Three retained its original name, *Ease of Use.* The remaining items placed in Factor Five measured similar characteristics, i.e. the design of the portal. Hence, Factor Five was named *Portal Design.* All of the items generated to measure *Convenience of Access* were loaded in Factor Four, retaining the original name, *Convenience of Access.*

ASSESSMENT OF RELIABILITY AND VALIDITY

Reliability

The Cronbach alpha was calculated to measure the reliability of the new scale. The 22-item scale had a reliability of 0.928, exceeding the minimum standard of 0.80 suggested for basic research (Nunnally, 1978). The reliability of each dimension was as follows: Factor *Usefulness* = 0.917; Factor *Confidentiality* = 0.893; Factor *Ease of Use* = 0.838; Factor *Convenience of Access* = 0.809; Factor *Portal Design* = 0.886. An investigation of corrected item-to-total for each of these 22 items revealed that only one item had a corrected item-to-total value below the cut-off value of 0.40 (Wang, 2003). It was decided to retain the item for two reasons. Firstly, its corrected item-to-total value is very close to the cut off value and its loading to the respective factor is high (see Table 1). Secondly, a minimum of three items is required to measure one factor (Cook et al., 1981).

Concurrent Validity

The aim of measuring concurrent validity in the current study is to explore whether the B2EPUS scale is indeed a measure of user satisfaction. In other words, concurrent validity measures the extent to which performance on the criterion measure can be estimated using the new scale (Martuza, 1977). Hence, in order to investigate a proper concurrent validity assessment, we need a criterion measure that is widely used so that can be regarded as a valid measure. In the current study, we adopted the criterion measure from Doll and Torkzadeh (1988). The criterion measure consists of two items namely *'In general, I am satisfied with the staff portal'* and *'In general, the staff portal is successful'* which measure overall user satisfaction and are thus appropriate for correlation with the B2EPUS scale. A positive relationship was expected between the total B2EPUS score and the valid criterion if the B2EPUS scale were to be capable of measuring user satisfaction construct. The finding indicated a positive correlation between the two scales ($r=0.793$, $n=145$, $p<0.01$). The high value of correlation indicates a strong relationship between these two scales and thus concurrent validity of the B2EPUS scale was achieved.

Convergent and Discriminant Validity

Following correlation matrix approach introduced by Doll and Torkzadeh (1988), we conducted the convergent and discriminant validity assessment. A scale is considered to have convergent validity when each item is highly correlated with other items designed to measure the same factor. Establishing whether the scale has convergent validity using this approach requires that the correlation between the items measuring the same dimension should be significantly different from zero and large enough to warrant further investigation. Appendix 1 shows the correlation matrix of the new scale. The smallest within-item correlations for the various items are *Usefulness* = 0.420, *Confidentiality* = 0.427, *Ease of Use* = 0.476, *Convenience of Access* = 0.497, *Portal Design* = 0.645. These correlations are significantly different from zero at $p < 0.05$ level and large enough to proceed with discriminant validity analysis.

Discriminant validity is indicated by low correlations between the measure of interest and other measures supposedly not measuring the same concept (Churchill, 1979). It is tested by counting the number of times that the item correlates higher with items of other factors than with items of its own theoretical factor (Doll & Torkzadeh, 1988). To claim that a scale has discriminant validity, Campbell and Fiske (1959) suggest that the count should be less than 50% of the potential comparisons. An examination of the correlation matrix in Appendix 1 reveals that there are 36 violations of the discriminant validity condition from 384 comparisons. The number of violations is clearly below 50% of the potential comparisons and hence, the discriminant validity of the new scale is supported.

DISCUSSIONS & CONCLUSION

After a series of refinements, the final 22-item B2E portal user satisfaction scale was demonstrated to have a good correspondence between the assignment of items to the dimensions and the factor structure and high internal consistency reliability for each of the dimensions. The empirical data also supported its content validity, concurrent validity, convergent and discriminant validity.

Figure 2 shows a comparison of the underlying dimensions between the most widely used satisfaction scales, the UIS (Ives, et al., 1983) and the EUCS (Doll & Torkzadeh, 1988) and the recently developed B2EPUS scale. It clearly shows that the UIS and EUCS scales are not entirely appropriate for measuring user satisfaction with web-based systems since there are other factors specifically associated with the web-based environment that were not captured in the earlier scales. *Confidentiality* and *Convenience of Access* are examples of user satisfaction dimensions, which are unique to the web-based systems.

The rigorous validation procedure allows the development of a scale for measuring user satisfaction with B2E portal. Nevertheless, this research contains limitations that could be addressed in future studies. Firstly, respondents to this research study were derived mainly from staff in the Australian higher educational sector and consequently, caution needs to be taken to generalise the findings. Secondly, this research study was conducted at one particular time only. Test-retest reliability investigation should be considered in future research to examine the stability of the construct over time. The evidence of the validity of the B2EPUS scale established in the current study should be treated as preliminary evidence that needs to be repeatedly investigated in other research contexts. That is, there exists a need for further validation of the scale. In fact, we have performed Confirmatory Factor Analysis to complete the research cycle of a scale development study. The findings strongly support the reliability

Figure 2. Comparison of underlying dimensions between UIS, EUCS, and B2EPUS

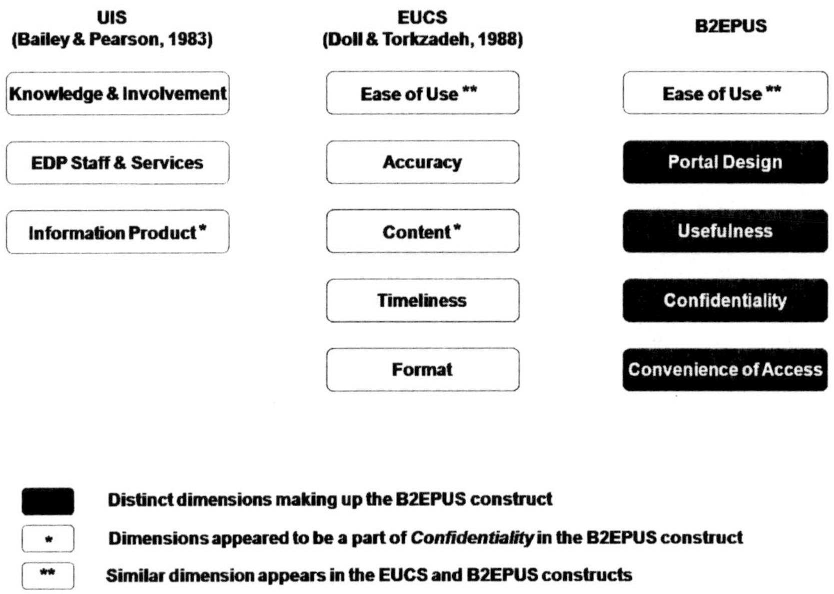

and validity of the B2EPUS scale (see Sugianto, Tojib, & Burstein, 2007 for details). Lastly, cross-cultural validation using different large samples from different industries would help to confirm and refine the current findings.

To the best of our knowledge, this was the first study that aimed to operationalise the construct of B2E portal user satisfaction into conceptually distinct indicators, which can be observed and assessed. The findings provide a better understanding of the multi-dimensionality of the B2E portal user satisfaction construct, which comprises five empirically distinguishable dimensions. In practice, the 22-item, five-factor scale is an accessible and easily administered measure of B2E portal user satisfaction, which can be used, in an organizational setting for various purposes. It can be used to compare user satisfaction with specific components (that is, *Convenience of Access*, *Usefulness*, *Ease of Use*, *Portal Design*, and *Confidentiality*) of the B2E portal systems. The scale can be adapted or supplemented to fit the specific research or practical needs of a particular environment. To conclude, considering the thoroughly developed B2EPUS scale has been proven to be reliable and valid, organizations may now use the scale to measure the extent to which their B2E portals deliver the intended benefits, offer insights on how to improve the portals, and help promote overall employee satisfaction and productivity.

REFERENCES

Abdinnour-Helm, S.F., Chaparro, B.S., & Farmer, S.M. (2005). Using the end user computing satisfaction instrument to measure satisfaction with a Web site. *Decision Sciences*, *36*(2), 341-363.

Allen, M.J. & Yen, W.M. (1979). *Introduction to measurement theory*. Monterey, CA: Brooks/Cole.

Bagozzi, R. P., & Baumgartner, H. (1994). The evaluation of structural equation models and hypothesis testing. In R.P. Bagozzi. (Ed.), *Principles of Marketing Research* (pp. 386-422). Cambridge: Blackwell.

Bailey, J. E., & Pearson, S. W. (1983). Development of a tool for measurement and analysing computer user satisfaction. *Management Science, 29*(5), 530-545.

Basselilier, G., Benbasat, I., & Reich, B.H. (2003). The Influence of business managers' IT competence on championing IT. *Information Systems Research, 14*(4), 317-336.

Bentler, P. (1995). *EQS structural equations program Manual*, Encino, CA: Multivariate Software.

Bergeron, F. (1986). Factors influencing the use of DP charge back information. *MIS Quarterly, 10*(3), 225-238.

Bhattacherjee, A. (2001). Understanding information systems continuance: An expectation-confirmation model. *MIS Quarterly, 25*(3), 351-370.

Calisir, F., & Calisir, F. (2004). The relation of interface usability characteristics, perceived usefulness, and perceived ease of use to end-user satisfaction with enterprise resource planning (ERP) systems. *Computers in Human Behaviour, 20*, 505-515.

Campbell, D.T., & Fiske, D.W. (1959). Convergent and discriminant validation by the multitrait-multimethod matrix. *Psychological Bulletin, 56*(2), 81-105.

Cedar Crestone (2007). *Cedar Crestone 2007-2008 HR Systems Survey: HR Technologies, Service Delivery Approaches, and Metrics 10th Annual Edition*, Cedar Crestone Inc.

Chen, L., Soliman, K. S., Mao, E., & Frolick, M. N. (2000). Measuring user satisfaction with data warehouses: An exploratory study. *Information & Management, 37*(3), 103-110.

Cho, N., & Park, S. (2001). Development of electronic commerce user-consumer satisfaction index (ECUSI) for Internet shopping. *Industrial Management & Data Systems, 101*(8), 400-405.

Churchill, G. A. (1979). A paradigm for developing better measures of marketing constructs. *Journal of Marketing Research, 19*(4), 491-504.

Comrey, A. L., & Lee, H. B. (1992). *A first course in factor analysis*. New Jersey: Lawrence Erlbaum.

Cook, J.D., Hepworth, S.J., Wall, T.D., & Warr, P.B. (1981). *The experience of work*. San Diego: Academic Press.

Davis, L. (1992). Instrument review: Getting the most from your panel of experts. *Applied Nursing Research, 5*, 104-107.

Deans, K. R., & Allmen, S. (2002). Poo poo portals at your peril. *Proceedings of the Eighth Australian World Wide Web Conference* (pp. 117-135). Queensland, Australia.

Doll, W. J., & Torkzadeh, G. (1988). The measurement of end-user computing satisfaction. *MIS Quarterly, 12*(2), 259-275.

Fehring, R. J. (1987). Methods to validate nursing diagnoses. *Heart and Lung, 16*(6), 625-629.

Galletta, D.F., & Lederer, A.L. (1989). Some cautions on the measurement of user information satisfaction. *Decision Sciences. 20*(3), 419-438.

Grant, J.S., & Davis, L.L. (1997). Focus on quantitative methods: Selection and use of content experts for instrument development. *Research in Nursing & Health, 20*, 269-274.

Grant, J., & Kinney, M. (1992). Using the delphi technique to examine the content validity of nursing diagnoses. *Nursing Diagnosis, 3*, 12-22.

Harrison, A.W., & Rainer Jr., R.K. (1996). A general measure of user computing satisfaction. *Computers in Human Behavior, 12*(1), 79-92.

Hinkin, T.R., & Schriesheim, C.A. (1989). Development and application of new scales to measure the French and Raven (1959) bases of social power. *Journal of Applied Psychology, 74*(4), 561-567.

Huang, J. H., Yang, C., Jin, B. H., & Chiu, H. (2004). Measuring satisfaction with business-to-employee systems. *Computers in Human Behaviour, 20*, 17-35.

Igbaria, M., Pavri, F.N., & Huff, S.L. (1989). Microcomputer applications: An empirical look at usage. *Information & Management, 16*(4), 187-196.

Ives, B., Olson, M. H., & Baroudi, J. J. (1983). The measurement of user information satisfaction. *Communications of the ACM, 26*(10), 785-793.

Jenster, P.V. (1987). Firm performance and monitoring of critical success factors in different strategic contexts. *Journal of Management Information Systems, 3*(3), 17-33.

Jones, M.C., & Beatty, R.C. (2001). User satisfaction with EDI: An empirical investigation. *Information Resources Management Journal, 14*(2), 17-26.

Katz, R. N. (2002). *Web portals and higher education: technologies to make IT personal.* San Francisco: Jossey-Bass.

Loiacono, E. T., Watson, R.T., & Goodhue, D.L. (2002). WebQual: A measure of Web site quality. In Evans, K.K. and Scheer, L.K. (Eds). *Marketing Theory and Applications, 13.* Chicago: American Marketing Association.

Lynn, M. (1986). Determination and quantification of content validity. *Nursing Research, 35*(6), 82-385.

Mahmood, Mo. A., & Medewitz, J.N. (1985). Impact of design methods on decision support systems success: An empirical assessment, *Information & Management, 9*(3), 137-151.

Martuza, V.R. (1977). *Applying norm-referenced and criterion-referenced measurement in education.* Boston: Allyn and Bacon, Inc.

Mawhinney, C. H., & Lederer, A. L. (1990). A study of personal computer utilization by managers. *Information & Management, 18*(5), 243-253.

Moore, G.C., & Benbasat, I. (1991). Development of an instrument to measure the perceptions of adopting an information technology innovation. *Information Systems Research, 2*(3), 192-222.

Muylle, S., Moenert, R., & Despontin, M.(2004). The conceptualization and empirical validation of Web site user satisfaction. *Information & Management, 41*, 543-560.

Nath, R. (1989). Are frequent users more satisfied. *Information Processing and Management, 25*(5), 557-562.

Nunnally, J. C. (1978). *Psychometric Theory.* (2nd ed.). New York: McGraw-Hill.

Ong, C. S., & Lai, J. Y. (2004). Developing an instrument for measuring user satisfaction with knowledge management systems. In *Proceedings of the 37th Hawaii International Conference on System Sciences*, Hawaii.

Pallant, J. (2005). *SPSS Surviving Manual.* (2nd Ed.) New South Wales: Allen & Unwin.

Srinivasan, A. (1985). Alternative measures of system effectiveness: associations and implications. *MIS Quarterly, 9*(3), 243-253.

Sugianto, L., & Tojib, D.R. (2006). Information at your fingertips: B2E portal as a strategic tool for todays workforce. In A Cox (Ed.) *Portals: people, processes, and technology*, London: Facet Publishing.

Sugianto, L., Tojib, D.R., & Burstein, F. (2007). A practical measure of employee satisfaction with B2E portals. In *Proceedings of the 28th International Conference on Information Systems*, Montreal, Canada.

Tabachnick, B. G., & Fidell, L. S. (2001). *Using multivariate statistics* (4th ed.). Boston: Allyn & Bacon.

Tojib, D. R. (2003). *The use of business to employee (B2E) portal in workplace: A preliminary investigation*. Unpublished honours thesis, Monash University, Victoria, Australia.

Vlahos, G. E., & Ferratt, T. W. (1995). Information technology use by managers in Greece to support decision making: amount, perceived value, and satisfaction. *Information & Management, 29*, 305-315.

Wang, Y. S. (2003). Assessment of learner satisfaction with asynchronous electronic learning systems. *Information & Management, 41*, 75-86.

Wang, Y-S., Tang, T-I., & Tang, J.E. (2001). An instrument for measuring customer satisfaction toward websites that market digital products and services. *Journal of Electronic Commerce Research, 2*(3), 89-102.

Xiao, L., & Dasgupta, S. (2002). Measurement of user satisfaction with Web-based IS: An empirical study. *Proceedings of the Eight Americas Conference on Information Systems* (pp. 1149-1156).

Yang, Z., Cai, S., Zhou, Z., & Zhou, N. (2005). Development and validation of an instrument to measure user perceived service quality of information presenting Web portals. *Information & Management, 42*, 575-589.

Yoo, B. & Donthu, N. (2001). Developing a scale to measure the perceived quality of an Internet shopping site (SiteQual). *Quarterly Journal of Electronic Commerce, 2*(1), 31-45.

Zimet, D., Dahlem, N.W., Zimet, S.G., & Farley, G.K. (1988). The multidimensional scale of perceived social support. *Journal of Personality Assessment, 52*(1), 30-41.

Zviran, M., & Erlich, Z. (2003). Measuring IS user satisfaction: Review & implications. *Communications of the Association for Information Systems, 12*, 81-103.

APPENDIX 1. CORRELATION MATRIX OF THE 22- ITEM SCALE

	UF1	UF2	UF3	UF4	UF5	UF6	CF1	CF2	CF3	CF4	CF5	CF6	EU1	EU2	EU3	EU4	CA1	CA2	CA3	DS1	DS2	DS3
UF1	1.00																					
UF2	0.87	1.00																				
UF3	0.86	0.88	1.00																			
UF4	0.56	0.58	0.50	1.00																		
UF5	0.65	0.67	0.61	0.68	1.00																	
UF6	0.65	0.63	0.64	0.42	0.53	1.00																
CF1	0.22	0.21	0.24	0.33	0.33	0.20	1.00															
CF2	0.19	0.22	0.22	0.39	0.33	0.22	0.86	1.00														
CF3	0.21	0.20	0.11	0.30	0.29	0.16	0.60	0.57	1.00													
CF4	0.24	0.22	0.16	0.31	0.33	0.24	0.61	0.60	0.92	1.00												
CF5	0.06	0.08	0.06	0.31	0.23	0.01	0.49	0.49	0.46	0.47	1.00											
CF6	0.27	0.25	0.18	0.34	0.37	0.22	0.47	0.47	0.65	0.66	0.43	1.00										
EU1	0.19	0.15	0.18	0.07	0.10	0.16	0.23	0.19	0.28	0.28	0.13	0.21	1.00									
EU2	0.40	0.33	0.33	0.28	0.30	0.35	0.21	0.22	0.32	0.36	0.22	0.36	0.51	1.00								
EU3	0.33	0.34	0.32	0.21	0.24	0.40	0.38	0.40	0.44	0.48	0.24	0.33	0.57	0.54	1.00							
EU4	0.21	0.18	0.15	0.24	0.25	0.21	0.40	0.47	0.48	0.54	0.47	0.52	0.48	0.69	0.60	1.00						
CA1	0.06	0.11	0.10	0.12	0.06	0.31	0.20	0.30	0.26	0.30	0.20	0.19	0.26	0.31	0.35	0.43	1.00					
CA2	0.23	0.22	0.25	0.21	0.32	0.40	0.37	0.46	0.37	0.45	0.32	0.41	0.42	0.48	0.51	0.65	0.60	1.00				
CA3	0.30	0.30	0.32	0.13	0.33	0.43	0.22	0.29	0.29	0.31	0.18	0.35	0.42	0.42	0.39	0.42	0.50	0.66	1.00			
DS1	0.43	0.45	0.43	0.28	0.46	0.42	0.34	0.41	0.43	0.41	0.31	0.38	0.27	0.47	0.38	0.39	0.21	0.40	0.43	1.00		
DS2	0.36	0.46	0.43	0.24	0.42	0.40	0.31	0.39	0.40	0.42	0.27	0.35	0.30	0.47	0.38	0.42	0.27	0.42	0.39	0.83	1.00	
DS3	0.51	0.53	0.49	0.37	0.45	0.52	0.30	0.34	0.48	0.53	0.22	0.47	0.41	0.60	0.58	0.47	0.30	0.45	0.45	0.69	0.65	1.00

Chapter VII
Contingencies in the KMS Design:
A Tentative Design Model

Peter Baloh
University of Ljubljana, Slovenia

ABSTRACT

Improving how knowledge is leveraged in organizations for improved business performance is currently considered as a major organizational change. Knowledge management (KM) projects are stigmatized as demanding, fuzzy, and complex, with questionable outcomes—more than 70% of them do not deliver what they promised. While most organizations have deployed knowledge management systems (KMSs), only a handful have been able to leverage these investments. Existing knowledge management (KM) research offered valuable insights on how to introduce KMSs in a sense of innovation-diffusion, yet little guidance has been offered to KMS developers who need to decide on functionalities of a tool they are to introduce in particular organizational setting. The goal of this chapter is to propose theoretical background for design of KMS that successfully support and enable new knowledge creation and existing knowledge utilization. By using principles of the design science, design profiles proposed build upon works from organization and IS sciences, primarily the Evolutionary Information-Processing Theory of Knowledge Creation (Li & Kettinger, 2006) and the Task Technology Fit Theory (Zigurs & Buckland, 1998), the latter being amended for particularities of the KM environment. Proposed fit profiles suggest that one-size-fits-all approaches do not work and that organizations must take, in contrast with extant literature, a segmented approach to KM activities and fitting technological support.

INTRODUCTION

The practice of knowledge management (KM) has become pervasive and ubiquitous across business environments as successfully utilizing existing and creating new knowledge improves decision making, accelerates learning, improves innovation assimilation, increases productivity and minimizes reinvention and duplication (see e.g. Wing & Chua, 2005).

Even though ability of organization to create and utilize its knowledge depends heavily on social factors, many researchers and practitioners (e.g. Alavi & Leidner, 2001; Fahey & Prusak, 1998; Vance & Eynon, 1998; Zack, 1999b) are convinced that KMSs (e.g. document management systems, content management system, groupware, knowledge maps…) can be an important enabler and facilitator of KM.

Some companies, such as Ford, Chevron, and Texas Instruments estimate that their KMSs have saved them millions of dollars (Bose, 2004). Meanwhile other articles report that 70% of all KMS fail to meet the KM objectives originally established for the system (Ambrosio, 2000; Malhotra, 2005; Rigby, Reichheld & Schefter, 2002). However the question still remains why so many technology-enabled initiatives have failed to deliver on the benefits of KM (Chua & Lam, 2005; Davenport & Glaser, 2002; Desouza, 2006; Desouza & Awazu, 2005; Stewart, 2002; Wing & Chua, 2005). The answer might very well be captured in this quote (Desouza, 2006):

Technology solutions to KM problems take a cookie cutter or standardized approach to the problem. This is quite similar to using a hammer to hit a nail and also to swat a fly. The technology is used indiscriminately, especially without regard to the type of knowledge being managed or the nature of work being conducted by the knowledge worker. This just will not work and benefits will remain elusive. – Chief Knowledge Officer, Financial Services Organization

Another example from a large financial UK institution, where KM practices in corporate credit risk analysis have been observed (Mondale, Scott & Venters, 2006) shows, that implementation of the KMS failed because its design was influenced by the theory of finance, which states that credit officers use standard financial models to focus on quantitative credit risk management mechanisms in attempt to eliminate complexity and uncertainty. In practice, credit officers balance their use financial data, models and systems with less formal processes of meaning making within a community of practice (Mondale, Scott & Venters, 2006). In other words, design of KMS failed to include facilities for enabling or supporting collaboration, as they implemented 'what they thought they were doing' instead of 'what they actually were doing' (p. 13).

Leaving project management and cultural issues aside and focusing on the core of IS innovation adoption, literature highlights the misfit of KMS that doesn't satisfy users' needs which originate in the business context (working practices) as one of the fundamental reasons for KMS failures (Cooper, 2003; Stenmark & Lindgren, 2004; Wing & Chua, 2005). Clearly, there is a need for a KMS design theory that would further knowledge applicable to productive application of IT in supporting and enabling activities that result in improved utilization of existing and creation of new knowledge. The goal of this chapter is to propose a theoretical background for design of KMS.

Findings presented here are based on existing understanding of KMS design (from the literature) and amended with first-hand experience gained from several research and consulting projects in the area of IT support for knowledge-rich processes (e.g.Awazu et al., 2006; Baloh, Sitar & Vasić, 2005; Wecht & Baloh, 2006).

THEORETICAL FOUNDATIONS

KM Implementation Guidelines: KM Strategies

New knowledge creation and existing knowledge utilization are enabled and supported by *KM strategies* which *define* activities, organizational structure and *IT support* for implementing KM (Hansen, Nohria & Tierney, 1999; Zack, 1999a). Setting up such a KM strategy that successfully leverages knowledge assets in a company has been a widely discussed issue both in theory and in practice. Lacking a firm advice on how companies should go about injecting KM activities into existing work, we cannot expect that technology that should leverage knowledge will be designed and deployed successfully.

Many researchers and management consultants developed various frameworks and models for successful KM. Usually they are called "KM strategies" (i.e. (Choi & Lee, 2002; Hansen, Nohria & Tierney, 1999; Koenig, 2004; Zack, 1999a)), "models" (Swan & Newell, 2000), "schools" (Earl, 2001), "paths" (Kelleher & Levene, 2001), "approaches" (Kankanhalli et al., 2003). However, even though these enablers are essential for a firm's KM capability, findings about how to employ them are still inconclusive (Desouza, 2006; Kankanhalli et al., 2003; Koenig, 2004; Tsui, 2005).

Literature review shows, that some sort of balanced approach to KM leads to best performance. Ribiere (2001) looked at the influence of organizational culture on the choice of a KM strategy. His study was conducted at the organizational level as well as at the unit level. The findings show that each company balance its KM initiative based on its organizational culture (defined by its level of organizational trust and solidarity). For the 100 companies surveyed it was clear that the KM strategy emphasis on was balanced, companies very rarely focused more than 70% of their effort on a unique approach (codification versus personalization) (Ribière & Román-Velázquez, 2005). But what is the right balance and which solution is good for who? As knowledge creation and utilization are highly business-context dependent, it is suggested (contrary to prevailing body of knowledge) that more than one strategy for injecting KM practices can and should exist in a company. Instead of proposing a company-wide approach, KM practices should be considered to be implemented on a more granular organizational level. Business processes seem the right choice as this is where employees perform their everyday tasks, applying existing or creating new knowledge.

KM Systems: Information Technology that Supports KM Practices

An emerging line of information systems (IT artifacts) targets professional and managerial activities by focusing on dealing with 'knowledge' and 'knowledge resources' (Alavi & Leidner, 1999). Depending on the stance on KM taken, various definitions and field of deployment of KMS exist. In example, researchers that view knowledge as an object, socially independent of a person/group (Popper, 1972; Zack, 1999b) argue for KMS that support 'hard' approach, storing knowledge in form of information in databases, document management systems, etc. Social constructivists rather see knowledge as "socially constructed" (Kuhn, 1962), leading to KMS to be associated with "connectivity", with functions that support and enable collaboration among employees. An artificial dichotomization (i.e. tacit vs. explicit knowledge, and four 'distinct' processes of Nonaka's SECI cycle) led to a reinforced misunderstanding about how and where knowledge can be used in organizations.

Generally speaking, KMS serve to inform the user and instigate learning via knowledge transfer and reuse (Cooper, 2003), the process by which an entity is able to locate and use shared knowledge (Alavi & Leidner, 2001). When analyzing IS support for emergent knowledge processes Markus, Majchrzak and Gasser acknowledge that one of the important reasons why knowledge workers are not supported appropriately is that there are too many different tools that are not integrated. Employees have access to expert systems, decisions support systems, discussion boards, content management systems, however, this results in a "tool glut", leading to aversion to use of such systems (2002, p. 185). Knowledge, as it is conceptualized by many organizations, is thus not connected to the work of the enterprise (Seeley, 2002): many knowledge initiatives very often avoided existing IT systems and introduced additional IT artifacts, regardless of the true knowledge needs of employees who were to use those applications KMS. With that, they designed a 'parallel universe' where knowledge is paramount; however, it is disconnected from their working practices (Smith & McKeen, 2004).

This is in obvious contradiction with the "process-value of IT" aspect (Barua, Kriebel & Mukhopadhyay, 1995; Barua & Mukhopadhyay, 2000; Sambamurthy, 2001; Tallon, Kraemer & Gurbaxani, 2000), which advises that just throwing technology in a process does not improve that process. *"Indeed, such indiscriminant applications of technology may actually reduce process performance,"* (Ray, Muhanna & Barney, 2005, p. 643). IT can only create business value through enabling and supporting working practices to be done in more efficient and effective manner. *"The best tools integrate seamlessly into the work process," (Cooper, 2003, p. 128).*

Net value of the IS adoption hence depends on how well the tool matches the needs of intended users, by not disrupting the natural flow of activity, by not altering roles of human actors significantly and by including all the important contextual information (Cooper, 2003). Thus, IT capabilities can only be leveraged and influence KM success when assimilated within business processes with already injected KM activities.

This influentially informs our theoretical background of KMS design, as one can see that first business context (process) needs to be analyzed, appropriate KM-related activities introduced, and only then KMS solutions considered and deployed. "Appropriate" KMS functionalities would consequently depend on knowledge needs of employees, which should be analyzed at the business process level, taking into account individual users, especially their working and knowledge practices, and the knowledge processes embedded in the process' task structures. In conclusion, there is a need to understand how the nature of knowledge needs impacts the KMS-design needed for supporting and enabling everyday knowledge-rich working practices.

IS Design: Why Task Technology Fit Theory?

The introduction of any organizational intervention—including new IT tool—into an existing environment has the potential to serve as a catalyst for change. Research in the IS field generated various approaches to IS development. Examples include

1. Generally applicable Socio-Technical Systems (Cherns, 1976; Clegg, 2000) and Participative design (Emery, 1993),
2. More specific IS design guidelines i.e. Task-Technology Fit Theory for Group Support Systems (Zigurs & Buckland, 1998) and conceptualized Microsoft Usability Guidelines for Web & Wireless Site Usability (Venkatesh & Ramesh, 2006), and
3. "Somewhere in between" IS success models such as DeLone & McLean's (DeLone & McLean, 1992) and TAM (Davis, Bagozzi & Warshaw, 1989), who have mostly been concerned with conceptual inquiry in utilization and use of IT artifacts and thus need deep considerations when mirroring their

findings in specific design guidelines for particular organizational setting.

There are three reasons why Task-Technology Fit Theory concept is suitable to theoretically ground the design of KMS in. Firstly, **TTFT fits the contingency perspective of this research.** It is the stance of this chapter that both KM strategy and IT support for KM practices are dependent on business context, particularly, knowledge needs of the employees who need to utilize and create existing knowledge within working practices. This is in line with contingency theory positing that there is no "single best" organization, leadership or decision-making style as they all depend upon various factors. Contingency theory found its way in the IS field through Task-Technology Fit Theory (TTFT) (Goodhue & Thompson, 1995), which argues that task-technology fit is important antecedent that can be used to predict or assess the success of IS: the greater the fit of IS design with task characteristics, the better the performance of the user. This concurs with 'dependability' of KM strategy on the organizational context and the process-value TTFT. Thus, it represents an ideal base for theoretical background of KMS design, where the concern is the fit between knowledge needs, technology characteristics, and KMS success.

Secondly, **KMS success measure is suitable to the TTFT concept in KMS environment**. Several adoption and success models have been proposed in the IS field, as measuring success is vital in any organizational setting to provide targets and basis for feedback on implementation, to defend and secure funding, to assess implementation success, and to develop guidelines for future implementations (Jennex, 2005; Kankanhalli & Tan, 2004; Turban & Aronson, 2001). Recent research in KMS field argues (Jennex, 2005) that KMSs are in essence similar to other types of IS, having value in efficient and effective use (cf. Markus & Keil, 1994) yet significantly different in a sense that they have the most value when providing *infrequently used* knowledge *when needed.* Instead of quantitative measures such as number of contributions or number of uses, measures of KMS success are rather related to intention to use (end users) and intention to contribute (contributory users), and can be assessed with the help of Perceived Benefit Model (Jennex, 2005; Thompson, Higgins & Howell, 1991).

KMS need to support practices of knowledge utilization and creation during everyday work, leading to the very importance of business context analysis. Borrowing from the task-technology concept (Zigurs & Buckland, 1998) where group decision context has been analyzed through task characteristics, we can analyze the context through analysis of characteristics of knowledge needs (instead of tasks). Namely, *'fit'* is defined as a number of ideal profiles, where task and technology contingencies are internally consistent and aligned (Zigurs & Buckland, 1998). Discovering such ideal fit-profiles between knowledge needs and technology will account for richer understanding of business context. The more technology is aligned to knowledge needs, the greater their intentions to contribute and use KMS will be. Thus, using TTFT concept aligns well with the KMS success measure, as is therefore used in our theoretical background we are proposing for KMS design.

Thirdly, **TTFT provides context-specific model of IS design.** The models of IS adoption and success mentioned above can be considered by researcher/practitioner when designing features of an IS. They originate in behavioral-science paradigm, as classified by Hevner et al. (2004), which seeks to develop and verify theories that explain or predict human or organizational behavior surrounding the analysis, design, implementation, management, and use of information systems.

Such theories ultimately inform researchers and practitioners of the interactions among people, technology, and organizations that must be managed if an information system is to achieve its stated purpose, namely improving the effectiveness and efficacy of an organization, (Hevner et al., 2004, p. 76).

With an exception of the Microsoft Usability Guidelines (Agarwal & Venkatesh, 2002; Venkatesh & Ramesh, 2006), mentioned studies are not rich in context. In example, Technology Acceptance model (Davis, Bagozzi & Warshaw, 1989; Venkatesh & Davis, 2000) became *"somewhat of a gold standard for understanding individual reactions to technology and user behaviour given its combination of parsimony, reliable and valid scales, and generalisability to a range of contexts and technologies,"* (Venkatesh & Ramesh, 2006, p. 201). Parsimony and context-less-ness was authors' original intention and that is what makes analysis of attitudes of users towards using particular IT with TAM applicable and generalizable across a range of areas. Similarly, Unified Theory of Acceptance and Use of Technology – UTAUT (Venkatesh et al., 2003) reviewed and integrated different models of IS adoption and use to provide a theory with roots in IS, psychology and sociology. Again, parsimony and generalisability across various domains were authors' main intentions (Venkatesh et al., 2003). Generic Task-technology Fit Theory (Goodhue & Thompson, 1995) is not much different in this aspect. Authors have focused on relevance of the task-technology concept, rather than on identification of combinations of particular tasks and technologies that fit together and affect utilization and individual performance.

However, in line with 'IS-research-crisis call for theories' where *IT* artifact is *more central* to the theory development (Benbasat & Zmud, 2003), more context-specific theories should be (and have been) developed as they *"provide greater explanatory power and more actionable guidance to practitioners"*, argue Venkatesh and Ramesh (2006, p. 201). Their *Web and Wireless Site Usability* study (Venkatesh & Ramesh, 2006) is one of such investigations where IS designers' needs are addressed explicitly as general cognitive perceptions about technology are (successfully) replaced with a (richer) model of perceptions of design attributes affecting web and wireless sites use. Zigurs and Buckland (1998) adopted general TTF theory and investigated *task-technology fit in specific, group support system environment.* It was more context-specific and actionable than general TTFT: it proposed explicit 'ideal fit profiles' of task characteristics and GSS dimensions, which positively influence group performance and technology use. Based upon general and specific TTFT for group support systems (Goodhue & Thompson, 1995; Zigurs & Buckland, 1998), Gebauer, Shaw and Gribbins (2005) proposed similar *task-technology fit for mobile systems environments* to provide specific environment IS-design theory.

To sum up, context-specific design-related theories are needed in the IS field to provide guidance to practitioners and greater explanatory power to researchers (Benbasat & Zmud, 2003). IT artifacts extend the boundaries of human problem solving and organizational capabilities; theories regarding their application and impact will follow their development and use (Hevner et al., 2004). Technology and behavior are inseparable and IS research should be evaluated through practical implications of combining *truth* (justified theory) with *utility* (effective artifacts) (Hevner et al., 2004).

Our theoretical grounding of KMS design borrows the concept of fit from the TTFT. However, as KMS design requires rich business context analysis of working practices (how and where knowledge gets created and utilized), *knowledge needs characteristics*, the necessary *KMS characteristics*, and the *fit* between them, *are assessed.* By defining characteristics of KMS which support particular working practices, this chapter proposes KMS-specific models of IS design.

TOWARDS THE THEORETICAL PROPOSITIONS OF THE KMS DESIGN

Theories of Task Technology Fit

Goodhue and Thompson (1995) analyzed the link between IS and individual performance in hope to improve prediction and management of IS success. They confirmed that task-technology fit has significant explanatory power: Task-Technology Fit Theory (TTFT) they developed holds that IT is more likely to have a positive impact on individual performance and be used when IT functionalities *match* the characteristics of tasks that the users must perform. Goodhue and Thompson's studies have mostly not been focused on development of specific guidelines that would identify relevant characteristics of task, technology and fit constructs, which would be of operationalizable help to IS designers.

Independently, yet almost as response to this deficiency, Zigurs and Buckland in their 1998 seminal article proposed a specific theory of task-technology fit in group support systems (GSS) environments based on task complexity and their relationships to relevant dimensions of GSS technology. Although the Goodhue and Thompson's (1995) model operates at the individual level of analysis, Zigurs and Buckland (1998) present an analogous model operating at the group level, propositioning that good fit results in good group performance. Their theory explicitly conceptualizes all the used constructs: **tasks, technology** and **concept of fit between tasks and technology**, defined as *"ideal profiles composed of an internally consistent set of task contingencies and GSS elements that affect the group performance,"* (p. 323).

Both streams of TTFT research informed this work' KMS design suggestions. Goodhue and Thompson's (1995) TTFT theory is followed when considering tasks, technology, performance and utilization as main constructs: adapting TTFT to KMS environment, 'knowledge needs' are proposed to substitute the 'task' construct, 'technology' characteristic construct is adapted to KMS specificity, and KMS-specific success measures are proposed as 'IS success' measure. Similar to work of Zigurs and Buckland (1998), 'knowledge needs/technology fit' is considered as pre-defined set of profiles, which are developed in three steps. In the first step, knowledge needs are defined. Secondly, KMS technology characteristics are described. Finally, knowledge needs/technology fit profiles are proposed. The figure below shows

Figure 1. Top-level conceptual framework of KMS design propositions

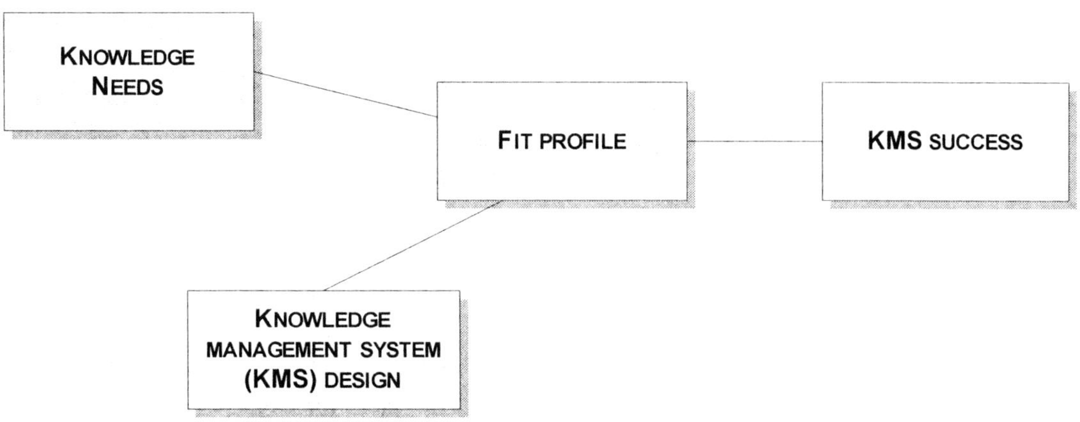

the top-level conceptual framework KMS design propositions.

After defining dimensions of 'KMS technology', analysis of 'knowledge needs' dimensions is made, and they are aligned to KMS dimensions. Accordingly, preliminary version of 'ideal fit profiles' of the KMS design is proposed after that.

KMS Technology Dimensions

Proposed characteristics for design of KMS in this work are conceptual and classified as "technical" and "human" oriented. They are considered through Nonaka's theory of dynamic knowledge creation (Nonaka & Takeuchi, 1995; Nonaka, Toyama & Konno, 2000), Hansen, Nohria and Tierney (HNT)'s discussion on KM strategies (Hansen, Nohria & Tierney, 1999) and Riempp's proposals in the field of Knowledge Management Systems (Riempp, 2004).

Two models of KMS, fulfilling different organizational needs, have been identified from existing research in the field of IS (Alavi & Leidner, 1999) and KM (Hansen, Nohria & Tierney, 1999; Zack, 1999b): machine-oriented and human-oriented designs.

Machine oriented design is a solution that draws upon HNT's *codification* strategy (Hansen, Nohria & Tierney, 1999), which is also known as "connect to information" strategy (McAdam & McCreedy, 1999). Technical model of KMS focuses on codification and storage facilities, where knowledge is stored in form of information in databases, documents in document management systems, etc., where it can be accessed by employees. Focus is on knowledge reuse "by providing access to codified expertise (Kankanhalli & Tan, 2004).

Electronic knowledge repositories (EKR) which to code and share explicit knowledge in a form of documents, manuals, technical specifications, blue prints, and other codified knowledge, represent this strategy (Alavi & Leidner, 2001; Kankanhalli & Tan, 2004). Distinct strengths of machine design includes being able to store large amounts of data that can be quickly transferred and seeked upon, not being subject to physical and emotional effects, and being able of complex computations (Desouza et al., 2004).

Human oriented design draws upon HNT's *personalization* strategy (Hansen, Nohria & Tierney, 1999), where underlying rationale is "connect to people" (McAdam & McCreedy, 1999) to enable transfer of tacit and explicit knowledge.

KMS thus connects people and channels individual expertise, facilitates conversation and helps in locating knowledge holders. KMS will have strong collaboration and experts-seeking functionalities. Examples of IT instantiations are forums for supporting communities of practice (see i.e. Baloh, Wecht & Desouza, 2006), wiki sites to conversationally craft a common body of knowledge (see i.e. Awazu et al., 2006), yellow pages, instant messaging solutions, etc. It has been noted that in order to access the knowledge in an organization that remains uncodified, mapping the internal expertise is useful (Kankanhalli & Tan, 2004; Ruggles, 1998). Distinct strengths of human design includes creativity, flexibility, and domain knowledge (Blattberg & Hoch, 1990; Desouza et al., 2004; van Bruggen, Smidts & Wierenga, 2001)

Such classification yielded results in previous studies; i.e. Desouza et al. (2004) applied human/machine strategy to the area of decision support systems. Authors showed that each strategy has both shortcomings and strengths in one context, however when context of application changes, shortcomings turn to strengths and strengths turn to shortcomings. It is thus necessary to understand what context we are trying to support and enable in order be able to make an informed decision for any of the strategies. This is in line with the previous discussion where it was argued that organizations need to take a segmented and focused approach to KM activities. Decision-making as the ultimate form of knowledge creation and utilization is done in particular business context, depending on individuals' working practices, which need to be analyzed to recognize employees' knowledge needs.

In order for KMS to actually support individual or collaborative knowledge creation, first the process of the latter needs to be understood. Process value of IT theory adopted in this research, namely argues that enterprise-level impact of IT can only be measured through intermediate (process level) contributions - IT is deployed to enable and support particular activities and processes and that impact of IT can only be assessed where effects are expected to be realized.

Such contextual framework was proposed by Li and Kettinger (2006) who clearly devised the knowledge creation and utilization process in six stages: problem recognition, goal setting, knowledge generation, tentative knowledge selection, knowledge retention, resource management. For each of the stages, authors suggested meta-description of organizational memory information system (OMIS) for those processes:

- The **problem recognition function**, which helps organizations to recognize organizational problems. It should also support decomposition of complex problems. In other words, how is the problem set?
- The **goal setting function,** which helps organizations to specify their goals, update the measure progress in the knowledge creation and update the problem status. In other words, how do we know how close to the goal we are?
- The **knowledge generation function** is highly important component as this is where tentative knowledge is created. OMIS needs to support and enable activities to recognize, capture, organize, distribute and combine existing internal and external knowledge to generate new knowledge.
- The **tentative knowledge selection function** should support specification and utilization of selection criteria to measure and select tentative knowledge. In other words, how is the tentative knowledge selected?
- The **knowledge retention function** performs functions of retaining selected knowledge variations as existing knowledge, which will be used (or not) in the next rounds of knowledge creation. In other words, how do we retain the knowledge created?
- The **resource management function** supports allocation of organizational resources during the knowledge creation and reconfiguration of organizational resources based on the new knowledge created. In other words, how are human resources, materials, and machines, allocated during the creation process?

In practical terms, decision maker (employee) seeks for a solution in existing knowledge space, which can either be in a form of individuals' memory, looking through documents and databases, either by asking "someone who knows the answer". As the problem/goal/resource combination has appeared before, solution is found in existing knowledge base and knowledge 'variation' (the same as 'old') is applied. Customer replies to employee that the problem is solved (positive feedback information), which means that there is no need for retention of 'new' knowledge. However if customer uses a new version of a product which responds to the solution differently as the old one did (feedback information is not positive), new tentative knowledge needs to be generated. That can be done i.e. by analyzing architecture of the product, consulting the engineers in manufacturing process etc., which, in case of complex problems, will maybe have to be solved by employees furnished with appropriate skills and capabilities, who will finally integrate the new knowledge in the OMIS – either in memory of individuals, improved manufacturing process, or as information - e.g. as an entry in an "intranet knowledge base".

With help of Li and Kettinger' knowledge creation theory (Li & Kettinger, 2006), furnished with understanding of information as opposed to knowledge, this research will assess KMS functionalities through the human/machine dimensions for each of the OMIS' six member functions.

Proposing the Ideal Fit Profiles

In this section, contingent factors of task domain, type of knowledge, and volatility of knowledge, which define knowledge needs of employees, are discussed. Suitable technological solutions, that is ideal fit profiles of KMS design, are proposed based on extant research and findings from several research and consulting projects undertaken in the past (e.g.Awazu et al., 2006; Baloh, Sitar & Vasić, 2005; Wecht & Baloh, 2006).

Task Domain

Knowledge workers engage with KMSs during the context of work assignments. The type of KMSs support needed to enable for effective and efficient completion of task assignments will heavily depend on the nature of the work and the kind of knowledge is involved in the task.

This contingency factor distinguishes focused and broad task domains. Focused tasks rely on functional knowledge embodied in a specific group of engineers, elemental technologies, information processing devices, databases and patents (Kusunoki, Nonaka & Nagata, 1998). Deep knowledge in particular area is required or knowledge that is high in specificity (Choudhury & Sampler, 1997; Pisano, 1994). Broad tasks necessitate working with employees from other processes within the organization through dynamic interaction, communication, coordination across different functional groups (Kusunoki, Nonaka & Nagata, 1998). The less variety of problems encountered in tasks requires specialization and decreases the need for inter-unit collaboration (von Krogh & Roos, 1995). Greater the variety of problems encountered in tasks, greater the need for working with other subunits within the organization.

According to Earl's (Earl, 2001) classification, technology school is perhaps the most relevant for focused tasks. It is plausible to say that using databases for transferring knowledge (hitherto in explicit form) might be appropriate for focused knowledge. Underlying argument here is that tacit and explicit knowledge are dependent qualities of knowledge: Polanyi (1966) argues that tacit knowing is needed to understand and interpret information received. Without some common knowledge base, transfer of knowledge will not happen. *"Only individuals with a requisite level of shared knowledge can truly exchange knowledge,"* (Alavi & Leidner, 2001, p. 112). When specialists share common knowledge base, knowledge is easier to transfer in explicit form, as they will easier comprehend the meaning articulated in codified form, as argued by (Zack, 1999b): complexity of connecting two individuals with close knowledge base is lesser than connecting two individuals with very different common knowledge. The more specialized knowledge two people have in common, the more sophisticated their communication can be: if two individuals with very different knowledge bases communicate, integration can occur only at a very basic knowledge (Grant, 1996b). Arguments above imply that technical design might hold promise for focused-task domain subunits that do not need much of integration with other subunits to perform their work.

From the opposite side of perspective constructivist theories of knowledge development holds their argument for social construction of knowledge: knowledge is not an object, socially independent of a person/group (as argued by many authors, starting from (Popper, 1972) and authors coming most notably from the IT literature (c.f. Alavi & Leidner, 2001)), but rather socially constructed (Kuhn, 1962). It is so because the transfer of "true belief", unbiased by personal interpretations between groups of people is questionable (Fiol, 1991; Weick, 1979). Therefore, social construction of knowledge is needed as all the background knowledge, necessary to understand new information, might be too much to transfer and digest in codified form. Rather the knowledge itself, when tasks are dependent on other individuals who have highly specific (but different base) knowledge, information about "who knows what" can be transferred, knowledge itself however, is exchanged face-to-face. Process of sharing deep knowledge is time consuming and expensive: It can't be systematized, so it can't be made efficient, meaning codified (Hansen, Nohria & Tierney, 1999). This holds promise for performing broad tasks where employees engage in collaboration across knowledge domains. In addition, it holds promise in the area where deep functional knowledge needs to be transferred in example from expert to novice, as performing focused tasks, learning must be individualized and personal. Internalization and externalization knowledge conversion modes of the whole Nonaka and Takeuchi' knowledge creation cycle (1995) are crucial for such tasks (Becerra-Fernandez & Sabherwal, 2001).

To sum up, performing focused task in particular subunit, technical design would be more appropriate, working on externalization knowledge conversion modes as a mean of articulating tacit knowledge in explicit form, and on internalization process as a mean of acquiring new tacit knowledge from explicit form. In focused tasks, internalization and externalization are enabled by common knowledge base. For example, engineers working on specific problems have deep expertise in their various domains. This deep knowledge is required for them to work on the specificities of their problem. Technology support is needed in the form of knowledge retrieval tools from databases. The specialist would need to scan and query large databases within his/her field for past knowledge about their problem. In most cases, the specialist will be able to formulate their queries with reasonable ease, as they understand their problems and the domain. The researchers are also able to receive automated updates, via email feeds, when new articles in their fields are loaded into the database, or when new articles cite articles that they have tagged as important.

Conversely, subunits where a task solving process requires individuals with different knowledge bases collaborating would benefit from pursuing human design of KMS. In example,. managers require broad knowledge. Their knowledge needs are integrative in nature – they need to know enough about the various domains they interact with to be able to build bridges and serve as an integrator across these domains. For the generalist rather than knowing everything within a domain, like the case of a specialist, the knowledge need is one of knowing who knows everything. KMS should be able to (1) provide updates of recent developments across a wide range of domains, and (2) provide a list of sources who can be contacted for further knowledge within each domain.

Type of Knowledge

This factor looks at two types of knowledge that are needed by particular person within tasks: "know what" and "know how". Employees in business process more often that not interact with different types of knowledge, thus requiring different types of KMSs' support. Processes are formed due to the degree of cohesion in internal activities of the tasks. Hence, people in those processes differentiate among themselves by the types of activities they conduct.

For example, the accounting department will be concerned with knowledge that is fairly declarative in nature (e.g. standards and procedures from the accounting boards), whereas the customer-service department will employ procedural or rule-based knowledge (e.g. how to fix a product bug). The differences in the knowledge needs will call for different types of KMSs.

Know-what is the base for content oriented tasks and has been associated with explicit (declarative) knowledge. If underlying knowledge bases are similar, technical design that focuses on knowledge codification, its transfer and retention in explicit form, would be suitable. Reading about success stories, reading manuals, documents, should facilitate tacit knowledge creation. Knowledge can be stored in databases, document management systems

Know-how (procedural knowledge) is the base for process oriented tasks, and has been associated with tacit knowledge of technological and scientific expertise, acquired through personal experience (Grant, 1996b; Pisano, 1994). Human design would be more appropriate for know-how as it is built by listening to stories, experimenting, trying, on-the-job training, brainstorming camps, communities of practice. Enabling and supporting person-to-person collaboration is the focus of the KMS. Additionally, storing experience might be possible in the instances when knowledge can be codified. Consider the example of Xerox and many other companies who store step-by-step procedural knowledge in knowledge-bases, where maintenance staff can pose questions and provide answers to a community of peers—engineers —across the globe.

Volatility of Knowledge

This contingency factor is concerned with "volatility" of knowledge: what is the volatility of knowledge in particular subunit or a process? The main question would be whether the knowledge used for performing tasks is rapidly changing, or can it be economically reused (stored and reapplied)?

Whether new/old knowledge is needed, depends on volatility of environment: if it is highly volatile, knowledge is time-sensitive and in order to "keep pace" with changing requests, has to be continuously created and continuously refreshed. If, on the other hand, knowledge is less time sensitive it can be reused over a relatively longer time span (Kankanhalli et al., 2003).

With repetition of similar tasks, complexity decreases (von Krogh & Roos, 1995) and knowledge gets acquired by performing parties. Only at this point knowledge can be converged to other employees as it can be explicated easier and organizationally legitimized and "approved" in a sense of either being articulated in a written form, code of practice, best practices, success stories, routines, standard operating procedures, emails, faxes, memos. Then, knowledge can be spatially detached from the source of its origin, argue von Krogh and Roos (1995), and at this point, technical design would be suitable. If problems are new or complex, tacit knowledge of one or more individuals might be needed for resolving particular task. Besides time sensitivity, knowledge goals are also different for each case. In the first case, the primary knowledge goal is knowledge augmentation (Markus, 2001), where employees augment the knowledge they already have by helping them acquire and deepen knowledge in particular domain. In the second case, the primary knowledge goal is knowledge substitution where existing knowledge of non-experts is substituted with experts' knowledge in order to uniformly direct the behavior of less-knowledgeable people (Markus, 2001).

These arguments suggest that for subunits with volatile knowledge requests human design orientation would be more appropriate, as it makes people to exchange knowledge in person. It is also less use to try storing such knowledge as it is less likely to be needed in the future. Rather than storing the knowledge itself, connectivity between people might be better to support, i.e. through enabling people to communicate and develop new knowledge. That would again imply use of human oriented design. In such environment, focus is on enabling person-to-person or group knowledge creation and transfer. Underlying philosophy is connectivity and collaboration:

make experts available for help and co-creating of new and innovative solutions.

In the case when standardized outputs are produced, knowledge can be retrieved from electronic databases and efficiently re-applied. Consider a case of consulting company where junior consultant is dispatched to perform due diligence project at a client company. When checking the health of financial statements and other areas of business operations, consultant uses relatively highly detailed instructions ("manual") on how to approach the audit in particular area (what is the procedure), how to perform different steps (what activities need to be performed, what needs to be checked and how, description of best practices and failure stories…). This implies that a combination of content and document management system, embedded in a workflow system, would be the focus of the KMS.

Proposed Profiles

Table 1 below recaps our propositions discussed above.

Even though informative, this table nevertheless inadequately addresses the needs of this research. Each process can be analyzed through each of distinct dimensions of knowledge needs (task domain, type of knowledge, volatility of knowledge). This gives out eight distinct knowledge needs profiles (Table 2) for which fitting KMS functionalities need to be selected in each of the six stages of Li and Kettinger's (2006) decision making framework. Using these eight profiles and based on the discussion so far, we can now propose fit profiles between the six KMS member functions and appropriate technology functionalities. Table 3 depicts these fit profiles per each of the possible fit profiles (KP1-KP8).

Table 1. KM specifities and main KMS characteristics per each knowledge needs dimension

Knowledge needs	KM specificities	Main KMS characteristics
Task domain		
Focused	Rely on deep functional knowledge; not much integration with others; specialization important; specialists share common knowledge base.	Machine design with strong knowledge codification and retrieval facilities
Broad	Employees work with different knowledge bases; breadth of knowledge and 'who knows what' is important.	Provide updates of developments across range of domains, provide 'who knows what' knowledge, connectivity
Type of knowledge		
Know-how	Knowledge is sticky and it is a base for process-oriented (procedural) tasks. Examples: Communities of practice, storytelling, experimenting, on-the-job-training.	Human design of KMS – enabling and supporting person to person collaboration.
Know-what	Associated with explicit declarative knowledge that can easily be codified.	Machine design of KMS.
Volatility of knowledge		
Highly volatile	No need to store knowledge as it is rapidly changing. If it is stored needs to be continuously refreshed. Knowledge gets exchanged in person. Focus on group or person-to-person knowledge creation when solving unique problems.	Human design enabling people to communicate and develop new knowledge. Tools to support group knowledge creation (i.e. wiki).
Less volatile	Knowledge can be reused over longer time-span. Focus on exploitation of once- organizationally-developed knowledge.	Machine design with knowledge storage and database retrieval facilities. Document management systems.

Table 2. Possible fit profiles

knowledge needs profile	task domain	knowledge volatility	type of knowledge
KP1	focused	low	informational
KP2	focused	low	procedural
KP3	focused	high	informational
KP4	focused	high	procedural
KP5	broad	low	informational
KP6	broad	low	procedural
KP7	broad	high	informational
KP8	broad	high	procedural

Table 3. Proposed knowledge needs-technology fit profiles

OMIS subsystem / Knowledge needs profile	Problem recognition	Goal setting	Generation of tentative knowledge	Selection of tentative knowledge	Retention of new knowledge	Resource management
KP1	M	M	M	M	M	M
KP2	M	M	M	M	M	M
KP3	M	M	M	M	M	M
KP4	M / H	M / H	H / M	H/ M	H/ M	H / M
KP5	M	M/ H	M	M	M	M/ H
KP6	M / H	M / H	H / M	M/ H	M	H
KP7	H	M	H	M	H / M	H / M
KP8	H	H	H	H	H	H/ M

For explanation of the Table 3, consider the example of the "KP1" working context. Employee in a customer support centre of software development company that replies to customer questions regarding installation of their software with particular existing hardware. Queries of this type are focused, standard, and require know-what type of knowledge. When a problem like this needs to be solved, problem recognition process can be operationalized in the 'machine' type of solution. In example, e-mail from a customer with description of the issue can trigger the need for a solution. Goal information – functioning software can also be explained in codified software. Problem and goal statements in a form of codified information triggers generation of tentative knowledge. Employee can check existing document-base for a particular solution using search functionality. Solution for exactly that type of problem is not found in the document base, but there is similar problem described in a document, which employee retrieves and creates a reply to the customer; the latter replies in a form of feedback information that this did not completely solve his problem as now another error message pops up. Employee searches again the document base and finds a solution for that error message. Feedback information from the customer is now positive and tentative knowledge is now selected. New knowledge about how to solve a particular problem—by combining two existing solutions, can now be codified in a new document and thus this knowledge can be stored in a form of document. Summary: For KP1 type of knowledge needs, 'machine' type of KMS is appropriate for each of the steps of the knowledge creation process. Similarly, other seven profiles can be explained.

CONCLUSION: FUTURE RESEARCH

While theory and practice have come a long way in terms of maturity of KM practices, concrete explanations as to how to deploy these practices successfully are lacking. We are also short of methods to clearly design and deploy the *right* technologies for the *right* kind of KM problem. Thus, the organizations needs to develop a competency in analyzing knowledge needs which influence the choice of KMS functionalities, managing multiple KMS and serving as a coordinator of these KMS, rather than taking the easy (and incorrect) road of trying to use a blanket one-size-fits-all approach as proposed by a large and influential body of literature. This thinking is in line with the one of Grant (1996a), who argued that the organization needs to be a coordinator of knowledge and expertise in the organization. Additionally, it needs to be a coordinator and integrator of KMS used throughout the company.

This chapter provides a theoretical background for a possible KMS design theory that would create and evaluate innovative IT artifacts that further productive application of IT for managerial and organizational purposes, in this instance, supporting and enabling creation of new and utilization of existing knowledge. The findings presented should be treated as propositions – due to the early stage of research, results are relatively of weak guidance to practitioners, however, are of strong importance to the research. Design science is inherently iterative, essentially a search process to discover an effective solution to a problem (Hevner et al., 2004). In this approach to building the model, Table 3 represents the first version of the knowledge needs-technology fit model for KMS, which has to be evaluated whether it yields utility in practice. Each of the four constructs of the future KMS design theory based on TTFT concepts need to be verified through a chosen success evaluation manner in the future research.

In the area of 'knowledge needs', future research is needed in the area of contingent factors. We have outlined three of them, however, some of the dimensions might have to be expanded (i.e. type of knowledge according to other, more detailed classifications rather only know how/know what). In the area of 'KMS technology', similar type of research is plausible.

'Fit profiles' need to be amended in exploratory case studies, and also appropriate KMS success measurement instrument needs to be adopted. Case studies do not necessarily provide sufficient basis for making general theoretical claims, but they do permit rigorous testing of propositions, looking for alternative explanations and seeking to explain why negative cases occur, thereby enabling the researcher to move towards the development of valid and well-grounded conclusions, and valid interpretation of occurring phenomena in the social world (Van Maanen, 1988). The outline of the constructs has been based on findings from previous research projects in the field of KM and KMS, and on the grounds of extant literature. Exploratory study will amend the dimensions of these constructs. Validity of the construct might need to be validated through a large-scale quantitative confirmatory study.

There are many potential contributions to the disciplines of information systems, knowledge management, management and organization science, and to KMS designers. First, while acknowledging importance of cultural, social and motivational factors, future research needs to addresses the technical aspects of KM by providing a design theory for KMS. As "design-type theory" (Gregor, 2006), such theory with explicitly outlined prescriptive statements for constructing an IT instantiation in particular organizational setting, would provide actionable guidelines to practitioners in the KMS-design field. With that, it is intended for the principal consumer of IS research – IT practice community (Benbasat & Zmud, 2003). Second, such theory can also be applied to analyze successful and unsuccessful KMS deployments retroactively. Third, based on the Task Technology Fit Theory (Dennis, Wixom & Vandenberg, 2001; Goodhue & Thompson, 1995; Zigurs & Buckland, 1998), such KMS design theory would complement other IS adoption/success theories some of which we already listed above: Technology Acceptance Model (Davis, Bagozzi & Warshaw, 1989; Venkatesh & Davis, 2000), DeLone & McLean's IS Success Model (DeLone & McLean, 1992; DeLone & McLean, 2003), Unified Theory of Acceptance and Use of Technology – UTAUT (Venkatesh et al., 2003), and Microsoft Usability Guidelines (Agarwal & Venkatesh, 2002; Venkatesh & Ramesh, 2006), by providing prescriptive guidelines on the fit in KMS environments, which is based on contingency factors of task characteristics (task domain), knowledge attributes (volatility of knowledge, type of knowledge), and their relationship to relevant KMS technology dimensions.

Such theory could be considered deterministic to the extent that it prescribes expectations about performance, based on characteristics of knowledge needs and technology. However, ideal task-technological fit profiles are part of greater social agenda (Barley, 1986; Orlikowski, 1992; Orlikowski & Barley, 2001; Zigurs & Buckland, 1998) in an organization and thus cannot be the sole guarantor for KMS success. Rather, such combinations for the latter are much more complex than the 'knowledge needs / KMS design' profiles. Having said that, the KMS design theory could be used as a starting point in technological context of the greater socio-technological complexity.

In the disciplines of management and organization science, there are again numerous potential contributions from this research. With a KMS design theory based on TTFT concepts, contingency theory could be used and applied in the area of KM, more specifically, design of IS that supports KM. KMS design theory would add to existing body of knowledge by researching KM strategies on more granular organizational level unlike most of prior research. Looking at business process level, tasks performed within could be viewed as an aggregate and are more homogenous (regarding their knowledge needs) than if looking at the whole company. Theory could be used to establish which KM strategies fit to particular business context, defined by particular knowledge needs of business processes, and with that, recognizing process- or even more granular-level-dependent KM strategies.

KMS designers would benefit significantly as such theory would have provided operationalizable advice in the area of KMS design, explicitly dealing with contingent factors of knowledge needs that fit particular KMS characteristics, and where degree of fit among 'knowledge needs / technology' affects success of KMS implementation. Secondly, KMS design theory could be applied both to identify promising areas for the application of available KMS, and to design information systems that successfully support KM activities. On the final note, KMS implementation is a problematic and ambiguous undertaking. Having a KMS design theory would inform how to go about the decision about type of KMS to be introduced, decreasing the chances of costly implementation failures.

REFERENCES

Agarwal, R., & Venkatesh, V. (2002). Assessing a firm's Web presence: A heuristic evaluation procedure for the measurement of usability. *Information Systems Research, 13*(2), 168-186.

Alavi, M., & Leidner, D. E. (1999). Knowledge management systems: Issues, challenges, and benefits. *Communications of the AIS, 1*(2es).

Alavi, M., & Leidner, D. E. (2001). Review: Knowledge management and knowledge management systems: Conceptual foundations and research issues. *MIS Quarterly, 25*(1), 107-136.

Ambrosio, J. (2000). Knowledge management mistakes: Experts reveal five pitfalls to avoid when starting down the knowledge management path. *Computerworld, 34*(27), 44.

Awazu, Y., Baloh, P., Desouza, K. C., Wecht, C. H., Kim, J. Y., & Jha, S. (2006). *Opening up innovation through information-communication technologies* (Technical Report No. #I4I-I3M-InnovICT-1): Institute for Innovation in Information Management, The Information School, University of Washington.

Baloh, P., Sitar, A. S., & Vasić, V. (2005). Practical approach to KM implementation: Industrial case of Slovenian company Gorenje d.d. In S. Hawamdeh (Ed.), *Knowledge management: Nurturing culture, innovation and technology. Proceedings of the 2005 International Conference on Knowledge Management, Charlotte, North Carolina, USA, October 27-28th, 2005* (pp. 689-690). New Jersey [etc.]: World Scientific.

Baloh, P., Wecht, C. H., & Desouza, K. C. (2006). Managing knowledge in fuzzy front end of open innovation. In *Proceedings of the MIC '06-7th International Conference of the Faculty of Management Koper, University of Primorska, November 23-25, 2006*. Portorož, Slovenia, 2006.

Barley, S. R. (1986). Technology as an occasion for structuring: Evidence from observations of CT scanners and the social order of radiology departments. *Adm Sci Q, 31*(1), 78-108.

Barua, A., Kriebel, C. H., & Mukhopadhyay, T. (1995). Information technologies and business value - An analytic and empirical-investigation. *Information Systems Research, 6*(1), 3-23.

Barua, A., & Mukhopadhyay, T. (2000). Information technology and business performance: past, present, and future. *Framing the Domains of IT Management: Projecting the Future Through the Past*, 65-84.

Becerra-Fernandez, I., & Sabherwal, R. (2001). Organization knowledge management: A contingency perspective. *Journal of Management Information Systems, 18*(1), 23-55.

Benbasat, I., & Zmud, R. W. (2003). The identity crisis within the IS discipline: Defining and communicating the discipline's core properties. *MIS Quarterly, 27*(2), 183-194.

Blattberg, R. C., & Hoch, S. J. (1990). Database models and managerial intuition: 50 Model+ 50 Manager. *Management Science, 36*(8), 887-899.

Bose, R. (2004). Knowledge management metrics. *Industrial Management & Data Systems, 6*(104), 457-468.

Cherns, A. (1976). The principles of sociotechnical design. *Human Relations, 29*(8), 783-792.

Choi, B., & Lee, H. (2002). Knowledge management strategy and its link to knowledge creation process. *Expert Systems with Applications, 23*(3), 173-187.

Choudhury, V., & Sampler, J. L. (1997). Information specificity and environmental scanning: An economic perspective. *MIS Quarterly, 21*(1), 25-53.

Chua, A., & Lam, W. (2005). Why KM projects fail: A multi-case analysis. *Journal of Knowledge Management, 9*(3), 6-17.

Clegg, C. W. (2000). Sociotechnical principles for system design. *Appl Ergon, 31*(5), 463-477.

Cooper, L. P. (2003). A research agenda to reduce risk in new product development through knowledge management: A practitioner perspective. *Journal of Engineering and Technology Management, 20*(1), 117-140.

Davenport, T., & Glaser, J. (2002). Just-in-time delivery comes to knowledge management. *Harvard Business Review, 80*(7), 5-9.

Davis, F. D., Bagozzi, R. P., & Warshaw, P. R. (1989). User acceptance of computer technology: A comparison of two theoretical models. *Management Science, 35*(8), 982-1003.

DeLone, W. H., & McLean, E. R. (1992). Information systems success: The quest for the dependent variable. *Information Systems Research,* (3), 60-95.

DeLone, W. H., & McLean, E. R. (2003). The DeLone and McLean model of information systems success: A ten-year update. *Journal of Management Information Systems, 19*(4), 9-30.

Dennis, A. R., Wixom, B. H., & Vandenberg, R. J. (2001). Understanding fit and appropriation effects in group support systems via meta-analysis. *MIS Quarterly, 25*(2), 167-193.

Desouza, K. C. (2006). *Knowledge management maturity model theoretical development and preliminary empirical testing.* Doctoral Thesis, University of Illinois at Chicago, Chicago, Illinois.

Desouza, K. C., & Awazu, Y. (2005). *Engaged knowledge management: Engagement with new realities.* Basingstoke: Palgrave Macmillan.

Desouza, K. C., Thomas, D., Yifeng, Z., & Awazu, Y. (2004). Information integrity in healthcare enterprises: strategies for mitigation of medical errors. *International Journal of Healthcare Technology & Management, 6*(2), 1-1.

Earl, M. (2001). Knowledge management strategies: Toward a taxonomy. *Journal of Management Information Systems, 18*(1), 215-233.

Emery, F. E. (1993). Characteristics of sociotechnical systems. In E. L. Trist & H. Murray (Eds.), *The Social Engagement of Social Science: A Tavistock Anthology vol. II. The Sociotechnical Perspective.* Philadelphia, PA: University of Pennsylvania Press.

Fahey, L., & Prusak, L. (1998). The eleven deadliest sins of knowledge management. *California Management Review, 40*(3), 265-276.

Fiol, C. M. (1991). Managing culture as a competitive resource - An identity-based view of sustainable competitive advantage. *Journal of Management, 17*(1), 191-211.

Gebauer, J., Shaw, M. J., & Gribbins, M. L. (2005). *Task-technology fit for mobile information systems. Working paper.* Retrieved September 1, 2006, from http://www.business.uiuc.edu/Working_Papers/papers/06-0107.pdf. Last Updated July 27, 2005.

Goodhue, D. L., & Thompson, R. L. (1995). Task-technology fit and individual-performance. *Mis Quarterly, 19*(2), 213-236.

Grant, R. M. (1996a). Prospering in dynamically-competitive environments: Organizational capability as knowledge integration. *Organization Science: A Journal of the Institute of Management Sciences, 7*(4), 375-387.

Grant, R. M. (1996b). Toward a knowledge-based theory of the firm. *Strategic Management Journal, 17(Winter96)*, 109-122.

Gregor, S. (2006). The nature of theory of information systems. *MIS Quarterly, 30*(3), 611-642.

Hansen, M. T., Nohria, N., & Tierney, T. (1999). What's your strategy for managing knowledge? *Harvard Business Review, 77*(2), 106-116.

Hevner, A. R., March, S. T., Park, J., & Ram, S. (2004). Design science research in information systems. *MIS Quarterly, 28*(1), 75-105.

Jennex, M. (2005). The issue of system use in knowledge management systems. In *Proceedings of the 38th Hawaii International Conference on System Sciences (HICSS-38 2005)* (pp. 9): IEEE Computer Society.

Kankanhalli, A., & Tan, B. C. Y. (2004). A review of metrics for knowledge management systems and knowledge management initiatives. In *Proceedings of the 37th Annual Hawaii International Conference on System Sciences (HICSS-37 2004)* (pp. 238-245): IEEE Computer Society.

Kankanhalli, A., Tanudidjaja, F., Sutanto, J., & Tan, B. C. Y. (2003). The role of IT in successful knowledge management initiatives. *Communications of the Acm, 46*(9), 69-73.

Kelleher, D., & Levene, S. (2001). *Knowledge management: A guide to good practice*. London: British Standards Institution.

Koenig, M. E. D. (2004). Knowledge management strategy: Codification Vs. Personalization (A False Dichotomy). In M. E. D. Koenig & T. Srikantaiah (Eds.), *Knowledge management lessons learned : what works and what doesn't* (pp. xxv, 595 p.). Medford, N.J.: Information Today.

Kuhn, T. S. (1962). *The structure of scientific revolutions*. Chicago; London: University of Chicago Press.

Kusunoki, K., Nonaka, I., & Nagata, A. (1998). Organizational capabilities in product development of Japanese firms: A conceptual framework and empirical findings. *Organization Science, 9*(6), 699-718.

Li, Y., & Kettinger, W. J. (2006). An evolutionary information-processing theory of knowledge creation. *Journal of the Association for Information Systems, 7*(9), 593-617.

Malhotra, Y. (2005). Integrating knowledge management technologies in organizational business processes: Getting real time enterprises to deliver real business performance. *Journal of Knowledge Management, 9*(1), 7-28.

Markus, M. L. (2001). Toward a theory of knowledge reuse: Types of knowledge reuse situations and factors in reuse success. *Journal of Management Information Systems, 18*(1), 57-93.

Markus, M. L., & Keil, M. (1994). If we build it, they will come: Designing information systems that people want to use. *Sloan Management Review* (Summer 1994), 11-25.

Markus, M. L., Majchrzak, A., & Gasser, L. (2002). A design theory for systems that support emergent knowledge processes. *MIS Quarterly, 26*(3), 179-212.

McAdam, R., & McCreedy, S. (1999). A critical review of knowledge management models. *The Learning Organization: An International Journal, 6*(3), 91-101.

Mondale, J., Scott, S., & Venters, W. (2006). Knowledge management as an image of the organization: Industry standards and processes of knowing in credit risk management. In *Proceedings of the 14. European Conference on Information Systems (ECIS), Gothenburg, Sweden, June 11-14 2006.*

Nonaka, I., & Takeuchi, H. (1995). *The knowledge-creating company: How Japanese companies create the dynamics of innovation*. New York: Oxford: Oxford University Press.

Nonaka, I., Toyama, R., & Konno, N. (2000). SECI, ba and leadership: A unified model of dynamic knowledge creation. *Long Range Planning, 33*(1), 5-34.

Orlikowski, W. J. (1992). The duality of technology: Rethinking the concept of technology in organizations. *Organization Science, 3*(3), 398-427.

Orlikowski, W. J., & Barley, S. R. (2001). Technology and institutions: What can research on information technology and research on organizations learn from each other. *MIS Quarterly, 25*(2), 145-165.

Pisano, G. P. (1994). Knowledge, integration, and the locus of learning - An empirical-analysis of process-development. *Strategic Management Journal, 15*, 85-100.

Polanyi, M. (1966). *The tacit dimension*: London: Routledge & Kegan Paul.

Popper, K. R. S. (1972). *Objective knowledge: An evolutionary approach*: Oxford: Clarendon Press.

Ray, G., Muhanna, W. A., & Barney, J. B. (2005). Information technology and the performance of the customer service process: A resource-based analysis. *MIS Quarterly, 29*(4), 625-651.

Riempp, G. (2004). *Integrierte wissensmanagement-systeme: Architektur und praktische anwendung; Mit 26 Tabellen*. Berlin [u.a.]: Springer.

Rigby, D., Reichheld, F., & Schefter, P. (2002). Avoid the four perils of CRM. *Harvard Bus Review, 80*(2), 101-109.

Ruggles, R. (1998). The State of the notion: Knowledge management in practice. *California Management Review, 40*(3), 80-89.

Sambamurthy, V. (2001). Editor's comments-research in information systems: What we haven't learned. *MIS Quarterly, 21*(4), v-xv.

Seeley, C. (2002). Igniting knowledge in your business processes. *KM Review, 5*(4), 12-15.

Smith, H. A., & McKeen, J. D. (2004). Developments in practice XII: Knowledge-enabling business processes. *Communications of AIS, 2004*(13), 25-38.

Stenmark, D., & Lindgren, R. (2004). Integrating knowledge management systems with everyday work: Design principles leveraging user practice. In *Proceedings of the 37th Hawaii International Conference on System Sciences (HICSS-37 2004)* (pp. 9): IEEE Computer Society.

Stewart, T. A. (2002). The case against knowledge management. *Business 2.0, 3*(1), 80.

Swan, J., & Newell, S. (2000). Linking knowledge management and innovation. In B. H. Hansen, M. Bichler & H. Mahrer (Eds.), *Proceedings of the 8th European Conference on Information Systems* (pp. 591-598). Vienna, Austria.

Tallon, P. P., Kraemer, K. L., & Gurbaxani, V. (2000). Executives' perceptions of the business value of information technology: A process-oriented approach. *Journal of Management Information Systems, 16*(4), 145-173.

Thompson, R. L., Higgins, C. A., & Howell, J. M. (1991). Personal computing: Toward a conceptual model of utilization. *MIS Quarterly, 15*(1), 125.

Tsui, E. (2005). The role of IT in KM: Where are we now and where are we heading? *Journal of Knowledge Management, 9*(1), 3-6.

Turban, E., & Aronson, J. E. (2001). *Decision Support Systems and Intelligent Systems* (6th ed.): Prentice Hall.

van Bruggen, G. H., Smidts, A., & Wierenga, B. (2001). The powerful triangle of marketing data, managerial judgment, and marketing management support systems. *European Journal of Marketing, 35*(7/8), 796-816.

Van Maanen, J. (1988). *Tales of the field: On writing ethnography*: University of Chicago Press.

Vance, D., & Eynon, J. (1998). On the requirements of knowledge-transfer using IS: A schema whereby such transfer is enhanced. In E. Hoadley & I. Benbasat (Eds.), *Proceedings of the 4th Americas Conference on Information Systems, Baltimore, 1998* (pp. 632-634).

Venkatesh, V., & Davis, F. D. (2000). A theoretical extension of the technology acceptance model:

Four longitudinal field studies. *Management Science, 45*(2), 186-204.

Venkatesh, V., Morris, M. G., Hall, M., Davis, G. B., & Davis, F. D. (2003). User acceptance of information technology: Toward a unified view. *MIS Quarterly, 27*(3), 425-478.

Venkatesh, V., & Ramesh, V. (2006). Web and wireless site usability: Understanding differences and modeling use. *MIS Quarterly.*

von Krogh, G., & Roos, J. (1995). A perspective on knowledge, competence and strategy. *Personnel Review, 24*(3), 56.

Wecht, C. H., & Baloh, P. (2006). Open Innovation: Can Modern ICT-Tools Facilitate Customer Integration Into the Fuzzy Front End of the Innovation Process? In *Proceedings of the 2nd European Conference on Management of Technology (EuroMOT), Birmingham, UK, September 10-12, 2006*: International Association for Management of Technology (IAMOT).

Weick, K. E. (1979). *The social psychology of organizing* (2nd ed. ed.). Reading, MA; London: Addison-Wesley.

Wing, L., & Chua, A. (2005). Knowledge management project abandonment: An exploratory examination of root causes. *Communications of AIS, 2005*(16), 723-743.

Zack, M. H. (1999a). Developing a knowledge strategy. *California Management Review, 41*(3), 125-+.

Zack, M. H. (1999b). Managing codified knowledge. *Sloan Management Review, 40*(4), 45-+.

Zigurs, I., & Buckland, B. K. (1998). A theory of task/technology fit and group support systems effectiveness. *MIS Quarterly, 22*(3), 313-334.

Chapter VIII
Users as Developers:
A Field Study of Call Centre Knowledge Work

Beryl Burns
University of Salford, UK

Ben Light
University of Salford, UK

ABSTRACT

We report the findings of a field study of the enactment of ICT supported knowledge work in a Human Resources contact centre, illustrating the negotiable boundary between what constitutes the developer and user. Drawing upon ideas from the social shaping of technology, we examine how discussions regarding producer-user relations require a degree of greater sophistication as we show how users develop technologies and work practices in-situ. In this case different forms of knowledge are practised to create and maintain a knowledge sharing system. We show how as staff simultaneously distance themselves from, and ally with, ICT supported encoded knowledge scripts, the system becomes materially important to the project of constructing the knowledge characteristic of professional identity. Our work implies that although much has been made of contextualising the user, as a user, further work is required to contextualise users as developers and moreover, developers as users.

INTRODUCTION

In this chapter we offer insights into how a group of users interact with what can be seen as a knowledge sharing system—an ICT-supported repository used by call centre staff who offer expert advice on employment issues as HR management workers. The study provides a case of ICT-enabled knowledge sharing, insights into complex knowledge work in what is often regarded as a highly standard, rules-based environment and, in particular, we emphasize the role of knowledge in systems development and use. To do this we focus upon the roles of users in the development, tailoring, and maintenance of "the knowledge" component of the system in everyday practice. Our research question was: How is knowledge made by professional users, and given the presence of ICTs in our field site, what is their role in this practice, if any? Drawing upon the social shaping tradition, which we shall expand upon later, we argue for the recognition of the, oft politically constructed and negotiable boundary between developers and users. In the remainder of this section, we briefly set our view of knowledge as a practice.

Although there are those who privilege ICTs as "the" mechanism for capturing, storing, and disseminating knowledge, this has been challenged as lacking insight into different kinds of knowledge and its provisional situated nature (Blackler, 1995; Fleck, 1997; Marshall & Brady, 2001; Sutton, 2001). For example, knowledge may be embodied (knowledge about how to do something, gained through doing), embedded (where routine arrangements are deployed), embrained (akin to the holding of conceptual skills and cognitive abilities), encultured (rooted in shared understandings), and encoded (conveyed by signs and symbols) (Blackler, 1995; Collins, 1993). However, it has been further argued that greater insights can be gained by studying the processes of knowledge construction, rather than trying to describe and define its different forms. Knowledge is mediated by various things, situated in a given time and place, provisional in that it is socially constructed, pragmatic in that it is purposive and object oriented, and contested as it has links with power and politics (Blackler, 1995). Blackler (1995) therefore recommends that we focus upon the systems through which people achieve their knowing, on the changes that are occurring within such systems, and on the process through which new knowledge may be generated. Indeed, it has been further argued that rather than simply define or describe knowledge networks, the challenge is to show how particular practices and discourses sustain networks of power-knowledge relations (Knights, Murray, & Willmott, 1993). For example, historically, task-continuous status organisations were prevalent where functional and hierarchical differentiation coincided. In this environment, positions were defined, by, among other things, knowledge ownership (Offe, 1976). But modern organisations are said to exhibit increased task-discontinuation structuring of status and the function of work performed (Hardy & Clegg, 1996). We therefore emphasize the need to go beyond simplistic notions of knowledge as a commodity to be extracted and transferred (Walsham, 2001). Knowledge may be used in innovation appropriation processes to provide access to other relevant knowledge and systems and as a political tool in support of particular interests (Hislop, Newell, Scarbrough, & Swan, 2000). Knowledge informs and justifies how we act, when it is taken as "truth," especially when it is understood as neutral and authoritative, then it is powerful (Alvesson & Willmott, 1996). As mentioned earlier, knowledge is situated and therefore it is necessary to understand that knowledge construction is somewhat predetermined by the fact of "growing up" in a society (Mannheim, 2004), in our case, an organisation. Thus we have to be careful to avoid an excessively volunteeristic account of knowledge work in which actors are depicted as autonomous agents who possess sufficient resources to make their network a reality (Knights et al., 1993). Indeed, as Orlikowski (2002, 2006) argues the role of material forms, systems, spaces, and infrastructures in everyday knowledgeable practice are important.

In this study we explore the ICT-related organisational practices in a call centre environment, "CarePoint," where complex forms of knowledge and processes for the construction of knowledge are used in practice. We examine how a group of staff maintain an important knowledge sharing system and how this facilitates the appropriation of it into the everyday practices. Through this we illustrate the links among: the development of an ICT system in use, different forms of knowledge, and the roles of users as developers. We point to the use of politicised knowledge in the creation of the negotiable boundary between development and use, and thus the naming of developers and users. In the next section we expand upon the call centre research tradition and its contested relationship to knowledge work. In particular we begin to raise questions regarding call centre working and knowledge where the employees are professionals. In the following section, we introduce a social shaping lens to highlight the negotiable boundary between development and use, and developers and users. Following this, we outline our approach to the field study and then we provide an interpretation of our findings. The findings are then discussed; we present our conclusions and some implications for research and practice.

CALL CENTRES AND KNOWLEDGE WORK

The use of call centres has grown over the past 20 years, predominantly in response to the needs for globalisation, the potential they offer for improving organisational efficiency, and the desire to become more customer facing in the light of the business hype surrounding customer relationship management (Light, 2003). Call centre workers often perform roles that, in more traditional organisational settings, would be performed by a number of people. Indeed, where outsourcing companies operate call centres, the workloads of employees can be distributed over a multi-organisational customer base in order that they are most efficiently utilised. In order to further maximise the efficiencies to be had from the call centre model, ICT-enabled surveillance is used extensively to monitor performance. For example, the automatic call distribution software used to allocate calls to agents is also used to measure performance in terms of, for instance, length of time taken to answer calls, calls lost, and revenue generated (Taylor, Mulvey, Hymann, & Bain, 2002). Conventional call centre agents have little or no officially sanctioned autonomy; scripts are built into ICTs by developers to guide them through their interactions in an efficient and standardized fashion. Yet, in recent years, there has been a significant drive towards the "professionalisation of call centre operations" with UK organisations such as the Call Centre Managers Association promoting "the profession" (www.ccma.org.uk) and the UK Customer Contact Association (www.cca.org.uk) offering call centre accreditation and portable qualification initiatives for call centre staff. Employers thus, often depict call centres as knowledge-intensive environments where skilled, semi-professional workers are employed for their interpersonal skills (Frenkel, Tam, Korczynski, & Shire, 1998). However, the majority of academic accounts of such environments stress that routinisation, repetition, and employee disempowerment are the prominent features, even though there are studies of employees "fighting back" (Menzies, 1995; Taylor & Bain, 1999; Taylor et al., 2002; Van Den Broek, Callaghan, & Thompson, 2002).

Given this context, how then might ICT-enabled knowledge work be performed in a call centre environment which is often organized and managed via scripts, within a unitary organisational frame of reference? More specifically, given our field site, call centre employees have to create, modify, and share knowledge as part of their professional contract—how does this happen in an environment where one would expect high levels of ICT-enabled scripting to take precedence? Moreover, given it has recently been argued that most ICT-based systems are still currently developed as static entities whose purpose is to model a dynamic world (Kanellis & Paul, 2005), and that knowledge is provisional. Then the use and development of ICTs in knowledge intensive environments needs further investigation in terms of the role of users and developers might have in making such systems work in practice. In the next section we introduce a social shaping of technology (SST) lens to assist with this.

USER-DEVELOPER RELATIONS: A VIEW FROM SST

The social shaping perspective we adopt to study knowledge construction seems particularly appropriate given its genealogy in the sociology of scientific knowledge (SSK). SSK proponents claim that the natural world has a small or non-existent role in the construction of scientific knowledge and emphasises the social influences upon science (Collins, 1993). However, as Orlikowski (2002, 2006) reminds us, it is important to take account of the materiality of knowledge. SST has therefore evolved over time to take account of materiality. Contemporary SST researchers reject technologically, and socially, deterministic accounts of the construction and appropriation of technologies recognising the mutually constitutive nature, and negotiable boundary between, society and technology (for overviews see Bijker & Law, 1994; Mackenzie & Wajcman, 1999; Pinch & Bijker, 1987; Sørensen, 2002). From this perspective technology applications do not have predictable outcomes. Instead, technologies are conceptualised as being shaped as they are designed and used. Therefore, while they may change situations, the technologies themselves may be subject to change resulting in intended and unintended consequences for socio-technical arrangements. However, as mentioned earlier, ICT-based systems are often reported as being delivered as complete solutions, which are sufficiently specified a priori (consider software vendor websites). The consequence of this is that many systems are still deemed failures at some point in time despite user involvement in system development and implementation (Cavaye, 1995). This has been termed the design fallacy—the presumption that the primary solution to meeting user needs is to build ever more extensive knowledge about the specific context and purposes of various users into technology design (Stewart & Williams, 2005). Stewart and Williams argue that the problem with this thinking is that it privileges prior design, it is unrealistic and unduly simplistic, it may not be effective in enhancing design/use, and it overlooks opportunities for intervention. Indeed, it has been argued within the field of IS, the reality of the situation is that organisational features are products of constant social negotiation and consensus building and this means we need to rethink how ICTs are developed (Truex III, Baskerville, & Klein, 1999).

A further issue is that users are often seen not considered in their context and instead are often thought of in systems development and use, as using a given ICT in isolation from other affiliations, identities, interactions, and environments (Lamb & Kling, 2003). Developers too, have also often been seen as objective experts whose sole aim in life is to build the best system possible for an undifferentiated group of users. However, it is now increasingly recognised that such views are simplistic and that development and use is loaded with power and politics on "both sides" (Franz & Robey, 1984; Markus, 1983; Markus & Bjørn-Andersen, 1987; Yourdon, 1986). Yet, the "two sides" of users and developers in ICT efforts are still a key feature of IS research. We understand that power is exercised by developers over users (Markus & Bjørn-Andersen, 1987), and that certain users may exercise power over developers (Howcroft & Light, 2006), but users as developers exercising power has received minimal attention. Even the long tradition of end-user computing still predominantly refers to users, not as developers, but as users who happen to develop ICT-based systems. Users are rarely discussed in terms of any role they may have as a developer and developers are similarly usually not seen as users (Friedman & Cornford, 1989). This is despite case studies of users as developers and their valuable role in product development (cf. Holmström, 2001). Consequently, questions remain about whom users and developers are.

We believe this is inextricably linked with the artificial, and arguably political, distinction made between the social and the technical often made in the field of IS, when such distinctions are clearly socially constructed and negotiable (Bloomfield & Vurdubakis, 1994).

Drawing upon SST we wish to expand upon this in terms of the roles associated with work deemed social and technical. In sum, we think the boundary between the usage and production (or development) of that deemed social and technical—the socio-technical—is also negotiable (Rohracher, 2005). Therefore, although discussions of producer-user relations have yielded many interesting and valuable insights, we think it might also be useful to recognise the ongoing work that "users" put into socio-technical systems in situ (Fleck, 1994; Rohracher, 2005; Stewart & Williams, 2005). Not only do they use such systems, they produce them in use too. This idea of ongoing development in use is well known within the body of work known as the SST (Fleck, 1994; Rohracher, 2005; Stewart & Williams, 2005). However, within IS the distinction between users and developers remains. In the next section we provide details of our approach to undertaking the field work.

RESEARCH APPROACH

This study is part of a wider programme of work looking at the deployment of ICTs in professionally populated environments. The research approach is interpretive and qualitative. The findings are based on primary and secondary data drawn from one of the cases. The case study research method is largely acknowledged and is frequently used to conduct qualitative data informed IS research (Klein & Myers, 1999; Orlikowski, 1996; Walsham, 1993; Walsham, 1995). A range of techniques associated with the case study method were used given our concern for the effects of the socio-technical arrangements within the company rather than the "technical" aspects alone (Myers, 1997).

CarePoint, the human resources contact centre we have studied, employs 78 staff, only five of whom are male; the average age of the staff is 34 years. During our time at CarePoint, we were given company identification badges and were allowed free access to the call centre when visiting. We were not accompanied by a member of staff, unless we requested this for introductory purposes. A cross section of 14 members of staff were interviewed and observed over a number of sessions during a six-month period—this included the call centre manager, policy makers, case workers, and a range of HR advisors. For this study, a number of relevant social groups have been identified by following the actors and historical snowballing in line with Bijker (1994). Moreover, as Bijker suggests "this is of course an ideal sketch as the researcher will have intuitive ideas about what set of relevant social groups is adequate for the analysis of a specific artefact" (p. 77). At the beginning of each interview we explained the study's purpose and made the participants aware that we intended to publish our findings. However, we assured each participant that they would be anonymised and we gave them the opportunity to decline participating in the interview and the research more generally. We also informed them that they could decline to answer any of our questions and that they could review their transcript (all participants reviewed their transcript and corrected any misinterpretations on our part). It was also made clear that we would only discuss any points of interest with their colleagues, if they agreed to this. No one declined to be interviewed, and everyone was happy to have their thoughts shared with their colleagues. The time scale of individual interviews varied between 0.5 and 1.5 hours. The interviews were semi-structured; they were recorded and transcribed providing 81 pages of transcription. Additionally, documentary evidence such as CarePoint policies was referred to as necessary, this was collected from the company intranet and it was given to us by various employees during the process of our investigation. We also drew upon numerous sessions of non-participant overt observation and photographic evidence that we collected. Analysis and data collection were simultaneously carried out. This began with the aim of discovering the nature of the contact centre agent's professional identity, their background and their working environment. Data was then collected, guided by Johnson's (2001) conceptual map of the characteristics of professions combined with a literature review of knowledge and ICT use in organisations. Johnson supplies us with a list of characteristics often related to professions: mastery of an esoteric body of knowledge; autonomy, formal organisation, and code of ethics; and a social function with occupational autonomy and knowledge being two defining features of this case study. The collected data was coded in relation to the literature and the features of the professions framework, and finally a subset of this related to knowledge was further unpacked eventually leading to the identification of the analysis reported here.

CONSTRUCTING KNOWLEDGE AT CAREPOINT

CarePoint is an internal HR contact centre within an international health care product firm. The staff prefer CarePoint to be referred to as a contact centre as opposed to a call centre so as to get away from the traditional portrayal of a call centre. This decision was made because the staff do more than enact standardised scripts; the calls require the practice of knowledge as related to various HR issues. The contact centre staff provide HR support and advice to the 55,000 employees employed by the company. This work is undertaken by HR professionals and involves advising on complex issues such as maternity leave, retirement policy, and staff recruitment. Employees will telephone the staff with a query, and it is their role to resolve it, escalating the caller to different levels of contact centre workers as necessary. The centre was set up in July 2003 with most of the staff recruited from within existing HR roles within the company. However, two staff with experience of call routing and working in a call centre environment were also recruited from elsewhere in the company. In both cases the existing knowledge base of the staff was seen as crucial to the creation of this new organisational function. The majority of the staff within CarePoint are chartered Institute of Personnel and Development (IPD) qualified or are undergoing one of their programmes of education. They regard themselves as professionals and consider HR to be a profession.

The ICT-based system supporting the function was created by integrating pieces of software from existing systems throughout the wider organisation. This comprises a generic help desk application, a call logging system based on a customer relationship management package module, and a bespoke staff scheduling system written in Microsoft Excel. In addition, call scripts are created in Microsoft Word using HTML and these are integrated with the customer relationship management module. The customer relationship management module offers the facility for linking to external documents via HTML. Thus the scripts which the staff use are readily accessible by clicking on a link on the screen layout and this takes them to the HR "e-manual." This manual contains two kinds of scripts. First, it is a script in terms of the wording staff might use when dealing with an employee or manager's query. It also acts as a potential script in terms of how they might progress that query. For example, an employee may call in and want to know how much holiday entitlement they have for a given year. The script always begins with a prompt for a series of security checks, they are asked for their staff number and name. Their details are then retrieved by the call centre staff member and they are asked to confirm their date of birth so as to ensure the advisor is speaking to the right person. Once the details have been provided and checked for authenticity, the call centre staff would then, in theory, open the e-manual and find the section on holiday entitlement. This would tell them to ask the member of staff how long they have been working at the company and what grade they are (as entitlements differ accordingly). The call centre staff member will then follow the appropriate link, based on the staff's grade, and the system will present them with the holiday entitlement information organised by number of years of service. They then report back to the employee what their entitlement is. They would also advise that this is exclusive of any UK bank holiday entitlement.

Who are the Users and What do they Use?

CarePoint staff are the first obvious group of users within our case. They operate the ICT-based systems to provide advice to the managers and employees who contact them for guidance. Such knowledge-based interactions are mediated a number of things. A key aspect is the call flow model as shown in Figure 1 and it can be seen that Level 0 introduces another set of primary users, the employees and managers of the organisation CarePoint services.

Level 0 is "self help," where managers and employees can solve minor problems and queries themselves by referring to the standard HR operating manual, Storenet and other "toolkits" such as Breakfast News (a daily bulletin) via the Intranet. CarePoint advisors are organised across the remaining "levels." Level 1, the advisory service, is split into two sections: Level 1a comprises a team of *frontline advisors* who should be able to answer a broad range of questions and they usually take the incoming calls (when the contact centre is busy, all staff, irrespective of level, take incoming calls directly). Level 1b comprises a team of *senior advisors* who have the ability to give more expert advice; a case that is referred to this level of staff should take no longer than 20 minutes to complete the call. If a call is more complex and exceeds 20 minutes then the customer is routed to the next level of agent, Level 2, comprising a team of *case workers*. Case workers deal with cases that exceed 20 minutes that need to be discussed in more detail, such as issues regarding discrimination or bullying. They make the choice on how to

Figure 1. CarePoint conceptual operating model

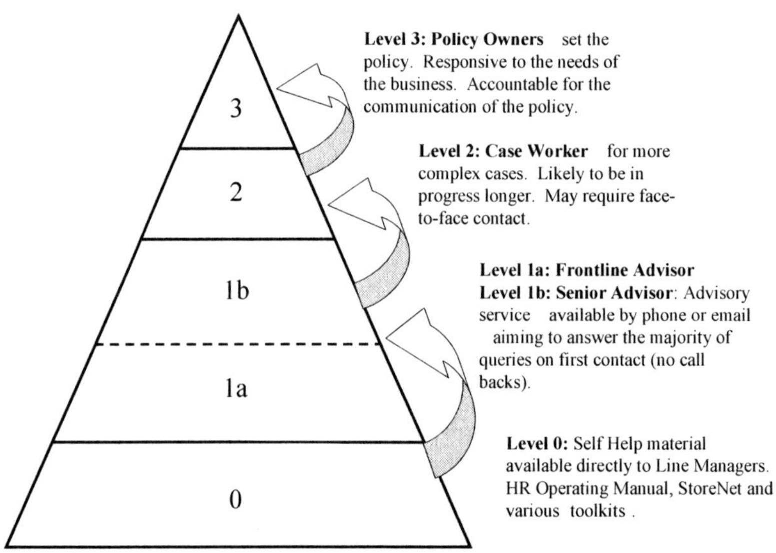

best support the line manager or employee, and if necessary, they visit them on site. Case workers also go out on site and deal with any new or extra businesses as they arise, the formation of new sites of operations for example. They also have particular responsibility for the development of the service; they look at how things can be done better. Level 3 hosts a team of *policy owners*; they deal with legal issues, set out various policies, are responsive to the needs of the business, and are accountable for the communication of changes in policy.

However, despite the internal structure of the contact centre appearing to be task-continuous, in this case the staff's relationship to other members of departments is quite discontinuous. Level 1 staff may deal with high ranking managers throughout the organisation because of their specialist encultured and embedded knowledge and by drawing upon the encoded knowledge in the scripting system. This meant call times could not be standardised. Staff may be dealing with a high ranking manager who does not want to be hurried. Moreover, some callers might have particularly difficult personal circumstances to deal with, for example bereavement or feelings of victimisation. Therefore, the staff we spoke with felt they would be neglecting their professional duties if they hurried people through the calls for the sake of improving call turnover statistics. This meant that when the call centre was set up, working arrangements had to be configured bearing this in mind, as one employee explained:

... and we came from HR, which could be very complicated and we needed to give them the benefit of our experience, so 15 minutes for the average call, and Janet [the call centre manager] wanted it cutting to 3 minutes and that's all you'll have and we had this ongoing debate, god you know there's no way we can share our marvellous knowledge in 3 minutes and if you look at the call stats now we actually do it in 7 minutes. (Adriana, Case Worker)

The uses of the "knowledge" component of the system, the scripts, are also deployed more flexibly than might be the case in a traditional call centre setting. Cathy, a senior advisor informed us that:

it would generally be the [level 1] advisors that would use the scripts, the stuff in scripts is fairly basic where a case is black or white, if it's a bit more grey and a bit more in depth then a bit more knowledge is needed

Even Level 1 advisors informed us that the scripts were of limited value:

I used the script regularly for the first two weeks while training, and sometimes I do still have to use them, but I have only been here three months, but because I come from an HR background I have prior knowledge of some issues, so that helps. (Peter, Junior Advisor)

We were told, and observed; that if the caller is a staff member then advisors stick rigidly to company policy on what is discussed referring the caller to the encoded knowledge in the scripts as necessary. However, if *line managers* call, advisors tend to draw upon encultured and embedded knowledge not recordable in the ICT-based scripts so they can give a more rounded discussion on the policy. The system is therefore used by more established members of staff in a supporting role if they are dealing with an unfamiliar case, with trainee staff using scripts in a supporting role, particularly in the first two weeks of training, and if they are new to the HR function. Case workers use the system to coach and develop advisors on legal issues and telephone manner and style; to learn from escalated and tribunal cases; to probe the case issues; and give guidance and advice. In addition all advisors are encouraged to use the system in their spare time to further enhance their knowledge and skills on various issues that they have yet to deal with. We noticed this happened on several occasions and when we asked the staff why they were doing it, they said it was part of their continuing professional development because the IPD was "strong on this."

A further distinguishing feature of this knowledge work environment is the approach to performance management. Data is logged and statistics are produced by senior call centre staff and used for reporting to the company on the number of calls taken, employer relations issues and trends, and to ensure each team is aware of their objectives. However the staff told us that this data was used to improve the service and their working conditions (for example by employing more staff as the number of calls increased) rather than to make them work any faster. We were also told that overt monitoring occasionally takes place for training purposes to ensure advisors are giving consistent advice on issues and in that respect all staff agreed that the system and monitoring taking place is beneficial to the company and the individual. As Peter, a junior HR advisor informed us:

We get messages to tell us that calls are monitored...people will always interpret things differently, possibly if you asked everyone here the same question, everyone will give you a slightly different answer, so it is important we all sing from the same hymn sheet when giving advice, but hopefully that's where our knowledge helps, it's a good thing.

The seating arrangements within CarePoint are very different to traditional call centre environments. Levels of staff are mixed together within the "pods" of the contact centre, whereas in more conventional call centre environments junior and senior staff tend to be segregated, as was the previous situation here. This action was taken to develop the group overall, thus reinforcing the group. So now if an advisor has a "tricky" call to deal with, they can put the caller on hold, ask for assistance from senior advisors and share that knowledge, therefore reducing their skills gap and instilling confidence:

If I get a strange call or something I can't deal with we have Senior Advisors seated alongside us, seating has been recently rearranged so that has helped a lot because we're taking more complex calls now and we can deal with them better because we're getting coaching from the team so we're up-skilling. (Heather, Junior HR Advisor)

Indeed, this is again an extension of recognising the knowledge base of the staff in configuring working practices. When the contact centre was first set up, staff had to be encouraged to move away from their comfort zone and develop their knowledge base further. As one manager commented:

The first challenge for me, as a line manager with my team of seven was because they all had their own jobs, they all stuck to their own jobs and they all had this long service, they weren't sharing their knowledge, well one of the first things I introduced them to was multi-skilling, so I made sure I had at least three people that could do each task. (Helen, Team Manager)

Development of the System in Use

So far, we have focussed upon the users in our study, now we move on to consider the developers of a particular part of it. Moreover, the users think this is the most important part—the ICT-based scripts that underpin the contact centres operations. As mentioned earlier, a central feature of traditional call centre working is the use of scripts for the purposes of maximising efficiency and reducing the need for skilled workers by imposing a standardised, prescriptive approach. Traditionally call centre scripts are produced by third party companies or in-house developers throughout the requirements gathering process of the development cycle, or in response to user needs. At CarePoint however, Jane, a senior advisor, told us that:

Senior Advisors construct the scripting, we have to think about how the questions will be asked in various ways, and look at how these can be answered. The HR manual is also reconstructed as and when necessary. When the script is written it goes to the Case Workers to check their interpretation of the Q and A to make sure it is all done properly, then it is saved.

In this case, users are also developers. As they are dealing with provisional knowledge, there is a need to constantly revise the scripts to build the base of encoded knowledge and it is a team effort to determine whether this happens or it just becomes encultured or embedded. As Helen a team manager commented:

It's a team effort though, we have meetings and if issues are raised regarding the scripts and they need to be altered, maybe a new query has been brought to light that we have no answer to, or maybe the answer to the question needed to be put into a different context, then it would be amended or written.

However, the professional identity of the group is also seen as something to be maintained. All codifiable forms of knowledge do not end up being encoded into the scripts. Case workers have monthly meetings where they share their learning and decide who this will be further shared with and how. The system may also not be developed further because of the temporally situated nature of the knowledge they deal with. For example, Heather, a junior HR advisor informed us that:

A useful tool we use is the Breakfast News that is communication between the teams about anything that needs to be shared...for instance we've got the Pope's funeral on Friday and we're starting to get calls about that, so everybody needs to give the same information...It doesn't go on the scripting because it's a one off and it's not in the HR manual. People are querying the one-minute silence. It's a lot easier to update people this way than to update the scripting.

Staff knowledge and expertise is a crucial element of working at CarePoint. Their professional identity is not only recognised by qualifications but in the main by an individual's knowledge and expertise. Jenny the talent management administrator told us that:

...there's a lot of people who have worked here for a long time and they've got a lot of knowledge about a lot of things and that's professional expertise, there's people here like that, their knowledge is so invaluable, so professional.

Such knowledge constructed through dealing with various cases is shared by pulling out "key learning" from the cases, selectively sharing the findings in regular meetings, deciding if there's anything they need to share with other staff (formally and informally), and recording it, if necessary, in the scripts. One advisor described the system as a "time saving tool" another as a "learning tool." This process of selectively sharing and encoding knowledge is also a powerful political tool. Advisors unanimously agreed that the system enhances their professionalism, finding out information via the intranet is easier, it makes it possible for individuals to have a broader span of what information is available concerning the various issues they are dealing with as Melanie, a senior advisor stated:

The system can pull up some wonderful information, it can tell you every minute of what's happening with a case...if you think about the responsibility we have got here, we support the whole of the company with HR issues, we are very powerful, the technology aids our power and knowledge...it enhances our professionalism, you know you could actually say that the technology in a way is your 'buddy' because the system holds all the answers for you to enable you to do your job properly.

Critically though, the staff of the contact centre control and enact the development of the scripts on their own terms, within the parameters of their role of course. In doing this they are able to present certain knowledge in an encoded format and label it as fairly basic "the black and white issues" that lay people might be able to understand if they write it correctly and if support is there from them to interpret it as necessary. However, they also get to choose which knowledge remains embedded and encultured and in turn this mystifies their professional position. Ironically, the ICTs that they value so much are also distanced in some circumstances as not being capable of doing what they do, even though they have developed them.

DISCUSSION

Our research question was: How is knowledge made by professional users, and given the presence of ICTs in our field site, what is their role in this practice, if any? In sum, we concur with prior studies which recognise the political nature of knowledge practice and the role of the material in this (Blackler, 1995; Hislop et al., 2000; Orlikowski, 2002, 2006). In this case, we see the prime materials as professional knowledge and ICTs. Thus, particular forms of knowledge and knowledge practices are implicated in the construction of further knowledge and knowledge practices and ICTs are enacted in a variety a ways in support of this project. Through these practices, the political boundary between users and developers, often constructed itself on the basis of claims to knowledge, is blurred as each becomes the other. We shall now explain this position further.

There are three main forms of knowledge that we focus upon in this study, encoded, embedded, and encultured. Encoded knowledge is that which is written into the scripts. Embedded and encultured knowledge is that retained by HR staff either because it is not possible to encode it or it is deemed not desirable to do so. Perhaps most starkly, we see that embedded and encultured knowledge is seen as securing the professional project. HR staff are very careful in determining which knowledge is encoded into the scripts—it is a political practice. Such knowledge is used as a source of power to retain and reinforce professional identity. The role of the ICT-based scripts in this practice is complex. Although, ICTs have been noted as heightening professional identities (Lamb & Kling, 2003) and we see this in our case, it is also clear that considerable work goes into development in situ to fend off any encroachment such systems might enable. Thus, the HR staff simultaneously see the technology of the scripts as, central to the professional project, what one advisor called, "a buddy" and also something that they do not need because what they know could not possibly be encoded. It is also used to reinforce their professional position through their control of its development to delegate certain tasks, which they see as lacking in professional status, to lay employees and managers. They are therefore exercising various forms of professional power in IS development (Markus & Bjørn-Andersen, 1987), as other "user/developer" studies have shown (Howcroft & Light, 2006). Their role becomes one of providing access to such "basic" knowledge, only stepping in when expert interpretation is required. This is interesting as Lamb and Kling (2003) state that, at the individual level—social actors exercise limited discretion in ICT choice and use, since, in organisational contexts; they articulate the preferences of a collection of actors. However, while we accept that this may be the case, knowledge is socio-technically structured (Mannheim, 2004), we also show that individuals can have a high level of discretion as to whether they use particular features of an ICT—scripts are optional and often downplayed in terms of their value. Therefore, the scripts are used as materials in knowledge practice in a variety of ways.

This usage of ICTs does, we argue, blur the line between development and use. We do recognize that many researchers may not see their creation, tailoring, and general maintenance of scripts as "proper development" but we take a different view. The line between development and use has to be seen as a politically constructed and negotiable knowledge boundary—developers are the experts who know how to develop, users do not. Yet, in our case, we have clear evidence that the HR staff have very definite, knowledgeable ideas about which practices can and should be encoded. Moreover, they are confident in implementing these developments using HTML code. Of course, knowledge is provisional, several years ago HTML would have been seen as the realm of the developer, now of course many would say that anyone can do it, or that "users" only have a rudimentary understanding of the language. The question is, does this matter? In this case, a group of users are shown to be developers of a knowledge management system that works for them in practice. Moreover, in this case, they are the developers of what they deem to be the most important part of the system. If the work scheduling system or call logging system failed, they could still answer the telephones and give advice using the scripts (accessed via the intranet). If the scripts were not in existence, they may (particularly new staff) encounter difficulties in service provision. The users are the developers of a mission critical system.

Finally, it is also worth discussing our findings in relation to the end-user computing literatures more generally. It is suggested that the process of developing an application not only predisposes an end user developer to be more satisfied with the application than they would be if it were developed by another end user, but it also leads them to perform better with the application than they would if it were developed by another end user (McGill, 2004). The implication of McGill's work is that performance relies on direct, hands-on use of the system. Our study also suggests that end users might perform better with an application they have had a hand in developing because they know when to use it and when not to. In terms of our case, we can think of performance in two ways. First, as related to answering an employees HR query—they can do it quickly without relying on the scripts. Second, in relation to performing the professional project—they can choose when and when not to encode particular knowledge practices. It has also been argued that use of user-developed applications requires substantial end user knowledge because of the lack of separation of data and processing that is commonly found (Hall, 1996; Ronen, Palley, & Lucas, 1989). However, in our study, this was not the case, the end-user-developed scripts were usable, and used by lay users (managers, employees, and new call centre recruits), because this application was developed for them as well as those who developed it. Again, this reinforces the user-developer's dexterity in systems analysis and design. However, of course the scripts are not always successfully navigated by users, and thus, like good developers, they engage in refinement of the system and user training on a continual basis. In this case, the socio-technical arrangements, as Truex III et al. (1999) suggest, are optimized for high maintenance.

CONCLUSION

Our study provides a case of users interactions with what can be viewed as a knowledge sharing system. Moreover, the case is of interest as a site of work relying on complex knowledge practices in what is often, despite managerial rhetorical efforts, a highly routine knowledge deprecating environment. We gain insights into a different call centre working model which rejects ICT process and socio-geographic configurations based on scientific management principles. Call centres are not, generally speaking, associated with knowledgeable autonomy, knowledge of the rules is required. Given the rise in the call for the professionalisation of call centre working, our site offers interesting insights for those looking to change working arrangements to give employees more credit for the ability to enact more complex knowledge practices. Moreover, we would suggest that further studies of professional user groups in other settings would yield interesting insights for studies of knowledge sharing and ICTs.

Our particular focus has been to shed light on a group of users, who traverse the politically constructed knowledge boundary into development, and how they get a knowledge sharing system to work for them in situ. In our case users develop a knowledge sharing system for use by themselves and others, in a variety of ways; as with other development processes, this is political. The implication of this for research and practice is the need to recognise the fact of ongoing ICT development, and the actors who perform this in situ. We would add this might be undertaken by users as developers. Moreover, our work also suggests the need for more attention to be paid to the role of developers as users. Given the rise of social software, open source communities, and the packaged software industry, a broader consideration of ICT developers as users is required. In each of these scenarios developers not only use the software as an application, but also as a system to fulfil a particular need—to sell to make a living for example. Much has been made of contextualising the user, further work is required to contextualise the developer as a user and understand the social actors in ICTs environments who straddle both politically constructed knowledge domains.

REFERENCES

Alvesson, M., & Willmott, H. (1996). *Making sense of management: A critical introduction.* London: Sage.

Bijker, W. E. (1994). The social construction of fluorescent lighting, or how and artefact was invented in its diffusion stage. In W. E. Bijker & J. Law (Eds), *Shaping technology/building society: Studies in sociotechnical change* (pp. 75-104). Cambridge, MA: MIT Press.

Bijker, W. E., & Law, J. (1994). *Shaping technology/building society: Studies in sociotechnical change.* Cambridge MA: MIT Press.

Blackler, F. (1995). Knowledge, knowledge work and organizations: An overview and interpretation. *Organization Studies, 16*(6), 1021-1046.

Bloomfield, B. P., & Vurdubakis, T. (1994). Boundary disputes: Negotiating the boundary between the technical and the social in the development of IT systems. *Information Technology and People, 7*(1), 9-24.

Cavaye, A. L. M. (1995). User participation in system development revisited. *Information and Management, 28*(5), 311-323.

Collins, H. (1993). The structure of knowledge. *Social Research, 60*(1), 95-116.

Fleck, J. (1994). Learning by trying: The implementation of configurational technology. *Research Policy, 23*(6), 637-652.

Fleck, J. (1997). Contingent knowledge and technology development. *Technology Analysis and Strategic Management, 9*(4), 383-397.

Franz, C. R., & Robey, D. (1984). An investigation of user-led system design: Rational and political perspectives. *Communications of the Association for Computing Machinery, 27*(12), 1202-1209.

Frenkel, S., Tam, M., Korczynski, M., & Shire, K. (1998). "Beyond bureaucracy" work organisation in call centres. *International Journal of Human Resource Management, 9*(6), 957-979.

Friedman, A. L., & Cornford, D. S. (1989). *Computer systems development: History, organization and implementation.* Chichester, England: John Wiley and Sons.

Hall, M. J. J. (1996). A risk and control oriented study of the practices of spreadsheet application developers. In *Proceedings of the 29th Hawaii International Conference on System Sciences* (pp. 364-373).

Hardy, C., & Clegg, S. R. (1996). Some dare call it power. In S. R. Clegg, C. Hardy, & W. R. Nord (Eds), *Handbook of organization studies* (pp. 622-641). London: Sage.

Hislop, D., Newell, S., Scarbrough, H., & Swan, J. (2000). Networks, knowledge and power: Decision making, politics and the process of innovation. *Technology Analysis and Strategic Management, 12*(3), 399-411.

Holmström, H. (2001). Virtual communities as platforms for product development: An interpretive case study of customer involvement in online game development. In S. Sarkar, V. C. Storey, & J. I. De Gross (Eds), *Proceedings of the 22nd International Conference on Information Systems Association for Information Systems*, New Orleans, LA (pp. 299-306).

Howcroft, D., & Light, B. (2006). Reflections on issues of power in packaged software selection. *Information Systems Journal, 16*(3), 215-235.

Johnson, D. G. (2001). *Computer ethics* (3rd ed.). Upper Saddle River, NJ: Prentice Hall.

Kanellis, P., & Paul, R. J. (2005). User behaving badly: Phenomena and paradoxes from an investigation into information systems fit. *Journal of Organizational and End User Computing, 17*(2), 64-91.

Klein, H. K., & Myers, M. D. (1999). A set of principles for conducting and evaluating interpretive field studies in information systems. *MIS Quarterley, 23*(1), 67-94.

Knights, D., Murray, F., & Willmott, H. (1993). Networking as knowledge work: A study of strategic inter-organizational development in the financial services industry. *Journal of Management Studies, 30*(6), 975-995.

Lamb, R., & Kling, R. (2003). Reconceptualizing users as social actors in information systems research. *MIS Quarterly, 27*(2), 197-235.

Light, B. (2003). A study of organisational experiences of CRM packaged software. *Business Process Management Journal, 9*(5), 603-616.

Mackenzie, D., & Wajcman, J. (Eds.). (1999). *The social shaping of technology* (2nd ed.). Maidenhead, England: Open University Press..

Mannheim, K. (2004). The sociology of knowledge and ideology. In C. Lemert (Ed.), *Social theory: The multicultural and classic readings* (3rd ed., pp. 213-216). Oxford, England: Westview Press.

Markus, M. L. (1983). Power, politics, and MIS implementation. *Communications of the Association for Computing Machinery, 26*(6), 430-444.

Markus, M. L., & Bjørn-Andersen, N. (1987). Power over users: Its exercise by system professionals. *Communications of the Association for Computing Machinery, 30*(6), 498-504.

Marshall, N., & Brady, T. (2001). Knowledge management and the politics of knowledge: Illustrations from complex products and systems. *European Journal of Information Systems, 10*(2), 99-112.

McGill, T. (2004). The effect of end user development on end user success. *Journal of Organizational and End User Computing, 16*(1), 41-58.

Menzies, H. (1995). *Whose brave new world? The information highway and the new economy,* Toronto, Canada: Between the Lines Press.

Myers, M. D. (1997). *Qualitative research in information systems.* Retrieved November 21, 1997, from http://www.auckland.ac.nz/msis/isworld/

Offe, C. (1976). *Industry and inequality.* London: Edward Arnold Publishers Limited. (Original work published in 1970)

Orlikowski, W. (2002). Knowing in practice: Enacting a collective capability in distributed organizing. *Organization Science, 13*(3), 249-273.

Orlikowski, W. (2006). Material knowing: The scaffolding of human knowledgeability. *European Journal of Information Systems, 15*(5), 460-466.

Orlikowski, W. J. (1996). Improvising organizational transformation over time: A situated change perspective. *Information Systems Research, 7*(1), 63-92.

Pinch, T., & Bijker, W. E. (1987). The social construction of facts and artifacts: Or how the sociology of science and the sociology of technology might benefit each other. In W. E. Bijker, T. P. Hughes, & T. Pinch (Eds), *The social construction of technological systems* (pp. 17-50). London: MIT Press.

Rohracher, H. (2005). From passive consumers to active participants: The diverse roles of users in innovation processes. In H. Rohracher (Ed.), *User involvement in innovation processes: Strategies and limitations form a socio-technical perspective* (pp. 9-35). Wien, Austria: Profil.

Ronen, B., Palley, M. A., & Lucas, H. C. (1989). Spreadsheet analysis and design. *Communications of the ACM, 32*(1), 84-93.

Sørensen, K. H. (2002). Social shaping on the move? On the policy relevance of the social shaping of technology perspective. In K. H. Sørensen & R. Williams (Eds.), *Shaping technology, guiding policy: Concepts, spaces and tools* (pp. 19-35). Cheltenham, England: Edward Elgar.

Stewart, J., & Williams, R. (2005). The wrong trousers? Beyond the design fallacy: Social learning and the user. In H. Rohracher (Ed.), *User involvement in innovation processes: Strategies and limitations from a socio-technical perspective* (pp. 39-71). Wien, Austria: Profil.

Sutton, D. C. (2001). What is knowledge and can it be managed? *European Journal of Information Systems, 10*(2), 80-88.

Taylor, P., & Bain, P. (1999). An assembly line in the head: Work and employee relations in the call centre. *Industrial Relations Journal, 30*(2), 101-117.

Taylor, P., Mulvey, G., Hymann, J., & Bain, P. (2002). Work organization, control and the experience of work in call centres. *Work, Employment and Society, 16*(1), 133-150.

Truex III, D. P., Baskerville, R., & Klein, H. (1999). Growing systems in emergent organizations. *Communications of the ACM, 42*(8), 117-123.

Van Den Broek, D., Callaghan, G., & Thompson, P. (2002). Call centres: Exploring the team paradox. In *Proceedings of the 6th International Workshop on Teamworking,* Malmo, Sweden.

Walsham, G. (1993). *Interpreting information systems in organizations.* Chichester, England: John Wiley & Sons.

Walsham, G. (1995). Interpretive case studies in IS research: Nature and method. *European Journal of Information Systems, 4*(2), 74-81.

Walsham, G. (2001). *Making a world of difference: IT in a global context.* Chichester, England: John Wiley and Sons Ltd.

Yourdon, E. (1986). *Managing the structured techniques: Strategies for software development in the 1990s* (3rd ed.). Englewood Cliffs, NJ: Yourdon Press.

This work was previously published in End-User Computing: Concepts, Methodologies, Tools, and Applications, edited by S. Clarke, copyright 2008 by Information Science Reference, formerly known as Idea Group Reference (an imprint of IGI Global).

Chapter IX
Two Experiments in Reducing Overconfidence in Spreadsheet Development

Raymond R. Panko
University of Hawai'i, USA

ABSTRACT

This chapter describes two experiments that examined overconfidence in spreadsheet development. Overconfidence has been seen widely in spreadsheet development and could account for the rarity of testing by end-user spreadsheet developers. The first experiment studied a new way of measuring overconfidence. It demonstrated that overconfidence really is strong among spreadsheet developers. The second experiment attempted to reduce overconfidence by telling subjects in the treatment group the percentage of students who made errors on the task in the past. This warning did reduce overconfidence, and it reduced errors somewhat, although not enough to make spreadsheet development safe.

INTRODUCTION

Spreadsheet development was one of the earliest end-user applications, along with word processing. Spreadsheet development continues to be among the most widely used computer applications in organizations (United States Bureau of the Census, 2003). Although many spreadsheets are small and simple throwaway calculations, surveys have shown that many spreadsheets are quite large (Cale, 1994; Cragg & King, 1993; Floyd, Walls, & Marr, 1995; Hall, 1996), complex (Hall, 1996), and very important to the firm (Chan & Storey, 1996; Gable, Yap, & Eng, 1991).

Unfortunately, there is growing evidence that inaccurate spreadsheets are commonplace. For instance, Table 1 shows that recent audits of 88 real-world spreadsheets have found errors in 94%; yet several studies only reported spreadsheets with

Table 1. Studies of spreadsheet errors

Study	Year	Number of Spreadsheets	Percent of Spreadsheets Containing at Least One Error	Formula Error Rate (FER): Percent of Cells Containing Errors
Field Audits				
Hicks	1995	1	100%	1.2%
Coopers & Lybrand (c)	1997	23	91%	
KPMG (b)	1998	22	91%	
Lukasic	1998	2	100%	2.2%, 2.5%
Butler	2000	7	86%	0.4%
Clermont, Hanin, & Mittermeier (a)	2002	3	100%	1.3%, 6.7%, 0.1%
Lawrence & Lee	2004	30	100%	Average of 6.9%
Total/Per Spreadsheet		88	94%	5.2%
Development Experiments				
Brown & Gould	1987	27	63%	
Olson & Nilsen (f,g)	1987-1988	14		21%
Lerch (f,g)	1988	21		9.3%
Hassinen (g)	1988	92	55%	4.3%
Panko & Halverson	1997	42	79%	5.6%
Panko & Halverson	1997	35	86%	4.6%
Teo & Tan	1997	168	42%	2.1%
Panko & Sprague (i)	1998	26	35%	2.1%
Panko & Sprague (j)	1998	17	24%	1.1%
Janvrin & Morrison (h)	2000	61		6.6%-9.6%
Janvrin & Morrison (h)	2000			8.4%-16.8%
Kreie (posttest)	2000	73	42%	2.5%

(a) Computed on basis of all non-empty cells instead of on the basis of formula cells. (b) Only spreadsheets with major errors were counted. (c) A dependent variable value was off by at least 5%. (d) Only errors large enough to demand additional tax payments were counted. (e) Only serious errors were counted. (f) Counted errors even if they were corrected by the developer. (g) CER is based only on formula cells. (h) CER was based only on high-risk formula cells. (i) MBA students with little or no development experience. (j) MBA students with at least 250 hours of spreadsheet development experience.
Source: Panko (http://panko.cba.hawaii.edu/ssr/). References to studies are given at the Web site.

serious errors. The implications of this ubiquity of errors are sobering.

As Table 1 shows, the field audits that measured the frequency of errors on a per-formula basis found an average formula error rate of 5.2%. This formula error rate explains why so many of the examined spreadsheets contained errors. Most large spreadsheets contain hundreds or thousands of formulas. Given the cell error rates found in field audits, the question is not whether large spreadsheets contain errors, but rather how many errors they contain and how serious these errors are.

These field audits and the experiments described later found three types of errors.

- Mechanical errors are mental/motor skill slips, such as typing the wrong number or pointing to the wrong cell when entering a formula.
- Logic errors are incorrect formulas caused by having the wrong algorithm or expressing the algorithm incorrectly.
- Finally, omission errors occur when the developer leaves something out of the model.

Although observed spreadsheet error rates are troubling, they should not be surprising. Human error research has shown consistently that for nontrivial cognitive actions, undetected and therefore uncorrected errors are always present in a few percent of all cognitive tasks (panko.cba.hawaii.edu/HumanErr). In software development, for instance, over 20 field studies have shown that about 2% to 5% of all lines of code will always be incorrect, even after a module is carefully developed (panko.cba.hawaii.edu/HumanErr/ProgNorm.htm).

In the face of such high error rates, software development projects usually devote about a third of their effort to postdevelopment error correction (Grady, 1995; Jones, 1998). Even after several rounds of postdevelopment testing, errors remain in 0.1% to 0.4% of all lines of code (panko.cba.hawaii.edu/HumanErr/ProgLate.htm).

The testing picture in spreadsheet development, however, is very different. Organizations rarely mandate that spreadsheets and other end-user applications be tested after development (Cale, 1994; Cragg & King, 1993; Floyd, Walls, & Marr, 1995; Galletta & Hufnagel, 1992; Hall, 1996; Speier & Brown, 1996), and individual developers rarely engage in systematic testing on their own spreadsheets after development (Cragg & King, 1993; Davies & Ikin, 1987, Hall, 1996; Schultheis & Sumner, 1994).

In the face of large error rates in spreadsheet development and other human cognitive domains, why is testing so rare in spreadsheet development? The answer may be that spreadsheet developers are overconfident of the accuracy of their spreadsheets. If they think that there are no errors or that errors are at least very unlikely, developers might not feel the need to do extensive testing. Rasmussen (1990) has noted that people use stopping rules to decide when to stop doing activities such as testing. If people are overconfident, they are likely to stop too early.

In the first known experiment to examine confidence in spreadsheet development, Brown and Gould (1987) had nine highly experienced spreadsheet developers each create three spreadsheets from word problems. Of the 27 spreadsheets developed, 63% contained errors, and all of the nine developers made at least one error. Yet, when subjects were asked to rate their confidence in the accuracy of their spreadsheets, their mean response was "quite confident." High confidence in the correctness of spreadsheets has also been seen in other spreadsheet experiments (Panko & Halverson, 1997), field audits (Davies & Ikin, 1987), and surveys (Floyd, Walls, & Marr, 1995).

However, these measurements of spreadsheet overconfidence used 5-point and 7-point Likert scales, which can be difficult to interpret. For instance, when developers in the Brown and Gould (1987) experiment rated themselves as "quite confident," this was still only 4 on a scale of 5, which perhaps indicates only moderate confidence.

Reithel, Nichols, and Robinson (1996) did an interesting experiment in which they showed students printouts of long and short spreadsheets that were either well formatted or poorly formatted. The long spreadsheets had 21 rows, while the short spreadsheets had only 9 rows. In both cases, the subjects only saw numbers, not the underlying formulas. The subjects had substantially more confidence in the correctness of long, well-formatted spreadsheets than in the correctness of the other three types of spreadsheets. In a personal communication with the author of this chapter in 1997, Reithel said that for the large fancy spreadsheet, 72% of the subjects gave the spreadsheet the highest confidence range (80%-100%), 18% chose the 60% to 70% range, and only 10% chose the 1%-60% range. For the other three conditions, about a third of the respondents chose each of these three ranges. This pattern of confidence is illogical because larger spreadsheets are more likely to have problems than smaller spreadsheets. The research reported in this chapter confirms that there is a strong illogical element in user confidence in spreadsheet accuracy.

TWO EXPERIMENTS

This chapter presents two experiments to shed light on overconfidence in spreadsheet development. The first—an exploratory study—develops a more easily interpreted measure of overconfidence than Likert scales to address whether the high levels of confidence seen in Likert scales really are as extreme as they seem to be. Specifically, each subject was asked to estimate the probability that he or she had made an error during the development of his or her spreadsheet. This is called the expected probability of error (EPE). The mean of these EPEs was compared with the actual percentage of incorrect spreadsheets.

The second experiment used a manipulation to see if feedback could reduce overconfidence and, hopefully, improve accuracy as a consequence of reduced overconfidence. Specifically, subjects in the treatment group were told the percentage of subjects who had produced incorrect spreadsheets from the task's word problem in the past, while subjects in the control group were not given this information. Both groups then built spreadsheets from the word problem.

OVERCONFIDENCE

In studying overconfidence, we can draw upon a broad literature. Overconfidence appears to be a strong general human tendency. Research has shown that most people believe that they are superior to most other people in many areas of life (Brown, 1990; Koriat, Lichtenstein, & Fischoff, 1980). Nor is overconfidence limited to personal life. Problem solvers and planners in industry also tend to overestimate their knowledge (Koriat, Lichtenstein, & Fischoff, 1980). Indeed, a survey of the overconfidence literature (Pulford & Colman, 1996) has shown that overconfidence is one of the most consistent findings in behavioral research.

Overconfidence can be dangerous. In error detection, as noted earlier, we have stopping rules that determine how far we will go to look for errors. If we are overconfident of our accuracy, we may stop looking for errors too soon. If spreadsheet developers are overconfident, this may lead them to stop error detection short of formal testing after development.

Fuller (1990) noted that engaging in risky behavior actually can be self-reinforcing. If we take risky actions when we drive, this rarely causes accidents, so we get little negative feedback to extinguish our behavior. At the same time, if we speed, we arrive earlier, and this reinforces our risky behavior. In spreadsheet development, developers who do not do comprehensive error checking are rewarded both by finishing faster and by avoiding onerous testing work.

Indeed, even if we find some errors as we work, this actually may reinforce risky behavior. In a simulation study of ship handling, Habberley, Shaddick, and Taylor (1986) observed that skilled watch officers consistently came hazardously close to other vessels. In addition, when risky behavior required error-avoiding actions, the watch officers experienced an increase in confidence in their "skills" because they had successfully avoided accidents. Similarly, in spreadsheet development, if we catch some errors as we work, we may believe that we are skilled in catching errors and so have no need for formal postdevelopment testing.

The most consistent finding within laboratory overconfidence research is the "hard-easy effect" (Clarke, 1960; Lichtenstein, Fischoff, & Philips, 1982; Plous, 1993; Pulford & Coleman, 1996; Wagenaar & Keren, 1986). Quite simply, overconfidence is much higher for hard tasks than for easy tasks. In studies that have probed this effect, subjects were given tasks of varying difficulty. These studies found that although accuracy fell in more difficult tasks, confidence levels fell only slightly, so that overconfidence increased. Task difficulty can be expressed in the percentage of people making errors. Given the high number of errors found in the spreadsheet audits and experiments shown in Table 1, spreadsheet development must be classified as a difficult task. Accordingly, we would expect to see substantial amounts of overconfidence in spreadsheet development.

Researchers have tried to reduce overconfidence using several procedural innovations. One study (Lichtenstein & Fischoff, 1980) found that systematic feedback was useful. Over a long series of trials, subjects were told whether or not they were correct for each question. Overconfidence decreased over the series. In another study (Arkes, Christensen, Lai, & Blumer, 1987), subjects had lower confidence (and therefore less overconfidence) when given feedback after five deceptively difficult problems. In addition, we know from Kasper's (1996) overview of decision support system (DSS) research that merely providing information is not enough—feedback on the correctness of decisions must be detailed and consistent. These studies collectively suggest that feedback about errors can reduce overconfidence.

Most laboratory studies, like the ones described in this chapter, use students as subjects. However, studies have shown that experts also tend to be overconfident in their professional fields (Shanteau & Phelps, 1977; Wagenaar & Keren, 1986). One puzzle from research on experts is that experts in some occupations are very well calibrated in confidence (Keren, 1992; Shanteau & Phelps, 1977; Wagenaar & Keren, 1986), while experts in other occupations are very poorly calibrated (Camerer & Johnson, 1991; Johnson, 1988; Shanteau & Phelps, 1977; Wagenaar & Keren, 1986). Shanteau (1992) analyzed situations in which experts were either well or poorly calibrated. He discovered that experts tend to be well calibrated if and only if they receive consistent and detailed feedback on their error rates. Wagenaar and Reason (1990) also emphasized the importance of experts comparing large numbers of predictions with actual outcomes in a systematic way if their confidence is to be calibrated. This need for analyzed feedback among professionals is reminiscent of results from laboratory research to reduce overconfidence noted earlier.

Note that experience is not enough. Many studies of experts looked at people with extensive experience. In many cases, however, these experts did not receive detailed and consistent feedback. For instance, blackjack dealers, who merely deal and have no need to analyze and reflect upon the outcome of each deal afterward, are not better calibrated than lay people at blackjack (Wagenaar & Keren, 1986). In contrast, expert bridge players get feedback with each hand and analyze that feedback in detail (Wagenaar & Keren, 1986). They are well calibrated in confidence.

As noted above, spreadsheet developers rarely test their spreadsheets in detail after development. With little systematic feedback because of the rarity of postdevelopment testing, it would be surprising if spreadsheet developers were well calibrated in their confidence. In contrast, one of the tenets of software code inspection is the reporting of results after each inspection (Fagan, 1976). Therefore, software developers, who do extensive postdevelopment testing, also get detailed feedback for analysis. Consequently, they have the motivation to continue doing extensive testing because of the errors this testing reveals.

Most overconfidence studies have looked at individuals. However, managers and professionals spend much of their time working in groups. Therefore, we would like to know whether groups, like individuals, are chronically overconfident. In fact, there is evidence that overconfidence does occur in groups (Ono & Davis, 1988; Sniezek & Henry, 1989; Plous, 1995). This is important because Nardi and Miller (1991) found that groupwork is common in spreadsheet development, although often in limited degrees, such as error checking and providing advice for difficult parts of a spreadsheet.

Although the overconfidence literature is largely empirical and is weak in theory, a number of research results suggest that overconfidence is an important issue for spreadsheet accuracy.

- First, the broad body of the literature has shown that overconfidence is almost universal, so we should expect to see it in spreadsheet development.
- Second, as noted earlier, overconfidence tends to result in risky behavior, such as not testing for errors.
- Third, error rates shown in Table 1 indicate that spreadsheet development is a difficult task, so in accordance with the hard-easy effect, we should expect substantial overconfidence in spreadsheet development.
- Fourth, even experts are poorly calibrated in confidence unless they do consistent and reflective analysis after each task, which is uncommon in spreadsheet development.
- Fifth, it may be possible to reduce overconfidence by providing feedback.
- Sixth, reducing overconfidence may reduce errors, although this link is not demonstrated explicitly in the overconfidence literature.

EXPERIMENT 1: ESTABLISHING THE PRESENCE OF OVERCONFIDENCE

In our first experiment, the goals were simple: to see if the high apparent confidence levels seen previously with Likert scale questions really indicate a very low perceived likelihood of making an error, and to see if the method for measuring confidence used in this study appears to be useful. We measured confidence after development, and we had no manipulation of confidence. The second experiment added a confidence manipulation and before-and-after confidence measures.

Sample

The sample consisted of upper-division undergraduate management information systems majors in the business school of a medium-size state university in the middle of the Pacific Ocean. All had taken a course that taught spreadsheet development and a subsequent course that used spreadsheets extensively, and all had taken two accounting courses.

Subjects engaged in the experiment to receive extra credit—one quarter of a letter grade. Over 80% of the students in the class participated. Accounting and finance students were excluded because of their specialized task knowledge. This left 80 participants. Subjects either worked alone or in groups of three (triads). Forty-five students worked in triads, while 35 worked alone.

Task (MicroSlo)

The task used in this study was the MicroSlo task, which required students to build a pro-forma income statement from a word problem. This task was selected because all subjects had taken one year of accounting and thus should have been able to do the task. The MicroSlo task is based on the Galumpke task, developed previously by Panko and Halverson (1997). MicroSlo is the Galumpke task minus a capital purchase subtask, which could not be handled by most students (Panko & Halverson, 1997).

Your task is to build a two-year pro-forma income statement for a company. The company sells microwave slow cookers for use in restaurants. The owner will draw a salary of $80,000 per year. There is also a manager of operations, who will draw a salary of $60,000 per year. The income tax rate is expected to be 25% in each of the two years. Each MicroSlo cooker will require $40 in materials costs and $25 in labor costs in the first year. These numbers are expected to change to $35 and $29 in the second year. The unit sale price is expected to be $200 in the first year and to grow by 10% in the second year. There will be three sales people. Their salary is expected to average $30,000 per person in the first year and $31,000 in the second. Factory rent will be $3,000 per month. The company expects to sell 3,000 MicroSlo cookers in the first year. In the second, it expects to sell 3,200.

Dependent Variables

After subjects had built the spreadsheet, they were asked to estimate the probability that they (or their triad) had made an error when building the spreadsheet. As noted earlier, this was the estimated probability of error (EPE). A higher EPE indicates lower confidence. Error likelihood estimates could vary from 0% to 100%.

Procedure

Subjects working alone used computers in a common room. They were monitored to prevent cheating. Triads worked in other rooms, one triad to a room. Each triad shared a single computer.

Counting Errors

To count errors, the author saved each subject's spreadsheet under a different file name and compared the spreadsheet to a standard solution. If the spreadsheet had the same bottom-line results (income after tax for each year) as the standard

Table 2. Overconfidence in monadic and triadic spreadsheet development

	Subjects working alone			Subjects working in triads		
	Self	Other working alone	All triads	Own triad	All working alone	Other triads
Number of subjects	35	35	35	45	44	44
Mean estimated probability of an error (EPE)	18%	22%	10%	13%	33%	21%
Median estimated probability of an error	10%	10%	5%	5%	25%	10%
Actual percent of spreadsheets with errors	86%	86%	27%	27%	86%	27%
Percent of subjects who were overconfident	100%	100%	97%	86%	100%	80%
p (based upon the sign test)	<0.000	<0.000	<0.000	<0.000	<0.000	<0.000

Note: The expected probability of error (EPE) can range from 0% (the subject thought that there was no probability of an error) to 100% (the subject thought that error was certain).

solution developed by the author, the spreadsheet was recorded as being correct. If the spreadsheet's results differed from the standard solution, it was corrected until it was the correct. The corrections were recorded as errors.

Only the cell containing a correction was counted as an error; subsequent cells that were incorrect as a result were not counted as errors. If the same correction had to be made in multiple years, only one error was counted.

The cell error rate (CER) for the spreadsheet was the number of errors counted by the researcher divided by the number of cells in the standard solution.

Results

As Table 2 shows, subjects were highly overconfident according to our scale for estimating errors. Subjects working alone thought, on average, that there was only an 18% probability that they would make an error. The median EPE, 10%, was even lower. In fact, 86% of them made errors. The subjects working alone were very poorly calibrated in confidence, meaning that their estimated probability of making an error was very different (much lower) than their actual frequency of error.

For triads, the average EPE for their own group's spreadsheet were "only" about 50% too low—13% instead of 27%. As research on the hard-easy effect would lead us to expect, the better calibration of triads was more due to better performance than to reductions in confidence. Expected error probabilities for individuals and other triads working at the task were also poorly calibrated.

To test whether miscalibration was statistically significant, we used the standard but imperfect method (Dunning, Griffin, Milojkovic, & Ross, 1990) of testing for a difference between the mean EPE and the mean of the actual error distribution for the sample.

The appropriate nonparametric statistical test is the sign test. If the expected probability of error is less than the mean error for all subjects in the group, the sign is a plus. If the EPE equals the mean, the sign is a zero. If the EPE is greater the mean, this is S-. (Which is called S+ or S- is irrelevant; we chose S+ to indicate overconfidence.) Table 2 shows that there was high overconfidence in all conditions. In all cases, the null hypothesis of no overconfidence was rejected with probabilities of less than 0.000.

Another way to see miscalibration in confidence is to look at the percentage of estimates that were below the group mean. Table 2 shows that in three of the six estimates, all subjects had an estimated probability of error lower than the group mean. The lowest percentage of individual scores under the group mean was 80%. In other words, overconfidence was pervasive.

Finally, in line with other research on overconfidence (Brown, 1990), subjects working alone felt that they did better than other subjects working alone. Subjects working in triads likewise thought that their triad did better than other triads.

Discussion

Experiment 1 showed that subjects were indeed overconfident, substantially underestimating the probability that they had made an error and also believing that their odds of making an error were less than those of others. Our subjects, in other words, exhibited classic overconfidence behavior.

In fact, this overconfidence was persistent. After the experiment, when one class was debriefed, they were shown the data and were amazed by the results. When subjects who had worked alone were asked to raise their hands if they thought they were one of the 18% that did the task correctly, well over half raised their hands.

EXPERIMENT 2: WILL WARNINGS HELP?

As noted earlier, feedback on error frequency may be able to reduce overconfidence (Arkes et al., 1987; Kasper, 1996; Liechtenstein & Fischoff, 1980). Ideally, people should receive feedback on their own development errors over a long series of trials. However, simply giving the person an indication of how many other people had made errors previously in comparable circumstances might help. To explore this conjecture, the second experiment had subjects develop two spreadsheets from word problems. The experiment attempted to manipulate their overconfidence by telling subjects in the treatment group the percentage of previous subjects that had made errors in each task.

RESEARCH MODEL

Figure 1 shows our research model for confidence and accuracy in spreadsheet development. This model is based on stages of development. For each stage, we are concerned with number of errors, degree of confidence, and the interaction of errors and confidence. In this experiment, we only tested a few aspects of the model.

Initial Confidence is Likely to be Overconfidence

In Experiment 2, we manipulate initial (predevelopment) confidence by telling the subjects in the treatment group the percentage of subjects in past experiments who made errors on the two tasks used in the experiment. This is a surrogate for the feedback that developers would have if they did postdevelopment testing and analyzed the results systematically, as is done in software code inspection (Fagan 1976). The treatment

Figure 1. Research model for confidence and performance in spreadsheet development

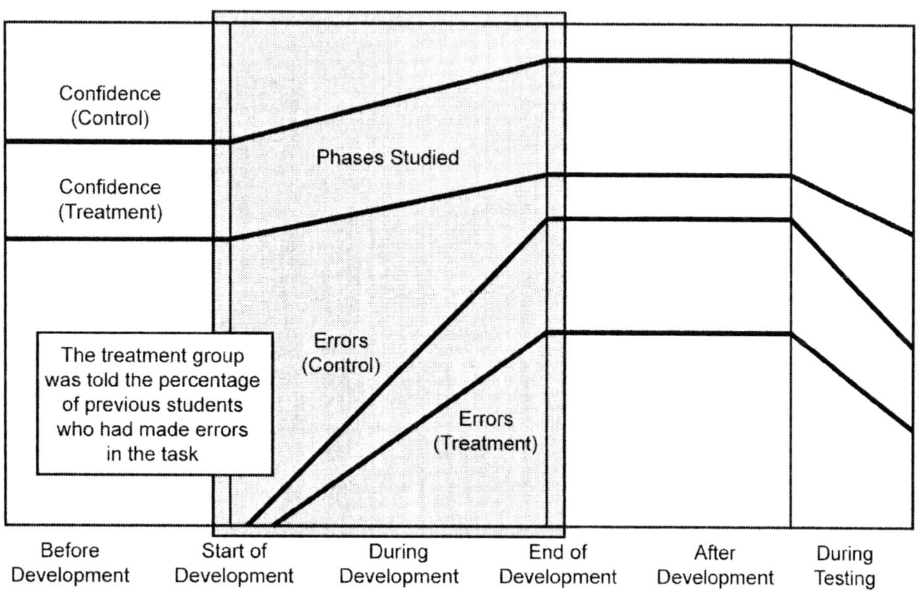

should reduce the initial confidence of subjects in the treatment group. This leads to our first hypothesis, H1:

H1: Initial confidence (after reading the problem statement but before developing the spreadsheet) should be lower for subjects who are told the percentage of past subjects who had created incorrect spreadsheets from this word problem.

Errors are Likely to Increase During Development

Quite simply, most subjects will make errors as they work. Allwood (1984) and others have systematically observed human problem solving. They have shown that people detect and correct many of the errors they make during development. Olson and Nilsen (1987-1988) specifically noted the presence of error detection and correction during spreadsheet development. However, most subjects will correct only some of their errors.

Confidence is Likely to Increase During Development

As noted above (Fuller, 1990; Habberley, Shaddick, & Taylor, 1986), when people detect and successfully correct errors, this may actually increase their confidence. This leads to hypothesis

H2: Confidence after development should be higher than confidence before development.

Feedback Manipulation Should Reduce Postdevelopment Confidence

We expect that the effect of the manipulation will last through the development stage. Consequently, we expect confidence after development to be lower for subjects who received the manipulation than for subjects in the control group.

H3: The manipulation should make confidence after development lower in the treatment group than in the control group.

Accuracy After Development

Finally, although the literature does not address the existence of a direct link between confidence and accuracy, we expect that reducing confidence by manipulation (or, in the real world, by systematically analyzing postdevelopment testing results) should increase accuracy. Otherwise, why bother? This leads to our final hypothesis:

H4: Subjects who receive the manipulation of being told the percentage of past students who have created incorrect spreadsheets from this task should have a larger percentage of accurate spreadsheets.

During and After Postdevelopment Testing

Post-development testing is rare in spreadsheet development, as noted earlier. However, if testing were done, the developer would be almost certain to find previously undetected errors in his or her "clean" spreadsheet. This would increase the accuracy of the spreadsheet. In addition, his or her confidence would fall as objective proof of errors would be seen repeatedly. If testing and analysis of the results is done systematically, confidence should be better calibrated, as indicated by the research on feedback and calibration discussed earlier. This study does not address the postdevelopment testing phase.

Sample

The second experiment used an entirely different sample, but its sample had the same background as the sample in the first experiment. These subjects also received extra credit equivalent to a quarter of a letter grade for participating in the study. Over 80% of the students in these classes participated. Sixty students participated in the experiment, but five data sets had to be discarded because the students failed to fill out the confidence survey information (three data sets) or because the files on disk could not be read (two data sets). This resulted in 55 useful data sets—27 in the control group and 28 in the treatment group. None of the students who participated in the first experiment also participated in the second experiment. (Subjects were drawn from the same course two semesters apart.)

Tasks

All students completed two tasks, which they performed in random order. One was the Kooker task, which was the MicroSlo task shown in the previous experiment with a capital purchase added. The other was the Wall task, developed by Panko and Sprague (1998). This task was purposely designed to be simple and to be almost free of domain knowledge requirements. Consequently, student domain knowledge should be unimportant.

You work for a wall-building company. You are to build a spreadsheet model to help you create a bid to build a wall. You will offer two options—lava rock or brick. Both walls will be built by crews of two. Crews will work three eight-hour days to build either type of wall. The wall will be 20 feet long, 6 feet tall, and 2 feet thick. Wages will be $10 per hour per person. You will have to add 20% to wages to cover fringe benefits. Lava rock will cost $3 per cubic foot. Brick will cost $2 per cubic foot. Your bid must add a profit margin of 30% to your expected cost.

Dependent Variables

Subjects were asked to estimate the probability that they would make an error building each of their spreadsheets after reading the task statement and after a warning (for the treatment group), but before doing the task. As noted earlier, we call this probability the EPE. Higher values for the estimated likelihood of making an error indicate lower confidence. Estimates could range from 0% to 100%.

Errors were counted in the same manner used in the first experiment. We decided not to use the number of errors as our accuracy measure because the error distribution was highly skewed with a strong zero and a long tail. We could not find a reasonable way to handle normalization problems or the treatment of many zero values. In addition, some spreadsheets were wildly incorrect. For instance, three subjects produced Kooker solutions that looked nothing like income statements. There was no way to count their number of errors, and excluding these spreadsheets presented conceptual difficulties as well. Consequently, we based hypothesis testing on whether a spreadsheet's bottom-line values were correct or not; this required a proportion test, as discussed later. It is also consistent with the concept of the expected probability of error measure.

Procedure

After reading the description of the task, subjects in the treatment group were told, in writing, the percentage of subjects who had made errors doing this task in the past (80% for Kooker and 40% for Wall). This information was written in boldface, and the experimenter emphasized verbally that some subjects had boldface information and should read it carefully, while other subjects did not have such information. After the experiment, a dozen subjects were asked if they had seen the boldface information and to characterize it. The seven who should have seen

Table 3. Results of the warning experiment

Variable	Control Group: No Warning	Treatment Group: Warning (80% Kooker) (40% Wall)	Hypothesis	z-Value	p
Number of Subjects	27	28			
Both Tasks					
Expected Error Probability (EPE), Before Averaged for Both Tasks	37%	56%	H1 (Wilcoxon Rank Sum Test)	2.370	0.009
Decline in EPE Pre-Post, Averaged for Both Tasks			H2 (Sign Test)	3.946	<0.000
EPE After, Averaged for Both Tasks	31%	43%	H3 (Wilcoxon Rank Sum Test)	1.795	0.036
Both Correct (Percentage incorrect on at least one)	2/27 (93%)	7/28 (75%)	H4 (proportion test)	1.763	0.039

Note: EPE is the expected error probability: the subject's expressed likelihood that he or she would or did commit an error during development.

the information in their packet all characterized it correctly as indicating how many people had made errors in the past. The five who could not have seen the information all said that they had not seen boldface information.

All subjects worked in a computer laboratory. They were monitored as they worked to ensure that there was no discussion and that students were not watching other screens.

Results

Table 3 shows the results for EPE and spreadsheet correctness. In the results, note that a higher EPE indicates less confidence than a lower EPE. The hypothesis testing was done on the average EPE of the subject across both tasks; the two EPEs were added and divided by two.

Overconfidence Before Development (H1)

For Hypothesis H1, the expected error probability was measured before development. For the Kooker task, the difference in average estimated probability of error before development was substantial: 40% in the control group versus 60% in the treatment group. For the Wall task, improvements were again encouraging. The two values were 34% to 53%, respectively. For the EPE averaged across the two tasks, the values for the two groups were 37% for the control group and 56% for the treatment group. As expected, EPE was higher in the treatment group than in the control group, indicating that the manipulation did reduce confidence.

According to the Wilcoxon Rank Sum Test (the nonparametric analog of the t-test), this difference in average EPE averaged across the two groups was significant at the 0.009 level (z=2.370).

Consequently, we reject the null hypothesis (that EPEs are the same across the control and treatment groups) and conclude that H1 (giving a warning reduces confidence before development) is supported. This is an important finding, because if it were not true, the entire experiment would be a failure.

Decline in Confidence After Development (H2)

As noted earlier, we expected that subjects would detect and correct some errors during development and that this would increase their overconfidence. According to H2, confidence should increase after development. All subjects were used in the test because the sample size was too small to study interaction effects.

The average estimated probability of error for the entire sample for the two tasks was 49% before development. This fell to 38% after development, indicating the expected increase in confidence.

The proper nonparametric test for paired data (before and after for individuals) is a sign test. There was a decrease in EPE in 31 of the cases after the models were built (an increase in confidence). Another 14 showed no change, and 6 cases saw an increase in EPE. This difference was highly significant ($z=3.946$, $p<0.000$). Consequently, we reject the null hypothesis and conclude that H2 (that confidence increases development) is supported.

Overconfidence After Development (H3)

H3 expects the manipulation to remain effective after development. As Table 3 shows, the estimated probability of error averaged over both tasks after development was 31% on average in the control group and 43% on average in the treatment group. This is in the right direction. Again, the appropriate test was the Wilcoxon Sum Rank Test. The difference was statistically significant ($z=1.795$, $p=0.036$). Again, we reject the null hypothesis and conclude that H3 is supported.

Correctness (H4)

The percentage of correct spreadsheets was taken as a proportion variable. A z-test for proportions was used to determine whether the difference in percentages between the experimental and control treatments was statistically significant. The dependent variable was whether or not both spreadsheets were correct.

In the control group, only 7% of all spreadsheets were correct. However, in the treatment group, 25% of the spreadsheets were correct. This difference was tested with the nonparametric proportion test. The difference, while not large enough to make spreadsheet development safe, was statistically significant at the 0.039 level ($z=-1.763$). We therefore reject the null hypothesis and conclude that H4 (a warning about the percentage of previous subjects who had done each task incorrectly increases accuracy) is supported.

DISCUSSION

As expected, our subjects in Experiment 2 were again overconfident. The percentage of incorrect spreadsheets in the control group and treatment group were 93% and 73%, respectively. Not one of the EPEs was anywhere near this high.

Second, as expected, confidence rose during the development phase. This provides support for the theory that finding errors during development increases confidence.

Third, the manipulation—giving information about the rates of incorrect spreadsheets for previous subjects—did tend to reduce confidence. The decrease was statistically significant, but confidence was not reduced to well-calibrated levels. The treatment merely reduced overconfidence somewhat.

Fourth, and perhaps most importantly, the manipulation increased the percentage of correct spreadsheets from 7% in the control group to 25% in the treatment group. This is encouraging because it represents more than a tripling in the percentage of spreadsheets that were correct. However, it certainly does not make spreadsheet development safe. Even in the treatment group, three out of four spreadsheets were incorrect.

Unfortunately, the second experiment could not measure whether subjects were more likely in the future to engage in systematic testing after development. Unless such long-term testing is done, error rates must be expected to remain unacceptably high.

CONCLUSION

As noted in the introduction, overconfidence is an almost universal human trait. The first experiment demonstrated that spreadsheet development is no exception. On average, subjects working alone rated the probability that they had made an error as only 18%. In fact, 86% made errors. Results for subjects working in triads were less extreme but still exhibited classic overconfidence patterns.

Can we reduce overconfidence, and if we do, will accuracy increase? The second experiment's response to these questions was a cautious "yes." Decreases in confidence before and after development were statistically significant. More importantly, reducing confidence increased accuracy. With a warning, the percentage of subjects getting both spreadsheets correct more than tripled, from 7% to 25%.

One point of caution in interpreting the results is that even with the improvements seen when warnings were given, subject accuracy still was too low to make spreadsheet development a safe activity. Unless feedback can decrease overconfidence sufficiently to motivate users to test their spreadsheets systematically after development, we are unlikely to get the order-of-magnitude error reduction needed to make spreadsheets even marginally safe.

Given the results of this experiment, a logical next step would be to follow the approaches of Arkes, et al. (1987) and Lichtenstein and Fischoff (1980) and give subjects a series of spreadsheet development tasks, providing them with detailed feedback on errors at each step.

In addition, it would be useful in future studies to see if wording could affect confidence estimates. In this experiment, we asked subjects to estimate the likelihood that they would make an error. It would be interesting to see if compatible data would result if we asked subjects to estimate the likelihood that their spreadsheet was correct.

More fundamentally, we need to understand how subjects think about accuracy and the multiplication of probabilities. Subjects were asked in the second experiment to estimate the likelihood that they would make an error in a formula cell on average, and to make an error estimate for the spreadsheet as a whole. For the Wall task, the average estimated probability of making an error in the spreadsheet for both groups was 32%; for making an error in any cell, the estimated probability was only 11%. This ratio is only 2.90. For the Microslo task, the percentages were 41% and 14%, for a ratio of 2.93.

These ratios make no sense. If there are N formula cells in a spreadsheet, and if the probability of making an error in a cell is e, the probability of an error in the spreadsheet as a whole, E, should be $E = 1 - (1+e)^N$ (Lorge & Solomon, 1955). Although this equation is only strictly true if all cells have the same probability of error, this equation should give us a good rough estimate of likely error rates.

The Wall task typically had 18 cells. Using the Lorge and Solomon formula with an 11% cell error rate for formula cells, the probability of making an error in the spreadsheet as a whole should be estimated as 88%, not 32%. Similarly, in the Kooker task, there were also about 18 formula cells, so with a formula cell error rate of 14%, the probability of making an error in the spreadsheet should have been estimated at 93%, not 41%.

This inability to realize that a long series of calculations should increase the likelihood of an error also seems to be indicated in the study we discussed earlier by Reithel, Nichols, and Robinson (1996). They had subjects look at spreadsheets that were short and poorly formatted, short and well formatted, long and poorly formatted, and long and well formatted. Subjects expressed substantially more confidence in the accuracy of the long and well-formatted spreadsheet than in the accuracy of other spreadsheets, despite the fact that longer spreadsheets should have more errors than shorter spreadsheets, given the Lorge and Solomon formula. The pattern seems to indicate a blindness in statistical thinking. We hope that future research can shed light on this interesting phenomenon.

One limitation in the experiment was the use of undergraduate subjects. Although all had previously taken a hands-on skills course and had undertaken at least two spreadsheet development homework assignments in their current class, none had extensive spreadsheet development experience at work. However, Panko and Sprague (1998) found almost identical error rates for the Wall task when the task was solved by undergraduate students, MBA students with little or no spreadsheet development experience at work, and MBA students with extensive spreadsheet development experience at work.

For the future, a major task for overconfidence research must be to go beyond its impressive body of empirical results and move to the creation of theory. Although we cannot offer a full theory, we offer several suggestions for directions such a theory may take.

Reason (1990) and Baars (1992) have argued that human cognition has two mechanisms that differ in important characteristics. First, we have an automatic cognition system that uses pattern matching. This automatic system is fast and effortless (Reason, 1990). Second, we have an attentional system that is linear, slow, and effortful (Reason, 1990). Postdevelopment testing, such as code inspection, appears to require a high degree of effort and tends to be unpleasant for people (Beck, 2000). Human beings appear to have a difficult time engaging the attentional system for more than brief periods of time. This may create strong resistance to formal testing.

In addition, there is a literature on denial, which focuses on illness and the fact that many people with terminal illnesses deny the seriousness of their condition or the need to take action. Apparently, what is very difficult and unpleasant to do also is difficult to contemplate. Although denial has only received intensive study in medical literature, it seems likely to occur whenever required actions are difficult or onerous. Given the effortful nature of spreadsheet testing, developers may be victims of denial, which may manifest itself in the form of overconfidence in accuracy so that extensive testing will not be needed.

Reinforcing the denial possibility for explaining overconfidence is the study by Reithel, Nichols, and Robinson (1996) mentioned earlier. For large, well-formatted spreadsheets, expressed confidence was much higher than for shorter spreadsheets. One explanation may be that subjects who see a spreadsheet too large to test easily and who get a cue from the formatting that the spreadsheet was well designed, turned off further judgmental processing.

Another possibility is that people actually are largely blind to risk, not simply poor at assessing risk (Naatanen & Summala, 1976). For example, Howarth (1988) observed drivers as their cars approached children who wanted to cross at an intersection. He found that fewer than 10% of all drivers took action, even these actions would have come too late if the children had started crossing the street.

As noted earlier, Fuller (1990) has suggested that we learn risky behavior because we rarely get negative consequences when we take risky action. Perhaps this means that we become blind to risk over time. Svenson (1977) studied drivers approaching narrow bends in a road. Unfamiliar drivers slowed down. Familiar drivers did not, and they approached at speeds that would have made accident avoidance impossible.

REFERENCES

Allwood, C. M. (1984). Error detection processes in statistical problem solving. *Cognitive Science, 8*(4), 413-437.

Arkes, H. R., Christensen, C., Lai, C., & Blumer, C. (1987). Two methods of reducing overconfidence. *Organizational Behavior and Human Decision Processes, 39*(1), 134-144.

Baars, B. J. (1992). A new ideomotor theory of voluntary control. In B. J. Baars (Ed.), *Experimental slips and human error* (pp. 93-120). New York: Plenum Press.

Beck, K. (2000). *Extreme programming explained: Embrace change*. Reading, Massachusetts: Addison Wesley.

Brown, I. D. (1990). Drivers' margins of safety considered as a focus for research on error. *Ergonomics, 33*(10/11), 1307-1314.

Brown, P. S., & Gould, J. D. (1987). An experimental study of people creating spreadsheets. *ACM Transactions on Office Information Systems, 5*(3), 258-272.

Butler, R. (2000). Is this spreadsheet a tax evader? How H. M. Customs & Excise test spreadsheet applications. In *Proceedings of the 33rd Hawaii International Conference on System Sciences*, CD-ROM. Los Alamitas, CA: IEEE Computer Society Press.

Cale, E. G., Jr. (1994). Quality issues for end-user developed software. *Journal of Systems Management, 45*(1), 36-39.

Camerer, C. F., & Johnson, E. C. (1991). The process-performance paradox in expert judgment: How can experts know so much and predict so badly? In K. A. Ericsson & J. Smith (Eds.), *Toward a general theory of expertise* (pp. 195-217). Cambridge, England: Cambridge University Press.

Chan, Y., & Storey, V. C. (1996). The use of spreadsheets in organizations: Determinants and consequences. *Information & Management, 31*(3), 119-134.

Clarke, F. R. (1990). Confidence ratings, second-choice responses, and confusion matrices in intelligibility tests. *Journal of the Acoustical Society of America, 32*(1), 35-46.

Clermont, M., Hanin, C. & Mittermeier, R. (2000). A spreadsheet auditing tool evaluated in an industrial context. In *Proceedings of the EuSpRIG 2000 Symposium, Spreadsheet Risks—the Hidden Corporate Gamble* (pp. 35-46). Greenwich, England: Greenwich University Press.

Cragg, P. G., & King, M. (1993). Spreadsheet modelling abuse: An opportunity for OR? *Journal of the Operational Research Society, 44*(8), 743-752.

Davies, N., & Ikin, C. (1987). Auditing spreadsheets. *Australian Accountant*, 54-56.

Dunning, D., Griffin, D. W., Milojkovic, J. D., & Ross, L. (1990). The overconfidence effect in social prediction. *Journal of Personality and Social Psychology, 58*(4), 568-581.

Fagan, M. E. (1976). Design and code inspections to reduce errors in program development. *IBM Systems Journal, 15*(3), 182-211.

Floyd, B. D., Walls, J., & Marr, K. (1995). Managing spreadsheet model development, *Journal of Systems Management, 46*(1), 38-43, 68.

Fuller, R. (1990). Learning to make errors: Evidence from a driving simulation. *Ergonomics, 33*(10/11), 1241-1250.

Gable, G., Yap, C. S., & Eng, M. N. (1991). Spreadsheet investment, criticality, and control. *Proceedings of the Twenty-Fourth Hawaii International Conference on System Sciences, Vol. III* (pp. 153-162). Los Alamitas, CA: IEEE Computer Society Press.

Galletta, D. F., Abraham, D., El Louadi, M., Lekse, W., Pollailis, Y. A., & Sampler, J. L. (1993). An empirical study of spreadsheet error-finding performance. *Journal of Accounting, Management, and Information Technology, 3*(2), 79-95.

Galletta, D. F., & Hufnagel, E. M. (1992). A model of end-user computing policy: Context, process, content and compliance. *Information and Management, 22,* 1-28.

Grady, R. B. (1995). Successfully applying software metrics. *Communications of the ACM, 38*(3), 18-25.

Habberley, J. S., Shaddick, C. A., & Taylor, D. H. (1986). *A behavioural study of the collision avoidance task in bridge watchkeeping.* Southampton, England: College of Marine Studies. Cited in Reason (1990).

Hall, M. J. J. (1996). A risk and control oriented study of the practices of spreadsheet application developers. *Proceedings of the Twenty-Ninth Hawaii International Conference on Systems Sciences, Vol. II* (pp. 364-373). Kihei, Hawaii. Los Alamitas, CA: IEEE Computer Society Press.

Howarth, C. I. The relationship between objective risk, subjective risk, and behavior. *Ergonomics, 31,* 527-535. Cited in Wagenaar & Reason, 1990.

Johnson, E. J. (1988). Expertise and decision making under uncertainty: Performance and process. In M. T. H. Chi, R. Glaser, & M. J. Farr (Eds.), *The nature of expertise* (pp. 209-228). Hillsdale, NJ: Lawrence Erlbaum Associates.

Jones, T. C. (1998). *Estimating software costs.* New York: McGraw-Hill.

Kasper, G. M. (1996). A theory of decision support system design for user calibration. *Information Systems Research, 7*(2), 215-232.

Keren, G. (1992). Improving decisions and judgments: The desirable versus the feasible. In G. Wright & F. Bolger (Eds.), *Expertise and decision support* (pp. 25-46). New York: Plenum Press.

Koriat, A., Lichtenstein, S., & Fischoff, B. (1980). Reasons for confidence. *Journal of Experimental Psychology: Human Learning and Memory, 6,* 107-117.

Lawrence, R. J., & Lee, J. (2004). *Financial modelling of project financing transactions.* Institute of Actuaries of Australia Financial Services Forum. The Institute of Actuaries of Australia, Level 7 Challis House 4 Martin Place, Sidney, NSW Australia 2000.

Lichtenstein, S., & Fischoff, B. (1980). Training for calibration. *Organizational Behavior and Human Performance, 26*(1), 149-171.

Lichtenstein, S., Fischoff, B., & Philips, L. D. (1982). Calibration of probabilities: The state of the art to 1980. In D. Kahneman, P. Slovic, & A. Tversky (Eds.), *Judgment under uncertainty: Heuristics and biases* (pp. 306-334). Cambridge, England: Cambridge University Press.

Lorge, I., & Solomon, H. (1995). Two models of group behavior in the solution of eureka-type problems. *Psychometrika, 20*(2), 139-148.

Lukasik, T. (1998). Personal communication with the author via e-mail.

Naatanen, R., & Summala, H. (1976). *Road user behavior and traffic accidents,* Amsterdam: North-Holland. Cited in Wagenaar & Reason, 1990.

Nardi, B. A., & Miller, J. R. (1991). Twinkling lights and nested loops: Distributed problem solving and spreadsheet development. *Interna-*

tional Journal of Man-Machine Studies, 34(2), 161-184.

Olson, J. R., & Nilsen, E. (1997-1998). Analysis of the cognition involved in spreadsheet interaction. *Human-Computer Interaction, 3*(4), 309-349.

Ono, K., & Davis, J. H. (1988). Individual judgment and group interaction: A variable perspective approach. *Organizational Behavior and Human Decision Processes, 41*(1), 211-232.

Panko, R. R., & Halverson, R. P., Jr. (1997). Are two heads better than one? (At reducing errors in spreadsheet modeling). *Office Systems Research Journal, 15*(1), 21-32.

Panko, R. R., & Sprague, R. H., Jr. (1998). Hitting the wall: Errors in developing and code inspecting a "simple" spreadsheet model. *Decision Support Systems, 22*(4), 337-353.

Plous, S. (1993). *The psychology of judgment and decision making.* Philadelphia: Temple University Press.

Plous, S. (1995). A comparison of strategies for reducing interval overconfidence in group judgments. *Journal of Applied Psychology, 80*(4), 443-454.

Pulford, B. D., & Coleman, A. M. (1996). Overconfidence, base rates and outcome positivity/negativity of predicted events. *British Journal of Psychology, 87*(3), 431-445.

Putnam, L. H. & Myers, W. (1992). *Measures for excellence: Reliable software on time, on budget.* Englewood Cliffs, NJ: Yourdon.

Rasmussen, J. (1990). The role of perception in organizing behaviour. *Ergonomics, 33*(10/11), 1185-1199.

Reason, J. (1990). *Human error.* Cambridge, England: Cambridge University Press.

Reithel, B. J., Nichols, D. L., & Robinson, R. K. (1996). An experimental investigation of the effects of size, format, and errors on spreadsheet reliability perception. *Journal of Computer Information Systems, 36*(3),54-64.

Schultheis, R., & Sumner, M. (1994). The relationship of application risks to application controls: A study of microcomputer-based spreadsheet applications. *Journal of End User Computing, 6*(2), 11-18.

Shanteau, J. (1992). The psychology of experts: An alternative view. In G. Wright & F. Bolger (Eds.), *Experts and decision support* (pp. 11-23). New York: Plenum Press.

Shanteau, J., & Phelps, R. H. (1997). Judgment and swine: Approaches and issues in applied judgement analysis. In M. F. Kaplan & S. Schwartz (Eds.), *Human judgment and decision processes in applied settings.* New York: Academic Press. Cited in Reason (1990).

Sniezek, J. A., & Henry, R. A. (1989). Accuracy and confidence in group judgment. *Organizational Behavior and Human Decision Processes, 43*(1), 1-28.

Speier, C. & Brown, C. V. (1996). Perceived risks and management actions: Differences in end-user application development across functional groups. In *Proceedings of the Twenty-Ninth Hawaii International Conference on System Sciences, Vol. II* (pp. 374-384). Kihei, Hawaii. Los Alamitas, CA: IEEE Computer Society Press.

Svensen, O. (1997). *Risks of road transportation from a psychological perspective: A pilot study, Report 3-77,* Project Risk Generation and Risk Assessment in a Social Perspective, Committee for Future-Oriented Research, Stockholm, Sweden. Cited in Fuller, 1990.

United States Bureau of the Census (2003). *Computer use in the United States, August 2003.* Department of Commerce, Washington, D.C.

Wagenaar, W. A., & Keren, G. B. (1996). Does the expert know? The reliability of predictions and confidence ratings of experts. In E. Hollnagel, G. Manici, & D. D. Woods (Eds.), *Intelligent decision support in process environments* (pp. 87-103). Berlin: Springer-Verlag.

Wagenaar, W. A., & Reason, J. T. (1990). Types and tokens in road accident causation. *Ergonomics, 33*(10/11), 1365-1375.

This work was previously published in End-User Computing: Concepts, Methodologies, Tools, and Applications, edited by S. Clarke, copyright 2008 by Information Science Reference, formerly known as Idea Group Reference (an imprint of IGI Global).

Chapter X
User Acceptance of Voice Recognition Technology:
An Empirical Extension of the Technology Acceptance Model

Steven John Simon
Mercer University, USA

David Paper
Utah State University, USA

ABSTRACT

Voice recognition technology-enabled devices possess extraordinary growth potential, yet some research indicates that organizations and consumers are resisting their adoption. This study investigates the implementation of a voice recognition device in the United States Navy. Grounded in the social psychology and information systems literature, the researchers adapted instruments and developed a tool to explain technology adoption in this environment. Using factor analysis and structural equation modeling, analysis of data from the 270 participants explained almost 90% of the variance in the model. This research adapts the technology acceptance model by adding elements of the theory of planned behavior, providing researchers and practitioners with a valuable instrument to predict technology adoption.

Supported by the development of easy-to-use and inexpensive technology, the adoption of voice recognition technology (VRT)-enabled devices by businesses and consumers is slated to grow 17% per year, becoming a $52 billion market by 2007 (Business Communications Company, 2003). Studies in hospital settings illustrate average productivity gains of 30% with the introduction of VRT devices (MedQuist, 2003). Additionally, organizations garner cost savings and the potential for improved security, while consumers are attracted to hands-free operation of personal devices including cellular phones and personal digital assistants. Yet, despite improvements in the technology itself and the potential productivity gains, studies suggest that organizations and consumers are resisting VRT adoption (Costanzo, 2003).

Social psychology and information systems research have extensively explored the adoption of innovations including information technology. A variety of tools have been developed to explain the adoption process, including the widely referenced theory of reasoned action (Fishbein & Ajzen, 1975), theory of planned behavior (Ajzen, 198; Fishbein & Ajzen, 1975), and technology acceptance model (Davis, 1986). This study marks the first time that the adoption of VRT has been explored with these research tools.

The adoption process is explored in this study during the initial implementation of a voice recognition device by the U.S. Navy. The next section describes the situation and the system to be implemented and is followed by a review of the social psychology and information systems literature. Then the research model and study's hypotheses are presented. A description of the data and its analysis is presented followed by the discussion of the findings. The chapter concludes with implications and suggestions for future research.

BACKGROUND

Coupling computer recognition of the human voice with a natural language processing system makes speech recognition by computers possible. By allowing data and commands to be entered into a computer without the need for typing, computer understanding of naturally spoken languages frees human hands for other tasks (Lai, 2000; Shneiderman, 2000). Speech recognition by computers can also increase the rate of data entry, improve spelling accuracy, permit remote access to databases utilizing wireless technology, and ease access to computer systems by those who lack typing skills (Boyce, 2002). The seamless integration of voice recognition technologies creates a human-machine interface that has been applied to consumer electronics, Internet appliances, telephones, automobiles, interactive toys, and industrial, medical, and home electronics and appliances (Soule, 2000). Applications of speech recognition technology are also being developed to improve access to higher education for people with disabilities (Goette, 2000; Leitch & Bain, 2000). Although speech recognition systems have existed for two decades, widespread use of this technology is a recent phenomenon.

Some of the most successful applications have been telephone based. Continuous speech recognition has been used to improve customer satisfaction and the quality of service on telephone systems (Charry, Pimentel, & Camargo, 2000; Goodliffe, 2000; Rolandi, 2000). Name-based dialing has become more ubiquitous, with phone control answer, hang-up, and call management (Gaddy, 2000a). These applications use intuitive human communication techniques to interact with electronic devices and systems (Shepard, 2000). BTexact Technologies, the Advanced Communications Technology Centre for British Telecommunications, uses the technology to provide automated directory assistance for 700 million calls each year at its UK bureau (Gorham & Graham, 2000). Haynes (2000) deployed a conversational interactive voice response system to demonstrate site-specific examples of how companies are leveraging their infrastructure investments, improving customer satisfaction, and receiving quick return on investments. Such applications demonstrate the use of speech recognition by business. The investigation of current customer needs and individual design options for accessing information utilizing speech recognition is key to gaining unique business advantages (Prizer, Thomas, & Suhm, 2000; Schalk, 2000). A long-awaited application of speech recognition, the automatic transcription of free-form dictation from professionals such as doctors and lawyers, lags behind other commercial applications (Stromberg, 2000).

Speech recognition systems are being tested for, or are already in use, in government and in private industry to the benefit of organizations and users. Speech technology has been applied to medical applications, particularly emergency medical care that depends on quick and accurate access of patient background information (Kundupoglu, 2000). The U.S. Defense Advance Research Projects Agency organized the Trauma Care Information Management System (TCIMS) Consortium to develop a prototype system for improving the timeliness, accuracy, and completeness of medical documentation. One outcome of TCIMS was the adoption of a speech-audio user interface for the prototype (Holtzman, 2000). The Federal Aviation Administration conducted a demonstration of how voice technology supports a facilities maintenance task. A voice-activated system proved to be less time consuming to use than the traditional paper manual approach, and study participants reported that the system was understandable, easy to control, and responsive to voice commands. Participants felt that the speech recognition system made the maintenance task easier to perform, was more efficient and effective than a paper manual, and would be better for handling large amounts of information (Mogford, Rosiles, Wagner, & Allendoerfer, 1997). Pilots must have good head/eye coordination when they shift their gaze between cockpit instruments and the outside environment. Boeing has investigated ways to free pilots from certain manual tasks and sharpen their focus on the flight environment. The latest solution includes the use of a rugged, lightweight, continuous-speech device that permits the operation of selected cockpit controls by voice commands alone. This technology is being applied in the noisy cockpit of the Joint Strike Fighter (Bokulich, 2000). Financial service organizations are beginning to explore the merits of speech recognition. Customers can use this technology to conduct "hands-free" banking. Such "customer-facing" applications are gaining popularity in the financial services industry (Costanzo, 2003).

Even though applications of speech recognition technology have been developed with increased frequency, the field is still in its infancy, and many limitations have yet to be resolved. For example, the success of speech recognition by desktop computers depends on the integration of speech technologies with the underlying processor and operating system and the complexity and availability of tools required to deploy a system. This limitation has had an impact on application development (Markowitz, 2000; Woo, 2000). Use of speech recognition technology in high-noise environments remains a challenge. For speech recognition systems to function properly, clean speech signals are required, with high signal-to-noise ratio and wide frequency response (Albers, 2000; Erten, Paoletti, & Salam, 2000; Sones, 2000; Wickstrom, 2000). One solution offered to reduce the negative consequences associated with noise-related errors is to implement a multimodal architecture. An example of a multimodal architecture is combining speech and pen input to reduce speech recognition errors (Oviatt, 2000). Oviatt suggests that "... multimodal systems will help stabilize error-prone recognition technologies, while also greatly expanding the accessibility of computing for everyday users and real-world environments" (p. 45). In conjunction with the multimodal architecture, a microphone system is critical in providing the required speech signal, and, therefore, has a direct effect on the accuracy of the speech recognition system (Andrea, 2000; Wenger, 2000).

Interference, changes in the user's voice, and additive noise—such as car engine noise, background chatter, and white noise—can reduce the accuracy of speech recognition systems. In military environments, additive noise and voice changes are common. For example, in military aviation, the stress resulting from low-level flying can cause a speaker's voice to change, reducing recognition accuracy (Christ, 1984). The change in voice modulation to compensate for environmental dynamics such as increased noise levels is called the Lombard effect (Oviatt, 2000). Oviatt suggests that speech recognition accuracy degrades when a system processes Lombard speech, and recommends that this effect should therefore be considered when designing such systems.

Despite the growth of voice recognition systems, we found only one study that examined the adoption of this innovative technology in organizations. This study explored the adoption of speech recognition technology within a sample of people with disabilities (Goette, 2000). Specifically, the Goette study focused on the perception of individuals with disabilities toward the adoption of a speech recognition technology designed specifically to facilitate a "hands-free" work environment. She found that the ability to use the technology for a trial period was the major factor influencing successful adoption. The next section describes the environment and system the U.S. Navy is implementing followed by an examination of the theoretical research.

NAVAL VOICE INTERACTIVE DEVICE (NVID)

Shipboard medical and engineering personnel regularly conduct comprehensive surveys to ensure the health and safety of the ship's crew, equipment, and working environment. Currently, surveillance data are collected and stored via manual data entry, a time-consuming process that involves typing handwritten survey findings into a word processor to produce a completed document. Typically, inspectors enter data and findings by hand onto paper forms and later transcribe these notes into a word processor to create a finished report. The process of manual note taking and entering data via keyboard into a computer database is time consuming, inefficient, and prone to error. To remedy these problems, the Naval Shipboard Information Program was developed, allowing data to be entered into portable laptop computers while a survey is conducted (Hermansen & Pugh, 1996). However, the cramped shipboard environment, the need for mobility by inspectors, and the inability to have both hands free to type during an inspection make the use of laptop computers during a walk-around survey difficult. Clearly, a hands-free, space-saving mode of data entry that would also enable examiners to access pertinent information during an inspection was desirable.

The Naval Voice Interactive Device (NVID) project was developed to replace existing, inefficient, repetitive survey procedures with a fully automated, voice-interactive system for voice-activated data input. In pursuit of this goal, the NVID team developed a lightweight, wearable, voice-interactive prototype capable of capturing, storing, processing, and forwarding data to a server for easy retrieval by users. The voice-interactive data input and output capability of NVID reduces obstacles to accurate and efficient data access and reduces the time required to complete inspections. NVID's voice-interactive technology allows a trainee to interact with a computerized system and still have hands and eyes free to manipulate materials and negotiate his or her environment (Ingram, 1991). The NVID has been designed to allow voice prompting by the survey program, as well as voice-activated, free-text dictation. An enhanced microphone system permits improved signal detection in noisy shipboard environments. All of these capabilities contribute to the improved efficiency and accuracy of the data collection and retrieval process by shipboard personnel. A comprehensive description of the NVID device and its development can be found in Paper, Rodger, and Simon (in press).

THEORETICAL BACKGROUND

The theory of reasoned action (TRA) (Fishbein & Ajzen, 1975) remains one of the most influential models of behavior in social psychology. The model assumes that behavior can be predicted on the basis of a person's behavioral intention. In turn, intention is determined by two components, one personal and the other social. The theory was designed to model how any specified behavior under volitional control is produced by beliefs, attitudes, and intentions toward that behavior. It further specifies that this intention is produced by both the individual's attitude toward performing that behavior, and the individual's perception of social pressures to perform that behavior, or the social norm. The TRA has been empirically tested in over a hundred studies, in a diverse range of fields including health, fitness, drug use, extraterrestrial beliefs, and social and consumer behaviors, and it has demonstrated strong support for the proposed determinants of intention (Sheppard, Hartwick, & Warshaw, 1988). Results of meta-analyses and quantitative reviews of TRA and its enhanced version theory of planned behavior (see below) indicate that attitude, subjective norm, and perceived behavioral control explained 40-50% of variance in intention, with intention explaining 19-38% of variance in behavior (Sutton, 1998).

The TRA assumes that human beings usually behave in a sensible manner so that they might obtain favorable outcomes and meet the expectations of others (Fishbein & Ajzen, 1975). According to the theory, behavioral intention is an immediate predecessor of a behavior and it is determined by attitude toward behavior and subjective norm. Attitude toward behavior is obtained by summing the products of behavioral beliefs to certain outcomes. Behavioral beliefs refer to the probability that a behavior leads to certain outcomes; the evaluation of outcomes is the extent to which the consequences of the behavior are favorable. Subjective norms are the summed products of normative beliefs and motivations to comply. Normative beliefs represent an individual's perceptions of a person's tendency to behave in a manner consistent with their reference group's belief.

Theory of planned behavior (TPB) was developed by Ajzen, who extended the TRA by introducing an additional construct—perceived behavioral control—to account for situations in which individuals lack complete control of the target behavior (Ajzen, 1985; Fishbein & Ajzen, 1975) (see Figure 1). Perceived behavioral control

Figure 1. Theory of Reasoned Action Plus Theory of Planned Behavior (TPB components inside box)

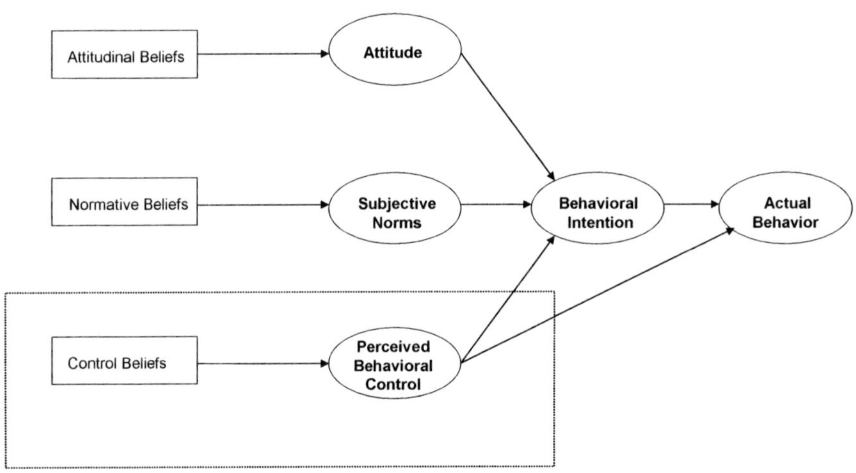

(PBC) is held to influence both intention and behavior. The rationale behind the addition of PBC is that it allows prediction of behaviors that are not completely under volitional control. While TRA can predict behaviors that are relatively straightforward, there are certain circumstances in which constraints on actions confounded its accuracy. The inclusion of PBC provides insights about the potential constraints on action by the actor, and is held to explain why intentions do not always predict behavior. A variety of studies have provided support for TPB (Ajzen, 1991; Blue, 1995; Conner & Sparks, 1996; Godin, 1993; Godin & Kok , 1996; Hausenblas, Carron, & Mack et al., 1997; Van den Putte, 1991; Armitage & Conner, 2001; Jonas & Doll, 1996; Manstead & Parker, 1995; Sparks, 1994).

From TRA, TPB retained the two other antecedents of intention: subjective norm and attitude toward behavior. Subjective norm refers to the individual's perceptions of general social pressure to perform a given behavior. If an individual perceives that significant others endorse the behavior, they are more likely to intend its performance. Attitude toward the behavior reflects the individual's global positive or negative evaluations of performing the behavior. Generally, the more favorable the attitude toward the behavior, the stronger the individual's intention to perform it (Fishbein & Ajzen, 1975).

The technology acceptance model (TAM) developed by Davis (1986) is grounded in both TRA and TPB. The goal of TAM is to "provide an explanation of the determinants of computer acceptance that is general, capable of explaining user behavior across a broad range of end-use computing technologies and user populations" (Davis, Bagozzi, & Warshaw, 1989, p. 985). This model (see Figure 2) seeks to identify a small number of fundamental variables that impact both behavioral intention to use computers and actual system use. The TAM has proven to be a robust model. The validity of TAM's predictive power was demonstrated by Mathieson (1991). He established that both TAM and TPB were good predictors of intention to use an information system, with TAM demonstrating a slight empirical advantage. He also commented that TAM was easier to apply while working with generalized users' opinions.

The original TAM posits that two particular beliefs, perceived usefulness and perceived ease of use, are critical for computing acceptance (Davis et al.,1989). An examination of the model indicates that computer use is determined by behavioral intention, with intention viewed as mutually determined by the individual's attitude and perceived usefulness of the computer system. Attitude, in this model, is influenced by perceived usefulness and perceived ease of use. During the

Figure 2. Technology Acceptance Model (TAM)

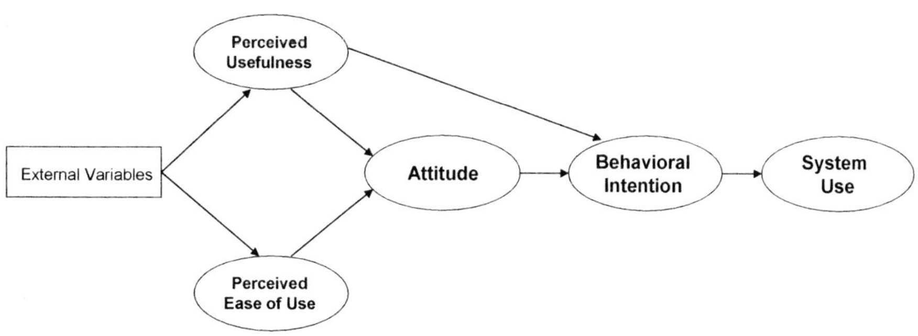

development of TAM, subjective norm—a critical element of TRA and TPA—was deleted due to theoretical and measurement issues (Davis et al., 1989), despite the psychology literature's finding that the construct was an important determinant of intention and behavior. A number of studies have found subjective norm to significantly contribute to the strength and predictability of the model (Mathieson, 1991; Hartwick & Barki, 1994; Taylor & Todd, 1995b; Gefen & Straub, 1997; Venkatesh & Morris, 2000l; Venkatesh & Davis, 2000).

Since its introduction, TAM has been modified and compared with other theories. Attitude has been removed as a means to simplify the model (Adams, Nelson, & Todd, 1992; Igbaria, Guimaraes, & Davis, 1995; Chau, 1996; Szajna, 1996). Other studies have added system use as a dependent variable, with subjects self-reporting their results (Davis et al., 1989; Adams et al., 1992; Igbaria et al., 1995). Additionally, behavioral intention has been used as a dependent variable (Davis, 1989; Chau, 1996). External variables have also been introduced into the model to explain situational factors that impact technology acceptance. In a study of microcomputer usage, Igbaria et al. (1995) added training, support, and organizational factors. Other studies (Thompson, Higgins, & Howell, 1991; Davis, 1993) removed behavioral intention from the model and directly link attitude to system use. These studies argue that the goal of the model is to determine factors of system use, not the intention to take part in future behavior. Results indicate that TAM is a valid model for predicting user acceptance of information technology. Perceived usefulness of a technology has been consistently identified as an integral predictor of attitude formation, while substantiation for perceived ease of use has been inconsistent or less significant. The leading explanation for the weak results for perceived ease of use as a predictor of attitude is that intention to use a technology may decrease as exposure to technology expands (Chau, 1996).

THIS STUDY

This is the second study in a two-phase examination of the development and introduction of NVID, an interactive voice operated device used in shipboard environments by the U.S. Navy. The first study (Paper, Rodger, & Simon, in press) explored the creation of a focus group, its requirements and issues for development, and the prototype device's initial testing. This study uses the information gathered during the first study to create and test an instrument that assists the Navy in the examination of intentions of sailors and their use of the device during its initial rollout. This phase of NVID's implementation was conducted with enlisted members—restricted to two job classification codes (healthcare specialists and machinists)—from the medical and engineering departments on a deployed (at sea) aircraft carrier. The NVID's use was voluntary since the Navy was still refining the device and did not feel that it was ready for full-scale adoption.

Sailors were introduced to NVID before the deployment. The device was demonstrated to participants from both departments, explained as experimental and voluntary, and presented as a means to improve their productivity and make their tasks easier. Training was then conducted on how to use the device and download the information into their department's computer, and then time was given to each sailor to practice and experiment with an actual unit. Additional opportunities for training and experimentation were provided before the ship deployed. Upon culmination of the department's training, sailors completed a questionnaire.

During the NVID implementation experiment, each sailor participating in the study was given a randomly generated code that he/she was required to input to activate the device. To use the NVID, a sailor would check it out from the department's office, input their assigned code, and then use the device while they were conducting their assigned tasks. When the sailor returned to the department's office, the NVID was connected to a workstation and the collected information, along with the user's code, was downloaded and saved for tracking. At the end of the implementation experiment, the researchers and the Navy were able to compile the number of times and when each sailor used the device as well as what tasks were performed. This provided a reliable means to determine system use without relying on self-reported measures that can be biased.

RESEARCH MODEL AND INSTRUMENT

The proposed modified TAM consists of the traditional TAM envisioned by Davis and the addition of subjective/social norm. The study's dependent variable remains system use for two reasons. The first reason is that the Navy was primarily interested in NVID's acceptance, and the second, to provide additional validity to TAM under this organizational situation. The extension of TAM is a result of the input from the focus groups during the first phase of this study. Members of these groups impressed upon the researchers that factors such as those used in the TAM instrument and the social norms measurements would impact the intention and ultimate decision to use NVID[1]. Subjective/social norm as indicated in earlier sections is derived from the referent TRA/TPB and was eliminated from TAM for theoretical and construct reasons.

The members of the focus group strongly suggested the inclusion of the additional factor because the Navy's goal was to implement NVID in the shipboard environment. A Navy ship's crew lives and works in very close quarters during deployment, when the ship sails from her home port for up to six months. Small ships have over 200 people, while an aircraft carrier has a crew of over 5,500. During the deployment, members of the crew become an extended family with sailors of specific departments, working, playing, eating, and sharing quarters together. The focus group felt that as a result of these "family" ties, there would be strong social pressure to adopt or reject the NVID. Hence, the inclusion of subjective/social norm, which as indicated earlier has strong validity in predicting intent, was recommended. Since sailors using NVID have similar educational and organizational backgrounds (the experiment was limited to two ratings—job classifiers—within two departments), our model purposefully excludes any external variables. Our research model is illustrated in Figure 3.

The instrument used to investigate technology adoption in this study is derived from well-

Figure 3. Adapted TAM Research Model

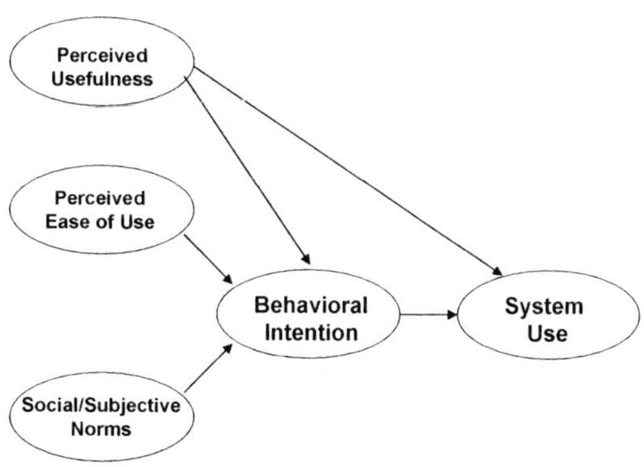

researched and validated measures. Nineteen items were culled from the literature and were formulated into question items using a 7-point Likert scale, with 1 designated "strongly agree" and 7 designated "strongly disagree." The researchers later added the use numbers collected from the ship's servers. The instrument was first validated by the focus group review panel plus a small number of potential users from departments not involved in NVID's test. They independently reviewed the instrument and suggested minor changes that were incorporated into the final design. The instrument was then balanced and randomized by the researchers.

The completed instrument was then evaluated for construct validity and reliability (Straub, 1989). Construct validity is concerned that the instrument actually operationalizes the constructs it purports to measure. Reliability is an index that refers to the respondents' ability to answer the same question in a similar manner over time. To test the instrument, a pretest was conducted with approximately 50 randomly selected groups of sailors from ship-based commands at a large Navy base. The results indicated acceptable Cronbach's alpha values (all above .80) and higher covariation among items for the same factors than among those for different factors. These results suggest that the instrument was of acceptable measurement reliability and had sufficient convergent and discriminate validity. Participants were instructed to consider their NVID experiences when completing this instrument. A copy of the final instrument is found in Table 1.

Table 1. Research Model Measurement Scales

Perceived usefulness
e1 Using computers enables me to accomplish tasks more quickly
e2 Using computers improves my performance
e3 Using computers improves my productivity
e4 Using computers enhances my effectiveness
e5 Using computers makes my job easier
e6 I find computers useful in my job
Perceived Ease of Use
pu1 Learning to use computers is easy for me
pu2 I find it easy to get computers to do what I want them to do
pu3 My interaction with computers is clear and understandable
pu4 I find computers flexible to interact with
pu5 It is easy for me to become skilled with computers
pu6 I find computers easy to use
Social/Subjective Norm
n1 People who are important to me use computers
n2 My friends use computers
n3 Using computers will help me get promoted
Behavioral Intention (Attitude)
b1 I intend to use computers when they are available
b2 To the extent possible, I would use computers to do various tasks
b3 To the extent possible, I would use computers frequently
b4 Using computers is a good idea
Note: All items were measured on a 7-point Likert scale, anchored with 1 = strongly disagree and 7 = strongly agree.

OPERATIONAL MEASURES OF THE STUDY VARIABLES AND HYPOTHESES

Perceived Usefulness

This factor is defined as "the prospective user's subjective probability that using a specific application system will increase his/her job performance within an organizational context" (Davis et al., 1989). User acceptance of computer systems is driven by perceived usefulness due to the reinforcement of value outcomes (Adams et al., 1992; Davis et al., 1989). The factor is similar to a component of relative advantage (Rogers, 1983) and identified by Dearing, Meyer, and Kazmierczak (1994) as the degree to which an innovation is more capable of achieving an ideal end-state. Six items were used to construct the perceived usefulness scale and are similar to those used in previous studies (Hu, Chau, Sheng, & Tam, 1999; Venkatesh & Morris, 2000; Al-Gahtani & King, 1999; Igbaria et al., 1995; Igbaria, Zinatelli, Cragg, & Cavaye, 1997).

H1a: Perceived usefulness is positively related to intentions toward NVID.
H1b: Perceived usefulness is positively related to NVID's use.

Perceived Ease of Use

Ease of use refers to the degree to which users expect the information system to be easy to understand and use (Davis et al., 1989). Some studies suggest that perceived ease of use will have a direct impact on attitude and behavioral intention and an indirect effect on perceived usefulness (Venkatesh & Morris, 2000). The direct effect should increase a user's acceptance of the system and lead to high levels of system use, while the indirect effect suggests that a system that is perceived as easier to use will be used more than one perceived harder to use (Davis et al., 1989). This study elects to retain only the direct effect given there is only a single system under testing with a key goal in NVID's development to create an easy-to-operate and user-friendly device. The items for this factor were also derived from previous studies (Hu et al., 1999; Venkatesh & Morris, 2000; Al-Gahtani & King, 1999; Igbaria et al., 1995; Igbaria et al., 1997). Participants were asked to indicate their agreement with six statements rated on a 7-point Likert scale.

H2: Perceived ease of use will positively influence users' attitudes toward NVID.

Subjective/Social Norm

Subjective norm relates to perceptions of general social pressure; the underlying normative beliefs are concerned with the likelihood that specific individuals or groups with whom the individual is motivated to comply will approve of the behavior. Subjective norm is considered to be a function of salient normative beliefs. Subjective norm was the last component added to the TRA and several authors have argued that it is the weakest component. As a result, studies have deliberately removed subjective norm from analysis, as was the case with the development of TAM. On the other hand, Trafimow and Finlay (1996) found evidence to suggest a distinction between individuals whose actions are driven primarily by attitudes and those who are driven by subjective norms. Their findings are confirmed across different behaviors where subjective norms have been found to be independently predictive of intentions (Venkatesh & Davis, 2000; Venkatesh & Morris, 2000; Conner & Sparks, 1996; Conner&& Armitage, 1998). These studies suggest that the explanation for the poor performance of the subjective norm component lies in its measurement with a number of studies using single, as opposed to more reliable multi-item, scales. The construct is composed of three adapted statements derived from previous studies (Venkatesh & Morris, 2000; Donald & Cooper, 2001).

H3: Social/subjective norms will positively influence attitude.

System Usage

The actual amount of usage by participants and the tasks they performed were logged by the NVID, uploaded to the department's server, and later collected by the researchers. Items collected included a date/time stamp, the number of times a sailor logged onto NVID, the particular tasks he/she performed, and the length of time used. Sailors were instructed to use only their assigned login code and when to log into and out of the NVID. Since sailors were required to check out the NVID unit, the researchers were confident that the user controls were sufficient, but realized that errors in length of time could still occur. Despite this limitation, the researchers believed that this system was superior to self-reported measures or user logs. Even though self-reported measures have been validated in past studies (Blair & Burton, 1987) their measures are not precise and are subject to user bias. Since each sailor had an equal opportunity to use NVID, the researchers created a metric of use by multiplying the number of times a user logged onto the device by the number of tasks he/she performed. The number of times the user logged on was matched to the date/time stamp to insure no errors in logging in. The length of time a sailor was logged on was not used since there was no control over an individual's routine.

DATA ANALYSIS AND RESULTS

A total of 270 sailors participated in the NVID trial. The participants had an average age of 20.4 years, an average enlistment time of 3.3 years, and an average time onboard the aircraft carrier of 17 months (standard shipboard assignment is 3 years).

Table 2. Principal components analysis rotated factor pattern (n = 270)

	Construct Loadings			
Item	Perc Ease of Use	Perc Usefulness	Beh Intention	Social/Subj Norms
e1	**82**	6	14	-1
e2	**90**	5	3	-1
e3	**82**	1	1	-4
e4	**81**	6	8	11
e5	**91**	9	12	1
e6	**93**	9	11	1
pu1	5	**89**	12	3
pu2	10	**85**	7	15
pu3	3	**81**	11	-3
pu4	4	**84**	7	16
pu5	8	**81**	17	-13
pu6	6	**85**	11	-5
b1	15	16	**85**	13
b2	7	12	**91**	4
b3	6	8	**87**	3
b4	10	17	**77**	20
n1	-8	2	20	**87**
n2	14	7	19	**69**

Table 3. Correlation matrix

Correlation Matrix

	Use	Behavior Intention	Social/Subj Norm	Ease of Use	Usefulness
Use	1.00000				
Behavior Intention	0.96046 <.0001	1.00000			
Social/Subj Norm	0.29153 <.0001	0.29339 <.0001	1.00000		
Ease of Use	0.15548 0.0105	0.19688 0.0011	0.05171 0.3974	1.00000	
Usefulness	0.21407 0.0004	0.26570 <.0001	0.10958 0.0722	0.15298 0.0118	1.00000

Reliability and construct validity of the instrument was evaluated using Cronbach's alpha with all values exceeding 0.80. Construct validity was evaluated by examining convergent and discriminant validity using both correlation analysis and exploratory factor analysis. Principal component factor analysis using a promax rotation procedure resulted in a 4-factor solution using the Eigen greater-than-one rule. The four factors yielded by the factor analysis procedure matched those proposed by the theoretical model. The analysis provided the researchers with sufficient justification to believe the ability of the items used to support this study's model. A copy of the factors, their item loadings, and Cronbach's alpha scores are found in Table 2, while the correlation matrix is present in Table 3.

TESTING THE MODEL

An examination of the proposed model was conducted using structural equation modeling (SEM) in the SAS statistical package, CALIS procedure, using maximum likelihood estimation. SEM has been found superior to other techniques including multiple regression analysis (Hankins, French, & Horne, 2000), and is used to test for whether the proposed model successfully accounts for the actual relationships observed in the sample. The proposed model was examined for overall goodness of fit and explanatory power, plus the individual causal links (paths) to test the study's hypotheses. The literature suggests seven indices be examined as a measure of the overall model's fit (Hoyle, 1995; Segars & Grover, 1993; Hatcher, 1994). All recommended indices were used in this study, including chi-square, goodness of fit index (GFI), adjusted goodness of fit index (AGFI), normalized fit index (NFI), nonnormalized fit index (NNFI), comparative fit index (CFI), and root mean square residual (RMSR). All indices provided justification to support the model's fit. The indices and their recommended values are presented in Table 4. Once the model's overall fit was satisfied, the individual constructs of the model were examined.

Table 4. Analysis of overall goodness of fit

Analysis of Overall Goodness of Fit

Indexes	Recommended value	Values from study
Chi-Square DF	$\leq$ 3.0	3.67
Goodness of Fit Index (GFI)	$\geq$ 0.90	0.99
Adjusted GFI (AGFI)	$\geq$ 0.80	0.96
Normalized Fit Index (NFI)	$\geq$ 0.90	0.99
Nonnormalized Fit Index	$\geq$ 0.90	0.98
Comparative Fit Index (CFI)	$\geq$ 0.90	0.99
Root Mean Square Residual (RMSR)	$\leq$ 0.10	0.07

Figure 4. Structural equation model

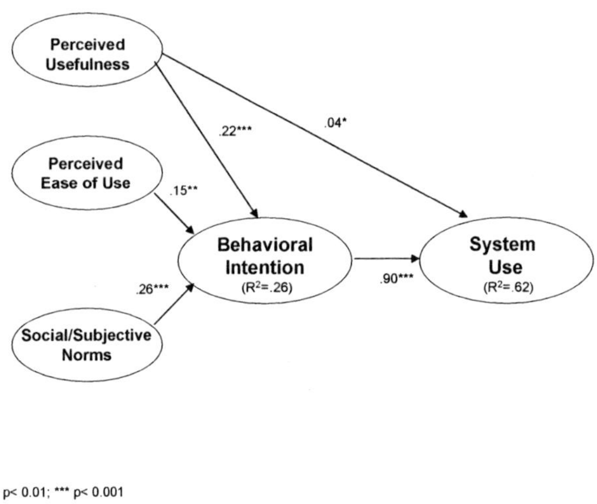

The constructs of interest in this study, behavior intention and system use, were examined using the adjusted R^2 for each dependent construct. Together the constructs explained 86% of the variance observed in the sailors' intention and use of the NVID device. As indicated in Figure 4, system use contributed the overwhelming amount of observed explanatory power of the model, with over 60%. During this phase of analysis the individual paths (hypotheses) were also examined. Using the standardized estimates, each path (hypothesis) was examined for its contribution to the model and level of statistical significance. All paths were found to be statistically significant, although the direct effect of perceived usefulness on system use was marginal. The strongest effect demonstrated was the .90 path coefficient between behavior intention and system use. This finding suggests that a one-unit increase in a sailor's intention results in a .90 unit in system use. The exogenous variables (perceived usefulness, ease of use, and social/subject norms) paths to behavior intention were retained during the analysis with statistically significant coefficients of .21, .16, and .26 respectively.

DISCUSSION

This study analyzed an adapted TAM using the introduction of a voice recognition device with a homogeneous sample of 270 deployed U.S. Navy sailors. The results suggest that the adapted TAM, using the additional factor social/subject norm, is a robust model with excellent ability to predict system use while performing almost exactly as predicted by the literature. Contributing to the success of the adapted TAM is the setting in which the experiment was conducted. A military environment is generally regarded as one in which activities such as computer usage is mandatory, yet this study was successfully carried out as a voluntary experiment, which suggests that user acceptance/usage should be high given either circumstance.

The exogenous variables—perceived usefulness, perceived ease of use, and social/subject norm—were all statistically significant predictors of attitude/behavior intention. The literature provided very strong support for the first two variables, although social/subject norm had received weaker support since it was dropped from the original TAM. Under the circumstance of this experiment we find that the variable provided similar if not stronger results than a recent study (Venkatesh & Davis, 2000) in which it was also included. Interestingly, both perceived usefulness and social/subject norm explained more variance (.22 and .26 respectively) than perceived ease of use (.15). The researchers suspect that given that all subjects in this experiment had ample computer experience and the NVID device was designed with ease of use in mind, the variable was considered less important. Additionally, the NVID was designed and tested specifically for shipboard use and to preserve the routine of its users. This may have contributed to the support for its usefulness, as time saved by using the device could be devoted to sailors' primary duties or advancement. The only surprise yielded by the study was the marginal direct influence of perceived usefulness on system use (.04).

The study's key finding is the adapted TAM's ability to predict sailors' intention and system use. The two factors (behavior intention and system use) predicted almost 90% of the model's explained variance, with system use contributing 62%. The findings clearly extend the literature's suggestions that attitude/behavior intention is an excellent predictor of system use (Venkatesh & Davis, 2000; Hu et al., 1999; Mathieson, 1991; Taylor & Todd, 1995; Szajna, 1996; Davis et al., 1989; Davis, 1986). Of some surprise was the relative weakness of attitude/behavior intention to explain more of the model's variance. The exogenous variables (perceived usefulness, perceived ease of use, and social/subject norm) accounted for only 26% of attitude/behavior intention, which is somewhat lower than in previous studies. Despite this less than expected result, the overall predictive power of the adapted TAM is excellent and should thereby incorporate the subject/social norm factor.

Although we provided evidence of the efficacy of the adapted TAM, the rapidly evolving nature of speech recognition technology (Boyce, 2002; Lai, 2001; Spring, 2003) may influence our model's predictive capability. As speech recognition technology becomes less error prone, more sophisticated, more powerful, and (hopefully) more user-friendly, perceptions of usefulness, ease of use, and social norm may fluctuate dramatically. As a result, their impact on intention and system use may also change. Given the potential rapid evolution of technology, future replications of this study may reveal that the adapted TAM is either more or less effective. However, we believe that the adapted TAM will prove even more effective in predicting system use because the future popularity of such technology hinges on its ability to attract new customers and retain current customers by providing a robust, reliable, and effective product (at least in terms of customer perceptions). A natural extension of the current study will be to replicate with the adapted TAM within the next 2 or 3 years.

The findings of this study make a contribution to theory by extending the exogenous variables of TAM with the inclusion and validation of subjective/social norms. The researchers felt that since the factor was part of the referent literature and the original TRA/TPB models, its inclusion was warranted. Moreover, the inclusion of subjective/social norms was empirically shown to be relevant with our sample. Since this study was conducted under somewhat unique conditions, we suggest that future research in other environmental settings also include subjective/social norms to further validate its applicability and explanatory ability. Additional research should also be conducted into the continued inclusion of perceived ease of use. Despite this study's finding of significance, its predictive power appears to be decreasing as this research stream continues. Information systems are more widely used than ever before and their design and user interfaces are created to improve and simplify user interaction. As a result, many users expect systems to be easy to use and simple in design and function. A trend is seen in the design and use of multifunction applications in today's mobile telephones, which contain planners once found in PDAs, Internet capabilities, and the original phone functions, all operated by voice or touch-screen. If this trend proves to be the future case and perceived ease of use is a less effective predictor of user intention and use, then the TAM model must be adapted accordingly. Either way, it appears that subjective/social norms will be an important influence on system use and hence our adaptation offers a contribution to the literature.

Future research should also continue to examine the external factors that impact a user's decision process to adopt and use an information system. While this study held the external factors constant, the researchers understand that in other settings these factors could significantly impact the intention and adoption process. Theoretical constructs should thereby extend the identification of these external factors and translate them into actionable steps implementers can take to improve and expedite the adoption/implementation process. For instance, if training and learning are critical external factors, researchers should identify how and by what steps these factors can be enhanced to correct deficiencies in deliverables. Additionally, researchers should examine TAM and adoption in a variety of multitiered settings. The vast majority of research samples encountered during this study, while drawn from different organizations, examined only one part or subsample of the organization's population. For experimental purposes this provides the researchers with stronger findings but is less applicable to the practitioner. Samples drawn from workers and managers throughout the organization could shed insight into variations required by the implementation and adoption process.

Limitations

The study set out to address the limitations of previous studies. The researchers were careful to incorporate into the work 1) a large sample of 270 participants, 2) a measurement scale that used at least three items per factor with Cronbach alphas of .80 or higher, 3) a structural equation model as an analysis technique, and 4) a direct measurement of system use to avoid self-reported system usage numbers. Yet despite this effort, the study did have some limitations that should be considered when interpreting the results. The sample in this study,

U.S. Navy sailors, was a relatively homogeneous group as indicated by the selection of the two ratings (job classifications) in two departments. The experiment was conducted under conditions that were somewhat unique when compared to other studies. Specifically, subjects were all living and working in close conditions, involved with testing a new computer-based system, and were voluntary users. The researchers, however, recognize that while the subjects had unique characteristics and participated under unique conditions, they are not so different from other sample groups. Any study participant that belongs to an organization is subject to a certain degree of social pressure, be it positive or negative. This is true for student samples as well as those in business settings. Further, more recent studies have investigated TAM using unique or newly introduced systems. For instance, Hu et al. (1999) examined TAM with physicians using telemedicine and Venkatesh and Davis (2000) used four situations with subjects examining new systems.

Given these disclaimers, the researchers realize that sailors deployed onboard a military ship possess unique characteristics and a unique culture when compared with other potential subjects. As a result, we suggest that this circumstance could potentially bias the results of this experiment (at least in terms of generalizability to less unique contexts). Despite voluntary participation, it is possible that sailors believed that they were under pressure from their departments to use the NVID and that in turn might have influenced and perhaps inflated their system usage statistics. However, this did not appear to be the case when reviewing the data gathered from a limited number of interviews at the conclusion of the experiment. Another potential limitation was that this study purposefully controlled for any external factors, such as training, computer experience, image, and so forth, which might have influenced the participants. This may not always be the case in the general population and should be taken into consideration.

CONCLUSION

This study yielded several practical findings. First, subjective/social norms proved a significant factor when investigating the adoption of an information system, at least within the conditions of this experiment. Second, assuming that these findings are generalizable to other organizational settings, subjective/social norm could have similar implications to adopters as perceived usefulness. That is, a socially acceptable technology may very well be similar to (or the same as) a perceptually acceptable one. If this is the case, information systems departments and those implementing the systems should be able to increase behavioral intention to use a system and ultimately system use itself by manipulating the organization's social environment. Third, social norms can be directly influenced by tweaking the reward system, technology training, and top management vision and participation. This suggestion is no different from actions undertaken by organizations to promote systems before implementation. For instance, a number of organizations involved in large-scale system implementation, such as ERP systems, actively "sell" the system to users through meetings, pep rallies, and promotional materials such as CDs and t-shirts. This in essence changes the organizational environment and creates social pressure for users to adopt the system. Once a critical mass of adopters is reached it becomes difficult for users to refuse or delay their adoption process.

Technology acceptance in organizations has been widely studied yet still continues to be a topic of interest to both researchers and practitioners. Despite gains made by TAM and other models, there is no "holy grail" that succinctly predicts and explains user intentions and use. Perhaps this is due to the complexity of the topic and the dynamic nature of organizations. However, by offering a model that explains almost 90% of the variance with regard to system use, we are confident that the adapted TAM provides strong inroads toward explaining the complicated relationship between intention and use. The combined organizational experiences of the researchers match the study results. In many organizational settings, we have found that social pressure is a great contributor to intention and system use. It is difficult, if not impossible, for organizational members to dismiss the allure of peer and managerial pressure if the adoption/implementation of a system is mandatory and made part of the organizational culture. We thereby believe that our work extends and advances previous theoretical gains on this critical issue.

REFERENCES

Adams, D. A., Nelson, R. R., & Todd, P. A. (1992). Perceived usefulness, ease of use, and usage of information technology: A replication. *MIS Quarterly, 16,* 227-242.

Ajzen, I. (1985). From intentions to action: A theory of planned behavior. In J. Kuhl & J. Beckman (Eds.), *Action control: From cognitions to behaviors.* New York: Springer.

Ajzen, I. (1991). Theory of planned behavior. *Organizational Behavior and Human Decision Processes, 50,* 179-211.

Albers, J. (2000). Successful speech applications in high noise environments. *SpeechTEK Proceedings* (pp. 147-154).

Al-Gahtani, S. S., & King, M. (1999). Attitudes, satisfaction and usage: Factors contributing to each in the acceptance of information technology. *Behaviour and Information Technology, 18,* 277-297.

Armitage, C. J., & Conner, M. (2001). Efficacy of the theory of planned behavior: A meta-analytic review. *British Journal of Social Psychology, 40,* 471-499.

Andrea, D. (2000). Improving the user interface: Digital far-field microphone technology. *SpeechTEK Proceedings*, 155-160.

Bagozzi, R. P. (1992). The self-regulation of attitudes, intentions, and behavior. *Social Psychology Quarterly, 55,* 178-204.

Blair, E., & Burton, S. (1987). Cognitive processes used by survey respondents to answer behavioral frequency questions. *Journal of Consumer Research, 14,* 280-288.

Blue, C. L. (1995). The predictive capacity of the theory of reasoned action and the theory of planned behavior in exercise research. *Research in Nursing and Health, 18,* 105-121.

Bokulich, F. (2000). JSF [Joint Strike Fighter] voice recognition. *Aerospace Engineering Online.* Retrieved from http://www.sae.org/aeromag/techupdate_5-00/03.htm

Boyce, C. (2002). Speech recognition market grows. *Research & Development, 44,* 17.

Business Communications Company, Inc. (2003). RG-268 Voice Recognition: Technologies, Markets and Opportunities. Retreived from http://www.bccresearch.com

Charry, M., Pimentel, H., & Camargo, R. (2000). User reactions in continuous speech recognition systems. *AVIOS Proceedings of The Speech Technology & Applications Expo*, (pp. 113-130).

Chau, P. Y. (1996). An empirical assessment of a modified technology acceptance model. *Journal of Management Information Systems, 13,* 185-204.

Christ, K. A. (1984). *Literature review of voice recognition and regeneration technology for Army helicopter applications* (Army Report No. HEL-TN-11-84). Aberdeen Proving Ground, MD: Human Engineering Lab.

Conner, M., & Armitage, C. J. (1998). Extending the theory of planned behavior: A review and avenues for further research. *Journal of Applied Social Psychology, 28,* 1429-1464.

Conner, M., & Sparks, P. (1996). The theory of planned behavior and health behaviors. In M. Conner & P. Norman (Eds.), *Predicating health behavior.* Buckingham: Open University Press.

Costanzo, C. (2003). Consumers still wary of voice recognition. *American Banker, 168,* 10-11.

Davis, F. D. (1986). *A technology acceptance model for empirically testing new end-user information systems: Theory and results.* Unpublished doctoral dissertation, Sloan School of Management, MIT.

Davis, F. D. (1993). User acceptance of information technology: System characteristics, User perceptions, and behavioral impacts. *International Journal of Man-Machine Studies, 38,* 475-487.

Davis, F. D., Bagozzi, R., & Warshaw, P. (1989). User acceptance of computer technology: A comparison of two theoretical models. *Management Science, 35,* 729-736.

Dearing, J. W., Meyer, G., & Kazmierczak, J. (1994). Portraying the new communication between university innovators and potential users. *Communication Science, 16,* 11-42.

Dishaw, M. T., & Strong, D. M. (1999). Extending the technology acceptance model with task-technology fot constructs. *Information and Management, 36,* 9-21.

Donald, I., & Cooper, S. R. (2001). A facet approach to extending the normative component of the theory of reasoned action. *British Journal of Social Psychology, 40,* 599-621.

Erten, G., Paoletti, D., & Salam, F. (2000). Speech recognition accuracy improvement in noisy environments using Clear Voice Capture (CVC) technology. *AVIOS Proceedings of The Speech Technology & Applications Expo* (pp. 193-198).

Fishbein, M., & Ajzen, I. (1975). *Belief, attitude, intention and behavior: An introduction to theory and research.* Reading: Addison-Wesley.

Gefen, D., & Straub, D. W. (1997). Gender differences in the perception and use of e-mail: An extension to the technology acceptance model. *MIS Quarterly, 21,* 389-400.

Gaddy, L. (2000). The future of speech I/O in mobile phones. *SpeechTEK Proceedings* (pp. 249-260).

Godin, G. (1993). The theories of reasoned action and planned behavior: Overview of findings, emergent research problems and usefulness for exercise promotion. *Journal of Applied Sport Psychology, 9,* 491-501.

Godin, G., & Kok, G. (1996). The theory of planned behavior: A review of its application to health-related behaviors. *American Journal of Health Promotion, 11,* 87-98.

Goette, T. (2000). Keys to the adoption and use of voice recognition technology in organizations. *Library Computing, 13,* 67-80.

Goodliffe, C. (2000). The telephone and the Internet. *AVIOS Proceedings of the Speech Technology & Applications Expo* (pp. 149-151).

Gorham, A., & Graham, J. (2000) Full automation of directory enquiries:A live customer trial in the United Kingdom. *AVIOS Proceedings of the Speech Technology & Applications Expo* (pp. 1-8).

Hankins, M., French, D., & Horne, R. (2000). Statistical guidelines for studies of the theory of reasoned action and the theory of planned behaviour. *Psychology and Health, 15,* 151-161.

Hartwick, J., & Barki, H. (1994). Explaining the role of user participation in information system use. *Management Science, 40,* 440-465.

Hatcher, L. (1994). *SAS system for factor analysis and structural equation modeling.* Cary: SAS.

Hausenblas, H. A., Carron, A. V., & Mack, D. E. (1997). Application of the theories of planned behavior and reasoned action to exercise behavior," *Journal of Sport and Exercise Psychology, 19,* 36-51.

Haynes, T. (2000). Conversational IVR: The future of speech recognition for the enterprise. *AVIOS Proceedings of the Speech Technology & Applications Expo*, (pp. 15-32).

Hermansen, L. A., & Pugh, W. M. (1996). *Conceptual design of an expert system for planning afloat industrial hygiene surveys* (Technical Report No. 96-5E). San Diego, CA: Naval Health Research Center.

Holtzman, T. (2000). Improving patient care through a speech-controlled emergency medical information system," *AVIOS Proceedings of the Speech Technology & Applications Expo*, (pp. 73-81).

Hoyle, R. H. (Ed.). (1995). *Structural equation modeling: Concepts, issues, and applications.* Thousand Oaks, CA: Sage.

Hu, P. J., Chau, P. Y., Sheng, O. R., & Tam, K. Y. (1999). Examining the technology acceptance model using physician acceptance of telemedicine technology. *Journal of Management Information Systems, 16,* 91-112.

Igbaria, M., Guimaraes, T., & Davis, G. B. (1995). Testing the determinants of microcomputer usage via a structural equation model. *Journal of Management Information Systems, 11,* 87-114.

Igbaria, M., Zinatelli, N., Cragg, P., & Cavaye, A. L. (1997). Personal computing acceptance factors in small firms: A structural equation model. *MIS Quarterly, 21,* 279-305.

Ingram, A. L. (1991). *Report of potential applications of voice technology to armor training* (Final Report: Sep 84-Mar 86). Cambridge: Scientific Systems Inc.

Jonas, K., & Doll, J. (1996). A critical evaluation of the theory of reasoned action and theory of planned behavior. *Zeitschrift fur Sozialpsycgologie, 27,* 19-31.

Kundupoglu, Y. (2000). Fundamentals for building a successful patent portfolio in the new millennium. *AVIOS Proceedings of the Speech Technology & Applications Expo*, (pp. 229-234).

Lai, J. (2000). Conversational interfaces. *Communications of the ACM, 43,* 24-27.

Lai, J. (2001). When computers speak, hear, and understand. *Communications of the ACM, 44,* 66-67.

Leitch, D., & Bain, K. (2000). Improving access for persons with disabilities in higher education using speech recognition technology. *AVIOS Proceedings of The Speech Technology & Applications Expo*, (pp. 83-86).

Mathieson, K. (1991). Predicting user intentions: Comparing the technology acceptance model with theory of planned behavior. *Information Systems Research, 2,* 173-191.

Manstead, A. S., & Parker, D. (1995). Evaluating and extending the theory of planned behavior. In *European Review of Social Psychology, 6,* 69-95.

Markowitz, J. (2000). The value of combining technologies. *AVIOS Proceedings of the Speech Technology & Applications Expo*, (pp. 199-206).

MedQuist (2003). http://www.rsleads.com/301ht-192

Mogford, R. M., Rosiles, A., Wagner, D., & Allendoerfer, K. R. (1997), *Voice technology study report* (Report No. DOT/FAA/CT-TN97/2). Atlantic City, NJ: FAA Technical Center.

Moore, G. C., & Benbassat, I. (1991). Development of an instrument to measure the perceptions of adopting an information technology innovation. *Information Systems Research, 2,* 192-222.

Nelson, R. R., & Cheney, P. H. (1987). Training end users: An exploratory study. *MIS Quarterly, 11,* 547-559.

Oviatt, S. (2000). Taming recognition errors with a multimodal interface. *Communications of the ACM, 43,* 45-51.

Paper, D., Rodger, J. A., & Simon, S. J. (in press). Voice says it all in the Navy. *Communications of the ACM.*

Prizer, B., Thomas, D., & Suhm, B. (2000). The business case for speech. *SpeechTEK Proceedings,* 15-26.

Rogers, E. M. (1983). *Diffusion of innovations* (4th ed.). New York: Free Press.

Rolandi, W. (2000). Speech recognition applications and user satisfaction in the imperfect world. *AVIOS Proceedings of the Speech Technology & Applications Expo,* (pp. 153-158).

Schalk, T. (2000). Design considerations for ASR telephony applications. *AVIOS Proceedings of the Speech Technology & Applications Expo,* (pp. 103-112).

Segars, A. H., & Grover, V. (1993). Re-examining perceived ease of use and usefulness: A confirmatory factor analysis. *MIS Quarterly, 17,* 517-525.

Shepard, D. (2000). Human user interface—HUI. *SpeechTEK Proceedings,* 262-270.

Sheppard, B. H., Hartwick, J., & Warshaw, P. R. (1988). The theory of reasoned action: A meta-analysis of past research with recommendations for modifications and future research. *Journal of Consumer Research, 15,* 325-343.

Shneiderman, B. (2000). The limits of speech recognition. *Communications of the ACM, 43,* 63-65.

Sones, R. (2000). Improving voice application performance in real-world environments. *SpeechTEK Proceedings,* 179-210.

Soule, E. (2000). Selecting the best embedded speech recognition solution. *SpeechTEK* Proceedings, 239-248.

Sparks, P. (1994). Attitudes towards food: Applying, assessing, and extending the theory of planned behavior. In D. Rutter & L. Quine (Eds.), *Social psychology and health: European perspectives.* Aldershot: Avery.

Spring, C. (2003). Productivity gains of speech-recognition technology. *Health Management Technology, 24,* 54-55.

Straub, D. W. (1989). Validating instruments in MIS research. *MIS Quarterly, 13,* 147-169.

Stromberg, A. (2000). Professional markets for speech recognition. *SpeechTEK Proceedings,* 101-124.

Sutton, S. (1998). Predicting and explaining intentions and behavior: How well are we doing? *Journal of Applied Social Psychology, 28,* 1317-1338.

Szajna, B. (1996). Empirical evaluation of the revised technology acceptance model. *Management Science, 42,* 85-92.

Taylor, S., & Todd, P. A. (1995a). Understanding information technology usage: A test of competing models. *Information Systems Research, 6,* 144-176.

Taylor, S., & Todd, P. A. (1995b). Assessing IT usage: The role of prior experience. *MIS Quarterly, 19,* 561-591.

Thompson, R. L., Higgins, C. A., & Howell, J. M. (1991). Personal computing: Toward a conceptual model of utilization. *MIS Quarterly, 15,* 125-143.

Trafimow, D., & Finlay, K. A. (1996). The importance of subject norms for a minority of people. *Personality and Social Psychology Bulletin, 22,* 820-828.

Van den Putte, B. (1991). *20 years of the theory of reasoned action: A meta-analysis.* Unpublished working paper, University of Amsterdam, The Netherlands.

Venkatesh, V., & Davis, F. D. (2000). A theoretical extension of the technology acceptance model: Four longitudinal field studies. *Management Science, 46,* 186-204.

Venkatesh, V., & Morris, M. G. (2000). Why don't men ever stop to ask for directions? Gender, social influence, and their role in technology acceptance and usage behavior. *MIS Quarterly, 24,* 115-139.

Wenger, M. (2000). "Noise rejection—The essence of good speech recognition. *AVIOS Proceedings of the Speech Technology & Applications Expo,* (pp. 51-63).

Wickstrom, T. (2000). Microphone voice recognition performance in noise: A proposed testing standard. *SpeechTEK Proceedings,* (pp. 211-219).

Woo, D. (2000). Desktop speech technology: A Macos perspective. *AVIOS Proceedings of the Speech Technology & Applications Expo,* (pp. 39-50).

This work was previously published in End-User Computing: Concepts, Methodologies, Tools, and Applications, edited by S. Clarke, copyright 2008 by Information Science Reference, formerly known as Idea Group Reference (an imprint of IGI Global).

Chapter XI
Educating Our Students in Computer Application Concepts:
A Case for Problem-Based Learning

Peter P. Mykytyn
Southern Illinois University, USA

ABSTRACT

Colleges of business have dealt with teaching computer literacy and advanced computer application concepts for many years, often with much difficulty. Traditional approaches to provide this type of instruction, that is, teaching tool-related features in a lecture in a computer lab, may not be the best medium for this type of material. Indeed, textbook publishers struggle as they attempt to compile and organize appropriate material. Faculty responsible for these courses often find it difficult to satisfy students. This chapter discusses problem-based learning (PBL) as an alternative approach to teaching computer application concepts, operationally defined herein as Microsoft Excel and Access, both very popular tools in use today. First PBL is identified in general, then we look at how it is developed and how it compares with more traditional instructional approaches. A scenario to be integrated into a semester-long course involving computer application concepts based on PBL is also presented. The chapter concludes with suggestions for research and concluding remarks.

INTRODUCTION

It probably would not surprise most Management Information Systems faculty and academics that both accredited and nonaccredited colleges of business continue to struggle with instruction in computer application concepts aimed at undergraduate students. A review of business school Web sites would indicate that a wide array of courses, course names, and course schedules exist: Introduction to Computers, Management Information Systems, and Microcomputer Applications might be some course names that could be found. Some of these courses might be sophomore level, whereas others might be found at the junior level. In other instances, more than one course might be found. One class might deal exclusively with computer application concepts, such as Microsoft Office, whereas another might relate more to MIS itself. Furthermore, the topic related to teaching MIS and application concepts is often raised in postings to ISWORLD (www.isworld.org). Thus, it is of little surprise that in the end, questions and uncertainties about these classes exist.

The focus of this perception-based chapter is toward teaching computer application concepts such as Microsoft Office. That is not to say that Microsoft Office is the sole way to teach these types of concepts. Indeed, some schools might teach other tools such as HTML, JAVA, Visual Basic, and so forth. In effect, however, my thoughts related to teaching these concepts are independent of the particular tool used in the classroom. At the same time, Microsoft Office, with particular emphasis on Excel and Access but with some instruction focusing perhaps on PowerPoint and FrontPage, seems to be the most prominent tool used today to provide students with basic computer application concepts.

With numerous surveys, questions posed on ISWORLD, questions raised by faculty at conferences, and continuing efforts by textbook publishers to develop the right set of books for these tools and applications, MIS faculty continue to struggle with this type of class. At the same time, students themselves raise objections. On the one hand, some students are already well skilled in these concepts as a result of taking similar classes in high school or at the community college level. Still others have worked with these tools professionally and do not see the need for taking another class that is perceived to have little to no value. In other instances where two required classes are taught, the overlap between the first and second class is so similar as to again provide seemingly little value to the student.

Instead of rehashing the same material over and over again, however, my thought is to suggest an alternative approach to teaching computer application concepts. The approach is called problem-based learning (PBL). This approach is by no means new or unique. However, it does seem to be somewhat unique to teaching computer application concepts; indeed, it appears to be quite unique in colleges of business. Ideally, readers of the chapter may question the approach presented leading them to investigate it more completely in terms of its applicability and use in this type of class. Additionally, new approaches such as this would lead to empirical research as well.

In the next section, a brief overview of PBL, its concepts, and how it might be applied to teaching computer application concepts is presented. A suggested research agenda follows. Conclusions are presented last.

PROBLEM-BASED LEARNING

In mid-January 2006 I searched one of the online database indices that our University subscribes to, EBSCO (www.ebsco.com/home/). I searched for the term "problem-based learning" and was rewarded with 1,125 hits. Choosing to refine the search somewhat, I again searched on "problem-based learning," this time as part of the title of articles. A total of 383 articles in that database contained "problem-based learning" as part of the title at the time of the search. This very unscientific sampling process indicated that the vast majority of articles are from the medical field: *Medical Education, Journal of Clinical Anesthesia, Medical Teacher*, and *Physiotherapy* are representative of the journals containing those articles. In fact, I continued with this process and noted that just one article found was related to colleges of business. That article appeared in the *International*

Journal of Technology Management, and it dealt with entrepreneurship. I was unable to identify any articles dealing with computer tools, such as Microsoft Office. To try to uncover anything of a PBL nature and information systems, I again searched EBSCO with "problem-based learning" appearing anywhere in the text of an article along with the following journals: *Journal of Management Information Systems, Information & Management, MIS Quarterly,* and *Decision Sciences*. No results were returned.

This very unscientific assessment does not come as a surprise, because the concepts of PBL do not appear to be mainstream as far as teaching computer and software concepts is concerned. One article dealing with PBL appeared in the spring 2003 issue of *Decision Sciences Journal of Innovative Education* (Kanet & Barut, 2003). It dealt with using PBL to teach production and operations management.

To begin with, it is perhaps easier to first indicate what PBL is not, or rather how it contrasts with a traditional approach to teaching. The opposite of PBL is subject-based learning. In this pedagogical environment, students are presented first with material that is specific to a discipline, such as medicine, nursing, geography; following this, students encounter or are presented with a scenario to which they apply what they have already been taught (Maskell and Grabau, 1998). In this situation, the instructor may present a lecture and take a very active part in the process. Following that, students, most likely acting alone, work out problems that may be very well structured and directly tied to the lecture presented earlier.

Problem-based learning is a process that is rooted in using a problem situation to direct and focus the learning activity. Boud and Feletti (1991) provide guidelines for PBL instruction:

Students are presented with an authentic problem, one that is based on a real-world situation. In its purest form, students work in groups to gather their thoughts and prior knowledge they may have and then seek to define the problem broadly.

Students develop questions that relate to problem aspects they don't understand; essentially, they understand what they know and do not know.

Students may rank order the issues or questions and may divide the process of obtaining answers among themselves.

When students reconvene, they present and summarize their findings as a group and decide how best to proceed.

Drennon (2005) indicates that the process outlined above provides students with a problem which then leads them to engage in a "self-directed, reiterative, reflective learning process" (Shepherd and Cosgriff, 1998, p. 50).

One of the major differences between a traditional approach to instruction, such as a standard lecture by the instructor, and PBL relates to the role of the class instructor. Indeed, much of the literature discussing PBL refers to the instructor as a "tutor."

The prime educative task of the tutor is to ensure that students make adequate progress towards formulating the problem, identifying what they need to learn in order to understand it better and deal with it...The tutor does this essentially by questioning, probing, encouraging critical reflection, suggesting and challenging in helpful ways—but only where necessary (Margetson, 1994, p. 14).

Dunlap (2005) iterates that a tutor would not provide direct instruction about the task facing the students or the direction in which the students should proceed. Rather, the instructor/tutor's role would be subtler, whereby he/she would guide students by asking metacognitive types of questions.

Other roles of a tutor would include generating and prioritizing ideas, helping decide what to do next, and determining whether and when a group reached consensus.

In summary, there are a number of pedagogical differences between a traditional approach to instruction and that based on problem-based learning. Shepherd and Cosgriff (1998) provide a synopsis of the major differences in these approaches; they are illustrated in Table 1. What may be one of the major differences between the two approaches, a difference that might surely need to be dealt with completely, is the change in the nature of authority in the classroom between the instructor and the students. Drennon (2005) summarizes it very clearly, indicating that PBL removes authority from the instructor and places it in the hands of the students, undoing students' "... submissiveness to the rules of the established order" (p. 389).

PROBLEM-BASED LEARNING APPLIED TO COMPUTER APPLICATION CONCEPTS

As taught in many colleges and universities, computer applications concepts often pertain to teaching Microsoft Office tools, especially Excel and Access. That is not to say that some institutions will not include Word, PowerPoint, FrontPage, or even Web tools such as HTML or JAVA. However, for the purposes of this chapter, it is assumed that Word and PowerPoint are elementary tools that are common knowledge today or are tools that students have learned on their own. This chapter also assumes that FrontPage, JAVA, and HTML are important tools today for the development of Web sites; however, the focus of these programs is quite specialized and might be better left to perhaps a Web development course. That said, it is certainly possible to apply PBL to instruction in these tools.

The discussion that follows below is one possible approach to incorporating PBL to teach computer application concepts. It is certainly not meant to imply that it is the only possible approach. By its very nature, PBL can be somewhat unstructured. In addition, the description below presumes that students are computer literate. That is, students would have already completed a literacy course that most The Association to Advance Collegiate Schools of Business (AACSB) accredited colleges of business include in their curricula. The course described here would be a 2nd level class, one that applies more advanced topics in Excel and Access.

The class is a three-credit-hour class taught at the junior level. It consists of two meetings per week: a 75-minute lecture on MIS concepts using one of many possible textbooks published for this purpose, such as texts by Turban, Leidner,

Table 1. Generalized differences between a traditional and a problem-oriented classroom (Adapted from Shepherd and Cosgriff [1998])

Traditional education	Problem-based learning
Curriculum as prescription	Curriculum as experience
Instructor-focused	Student-focused
Teaching as transmitting	Teaching as facilitating
Learning as receiving	Learning as constructing
Rigid environment	Flexible environment
Replicate and apply received knowledge	Construct and synthesize knowledge
Emphasis on a best solution	Emphasis on alternative acceptable solutions
Evaluation of result	Evaluation of result and learning process

McLean, and Wetherbe (2006) and Laudon and Laudon (2005). The other 75 minutes are devoted to advanced application concepts involving Excel and Access. The latter portion of the class would normally be taught in a computer lab with each student working independently at a computer workstation. Because the class focuses on more advanced concepts in Excel and Access, there are numerous scenarios that an instructor might incorporate in order to achieve the learning objectives. The scenario suggested here relates to 401(k) investments.

The scenario indicates that the student has just moved from one job to another and wishes to "roll over" $100,000 accumulated in a 401(k) program begun with the first employer. There are numerous mutual funds that the new employer will allow employees to choose from; these options would also be open to the rollover of the $100,000. For practical purposes, the number of funds suggested for this exercise is in excess of 150, and the student must select at least three funds at the beginning of the project.

In addition, each month the student will invest $1,000 of his/her money into the new employer's 401(k) program, and the new employer will contribute an additional $500, thereby providing the employee with a total new monthly investment of $1,500. The student is free to invest the $1,500 each month in one or more of the three funds selected at the beginning of the project. Or the student may select new funds for which to invest. Finally, the student is not restricted to keeping the same three funds that he/she started with at the beginning of the project. That is, the student may rollover money from one or more funds into one or more new funds. The goal of the project, in addition to having the student learn new concepts in Excel and Access, is to see which student in the class accumulates the most money at the end of the semester. An instructor may choose to reward the winning student (or perhaps the first three students) with some sort of award, such as actual prize money or extra points or credit for the class.

It should be stated at the outset that PBL exercises are reportedly based on collaboration with peers, that is, students who work together in groups. The mutual fund exercise described above could be designed for persons to work alone. That said, it might be argued that some of the benefits from PBL as suggested earlier by Drennon (2005) would be lost. They include allowing students to better organize ideas, defining the problem better, and deciding as a group on how best to proceed (Drennon, 2005). However, it is possible that the main focus of PBL might still be accomplished by having students work alone. In fact, one PBL study implied that, except for the nature of the problem students were working on, students could accomplish a PBL-based task by working alone (Lam, 2004). Harland (1998) pointed out too that students working in teams or groups can encounter difficulties. These include differential workloads and failing to delegate properly. Thus, team-oriented activities are not necessarily perfect.

One of the strengths of the project described here is its applicability to students enrolled in colleges of business administration and the real world nature of the issues that these students would encounter, perhaps sooner than expected but definitely not long after they complete their undergraduate studies. This scenario can serve a much greater good than many projects described in textbooks because of the personal association that students can have with the topic, and such an orientation is a primary objective of PBL-based instruction. Its applicability to students is enhanced too because it is designed as a semester-long project.

There are numerous parts to this problem that contribute to its somewhat unstructured nature, another characteristic of PBL-centered problems. These include the necessity of researching mutual funds in terms of fund objectives, types of investments a fund focuses on—stocks, bonds, international—and so forth. Mutual funds also assess various fees, such as management fees, and not all funds charge the same for them. Of course, the nature of returns on investments would be a primary concern of anyone. Additionally, capital gains and dividend histories would be relevant to long-term growth. The issues just listed are not necessarily exclusive of others that might be interjected into the project, but they are all key to mutual fund investing. More importantly, these criteria lend themselves very well to the use of Excel and Access.

There are a number of advanced concepts in Excel and Access that students could investigate and integrate into the project. For Excel, these include numerous analysis tools, such as sensitivity analysis, goal seeking, descriptive statistics, presentation tools such as histograms and pivot tables, ranks and percentiles, and exponential smoothing. In Access, students would be exposed to report writing and query processing as well as perhaps more advanced topics in modeling and database design.

The role of the instructor in this scenario would be consistent with the PBL approach to instruction. Recall that PBL approaches usually refer to the instructor as a tutor. The instructor would provide the background information to the students, and that information is consistent with the information provided above, that is, rolling over $100,000, identifying new mutual funds, and so forth. The instructor would not provide direct instruction in Excel and Access as needed to deal with the relevant questions about fund types, expenses and fees, returns, capital gains and dividends, gains and losses, and so on. Instead, consistent with Dunlap's (2005) suggestions, the instructor would cleverly guide students and their thinking by asking appropriate questions. These could include:

- What might be your long-term objectives with these investments?
- What information might you need to more appropriately and clearly make your investment decisions?
- Have you attempted to summarize what you know or do not know up to now?
- What type of analysis is necessary to manage your investment decisions? How would Excel and Access contribute to facilitating the analysis?

The role of the instructor/tutor may also be thought of as a facilitator. In this regard, the instructor would assist students in generating and prioritizing ideas, and expanding their knowledge beyond what they might find in a course text on Excel and Access, perhaps using numerous Web-based resources. Since personal investments centered on 401(k) programs might be new to these students, the instructor would encourage students to do additional research on the topic. It is also important for the instructor to monitor the student's progress to make sure they are "keeping on track." This can be accomplished by appropriate feedback in the formal class period as well as through the use of e-mail, chat rooms, and the like.

Harland (1998) discussed another potential problem that could occur in the scenario described here: students' lack of knowledge in basic Excel and Access skills. All students completing this class would have completed a computer literacy class as a prerequisite. However, it is a frequently discussed fact that literacy courses can and do differ considerably. In addition, students may have completed a literacy class years earlier, may have used different tools besides Excel and Access, and may have received varying levels of knowledge and instruction in the literacy class. As such, not all students are similarly prepared to enter the advanced class. Varied backgrounds such as these can necessitate that the instructor ensures that all students are guided properly, that appropriate Excel and Access tools that might be integral to the project are identified to the students who would then do additional research on them.

Potential Benefits of PBL

A number of benefits can be achieved by using a PBL-based approach to teaching computer application concepts. First, PBL is student-centered. In the current scenario, students are provided with an overview of the topic and problem area and are guided to develop an understanding of the problem. Students are not force-fed instruction about this or that command. The models and types of analyses they develop to have a better understanding of the mutual fund problem are their own ideas, as opposed to traditional approaches that may tell the student to "do this, then do this next, and finally conclude by doing this." Students may be able to draw upon prior learning much better. A PBL approach may help the students to understand the relevance of earlier material better and allow them to grasp the fact that there is indeed meaning behind coursework. In addition, the nature of the problem presented here, that is, rolling over mutual funds, is very relevant to students enrolled in colleges of business. Many will not seek employment as financial analysts, but most will indeed take a personal interest in their own investment portfolios. Steinemann (2003) refers to this latter point as applicability because students are solving real problems. In general, Steinemann (2003) indicated that PBL provides instruction for students that, in addition to applicability, is more active because students are directing themselves; students are more motivated because they are more interested in a subject when it is more personally oriented; and students can develop professional skills that they can take with them long after completing a semester-long course. The nature of the problem presented here can provide these benefits.

RESEARCH AGENDA

The use of PBL as a method of instruction involving many courses in colleges of business is an approach that needs formal research. As stated previously, there is little evidence found in the literature that PBL is common in business classes. The assessment of learning is obvious. Do students learn better in PBL environments than in traditional, lecture-based classes? Longitudinal studies over the length of a quarter or semester can help to answer this question. Dunlap (2005) found that self-efficacy improved in students enrolled in a computer science software engineering capstone course as a result of incorporating PBL as the method of instruction. Could the same results be found in teaching computer application concepts?

Another important research question relates to just how students go about the process of learning when following PBL. Closely related to learning as an outcome variable, understanding the process is also important. What did students do? What sources did they seek out to assist them in the problem? What processes did they follow when developing any models they use to track mutual fund performance?

Of paramount importance too is the comparison of group-oriented PBL instruction versus instruction that is individually based. As stated previously, PBL is essentially always oriented toward individuals working together in groups. As proposed above, it is suggested that students can work alone and achieve success. This should be investigated.

Is there a need for research on just how a PBL method of instruction in a given class affects learning in later related classes? Are students better prepared to take later, more advanced courses if earlier classes are taught following PBL?

Finally, it is necessary to determine the willingness of faculty to reorient their approach to teaching this type of material. Sometimes faculty may be resistant to changes, so it is necessary to ascertain what motivators might serve to have faculty institute significant changes in instructional methods.

CONCLUSION

Colleges of business continue to struggle with teaching computer application concepts. As recent as January 2006, an online survey posted to ISWORLD dealt with topics, tools, and concepts associated with this type of class. Schools seeking accreditation, or those working to maintain accreditation, are also faced with numerous issues surrounding curriculum and curriculum management and revision. It therefore follows that this type of class needs attention.

PBL has been used successfully in many disciplines, such as medicine, engineering, computer science, and geography. It follows that it might also be an approach that could meet with much success in business schools as well. In particular, Computer Application Concepts is a business course that is taught in almost all colleges of business. It is a course that is often difficult to teach and manage. PBL is an approach that should be investigated more completely as a possible solution to better instruction and better student learning.

REFERENCES

Boud, D., & Feletti, G. (1991). *The challenge of problem based learning.* New York: St. Martin's Press.

Drennon, C. (2005). Teaching geographic information systems in a problem-based learning environment. *Journal of Geography in Higher Education, 29*(3), 385-402.

Dunlap, J. (2005). Problem-based learning and self-efficacy: How a capstone course prepares students for a profession. *Educational Technology Research & Development, 53*(1), 65-85.

Harland, T. (1998). Moving towards problem-based learning. *Teaching in Higher Education, 3*(2), 219-230.

Kanet, J., & Barut, M. (2003). Problem-based learning for production and operations management. *Decision Sciences Journal of Innovative Education, 1*(1), 99-118.

Lam, D. (2004). Problem-based learning: An integration of theory and field. *Journal of Social Work Education, 40*(3), 371-389.

Laudon, K., & Laudon, J. (2005). *Essentials of management information systems: Managing the digital firm and student multimedia edition package* (6th ed.). Upper Saddle River, NJ: Prentice Hall.

Margetson, D. (1994). Current educational reform and the significance of problem-based learning. *Studies in Higher Education, 19*(1), 5-19.

Maskell, D., & Grabau, P. (1998). A multidisciplinary cooperative problem-based learning approach to embedded systems design. *IEEE Transactions on Education, 41*(2), 101-103.

Shepherd, A., & Cosgriff, B. (1998). Problem-based learning: A bridge between planning education and planning practice. *Journal of Planning Education and Research, 17,* 348-357.

Steinemann, A. (2003). Implementing sustainable development through problem-based learning: Pedagogy and practice. *Journal of Professional Issues in Engineering and Education and Practice, 129*(4), 216-224.

Turban, E., Leidner, D., McLean, E., & Wetherbe, J. (2006). *Information technology for management.* Hoboken, NJ: John Wiley & Sons.

This work was previously published in End-User Computing: Concepts, Methodologies, Tools, and Applications, edited by S. Clarke, copyright 2008 by Information Science Reference, formerly known as Idea Group Reference (an imprint of IGI Global).

Chapter XII
Covert End User Development: A Study of Success

Elaine H. Ferneley
University of Salford, UK

ABSTRACT

End user development (EUD) of system applications is typically undertaken by end users for their own, or closely aligned colleagues, business needs. EUD studies have focused on activity that is small scale, is undertaken with management consent and will ultimately be brought into alignment with the organisation's software development strategy. However, due to the increase pace of today's organisations EUD activity increasing takes place without the full knowledge or consent of management, such developments can be defined as covert rather than subversive, they emerge in response to the dynamic environments in which today's organisations operate. This chapter reports on a covert EUD project where a wide group of internal and external stakeholders worked collaboratively to drive an organisation's software development strategy. The research highlights the future inevitability of external stakeholders engaging in end user development as, with the emergence of wiki and blog-like environments, the boundaries of organisations' technological artifacts become increasingly hard to define.

INTRODUCTION

In today's environment of rapid business change facilitated by users with increased technical capabilities, there is a tacit understanding that end user development (EUD) activity is inevitable—development tools are more accessible, and end users are now technologically mature and expected to be proactive in their use of technology to enable enactment of their employment roles (Jawahar & Elango, 2001; Nelson & Todd, 1999). As the end user takes control of the development effort and develops systems with little or no input from information technology (IT) specialists so the ultimate level of end user involvement has arrived—the end user is no longer simply consulted, they have assumed the roles of the designer, developer and tester, they are the IT specialist for their software requirement (Cheney, Mann, & Amoroso, 1986; McGill, 2004).

To date, studies have focused on EUD that management is fully aware of and endorses, the assumption is that EUD activity is small scale and that it will ultimately be brought into the organisation's software development strategy. However, due to the increased pace of today's organisations, EUD activity increasingly takes place without the full knowledge or consent of management. Such developments can be defined as covert rather than subversive, and it can be argued that they emerge in response to the dynamic environments in which today's organisations operate (Nelson & Todd, 1999; Ouellette, 1999; McLean, Kapperlman, & Thompson., 1993).

This chapter reports on a field study on the effects of covert EUD activity in a publishing company. The chapter aims to enhance our understanding of covert EUD activity using an interpretive approach. We draw on the literature on the social construction of technology (SCOT) and apply this to covert EUD activity identifying a technology "path" (MacKenzie & Wajcman, 1985). The "path" may be born from an individual vision, but the multifaceted nature of technology requires disparate actors to contribute to technology success. Whilst the chapter does not purport to offer definitive solutions, the experiences reported suggest valuable lessons for organisations faced with the challenge of managing the dichotomous relationship of encouraging worker proactivity manifested in EUD whilst controlling maverick EUD activity.

LITERATURE REVIEW

Authors have begun to recognise the futility of attempting to align business strategy and technological infrastructures and have acknowledged that technological "drift" is inevitable, (Ciborra et al., 2000; Sauer & Burn, 1997; Ciborra, 1994; Orlikowski, 1996). This process of "drift" is largely assumed to be an overt process, management being aware that it is happening and either attempting realignment (usually futilely) or allowing the technology to develop a certain momentum of its own (for examples see Kanellis & Paul, 2005; Hanseth & Braa, 1998; Rolland & Monteiro, 2002). What is less frequently considered is the notion of, and rationale for, covert IT implementations that result in "drift," and the literature that does exist is primarily concerned with covert activity with the intention of sabotage (for examples see Gordon, 1996; Conti, 2005; Verton, 2001; Graham, 2004).

Such covert activity, whether for altruistic or subversive purposes, necessitates a degree of improvisation—using current resources to create new forms and order from tools and materials at hand, such an approach has been defined by anthropologists as "bricolage" (Levi-Strauss, 1966). When considering information systems bricolage, "materials at hand" are usually considered to be information technology hardware and software artefacts. However, it has also been suggested that the use of networking with preexisting professional and personal contacts is also a form of "network bricolage" (Mintzberg, 1994; Moorman & Miner, 1998; Baker, Miner, & Eesley, 2003).

RESEARCH DESIGN

To examine covert EUD activity within an organisation from multiple stakeholder perspectives requires an understanding of the social and contextual relationships that influence the organisation; there can be no single explanation of success. Our epistemological assumptions are that no individual account of social reality can be proven correct. Therefore, the research method employed has been interpretivist, with the aim being to understand the perspectives of the various stakeholders and the historical and socially situated contexts in which they reside (Hirschheim, Klein, & Lyytinen, 1996). The opportunity to gain access to a covert end user developer group emerged during the course of a wider longitudinal study that was undertaken by the author and a postgraduate student that examined the effect on various stakeholders of implementing IT solutions in small- to medium-sized enterprises (SMEs). The postgraduate student was employed by one of the studied SMEs (PublishCo) and became one of the covert EUDs, simultaneously exploring the dynamics and rationale for EUD whilst repeatedly intervening and stimulating change. Therefore, the research employed participatory action research: "...members of the organisation we study are actively engaged in the quest for information and ideas to guide their future actions" (Whyte, Greenwood, & Lazes, 1991). The research largely complied with the major characteristics of information systems action research as identified by Baskerville (1999) and Baskerville and Wood-Harper (1996). However, there is one area in which the researchers did not fully conform to the established principles for action research: informed consent. Whilst formal consent from PublishCo's management to study the effects of the introduction of IT into the organisation was gained, as the research evolved into a more specific study on covert EUD the researchers were faced with a moral dilemma—whether to inform management of the covert activity or whether to abuse management confidence and collude rather than confess. During the rest of this chapter we will discuss how this dilemma was accommodated, and it is up to the reader to decide if we acted morally.

Field Study: PublishCo

PublishCo is a small publishing house established in 1895, and it is involved in the publication of niche specialist magazines for the pet owning, breeding, and exhibiting community. Circulation is global, although the customer base at the start of the fieldwork was largely European. Over the course of 4 years data has been collected from multiple levels and perspectives across the organisation, from technology suppliers and from customers. The findings are the results from 4 years of field diaries, over 200 hours of transcribed interviews and numerous hours of observation.

This chapter reports on the covert development of a Web-based portal that has resulted in system integration across the company, and has contributed to increasing PublishCo's revenue by approximately 60%, improved the efficiency of the company's processes, built better customer relations, and simultaneously improved the working lives and morale of its employees. A summary of the various stakeholders' roles and perspectives is presented in Table 1.

IMPLICATIONS OF COVERT END USER DEVELOPMENT

Within the limitations of this field study, by considering the rationale and world views of these disparate groups who have all embraced EUD as a means of moving down a technology "path," a model for harnessing end user development activity emerges. The internal covert developers viewed the technology as emancipatory. They saw the technology as an opportunity to change the organisation's business model and introduce a new revenue stream, thereby potentially increasing sales turnover and hence commission. However, in the portal development the technical artefacts themselves became the technology path driver, and as the end user developers gained enthusiasm and technical knowledge they introduced more, disparate functionality into the portal, and the technology developed a momentum of its own. These disparate elements now define the portal—its quirky, eclectic style clearly differentiates it from the competition and have made it the market leader. The IT consultants also assumed the role of covert developers. The usual assumption is that IT consultants view technical implementations as a source of income, yet in

Table 1. Emerging characteristics, roles and influences for stakeholders in covert end user development

Stakeholder Groups	**Predevelopment**	**System Development**	**System Implementation**
Internal End User Developer	**Characteristics** Entrepreneurial Personal or financial reward **Role/influences** Participative approach Enrolling management Revision of organisation and IS, based on previous external experience, current education, etc. **Quote** *"Web publishing is the way forward for all niche magazines"* (Emp1). *"We've only got a single webpage with our contact details... it's too dated"* (Emp2)	**Characteristics** Risk taking. **Role/influences** External stakeholders—developer /software vendor, customers, competitors Access to technology—beta-software, packages External promotional events and training **Quote** *"There was so much to implement and limited resources so illegal software was the only solution"* (Emp3)	**Characteristics** Efficiency gains Technical rationalisation (networking, desktop, etc.) **Role / influences** Recognition of need to manage & integrate technology Upgrade to reflect changing use **Quote** *'as functionality was released customers started using it straight away, it was obvious immediately that the decision to move online was the right one ... we're always adding new things, it changes almost weekly* (Emp2)
Management	May not be aware EUDs occurring **Characteristics** Willingness to listen to strategic technologist **Role/influences** Trust change agent—internal or external Change-friendly climate Encourage autonomous "can do" employees **Quote** *"We were really wary of the developing a comprehensive website, especially as the dotcom catastrophe was just happening"* (Man2)	**Characteristics** Radical change (large or small scale) Business-led development **Role/influences** Acceptance of new direction Commit to technology education Technology as +ve lever for social change within the organisation **Quote** *"We can not envisage our customers using a website, their largely older, we don't expect them to be interested in an online site ... we regard the development as a necessary commercial distraction"* (Man1)	**Characteristics** Financial commitment Technology management **Role / influences** Availability of financial support for ongoing implementation Ensure legality of EUD systems Vision alignment with business, organization and technology strategies **Quote** *'the web site has made a significant difference to our business, the discussion forums keep all of us abreast of what's going on in* [the field] *and we're* [management] *now actively contributing to those discussions'* (Man2)
External IT Agencies	**Characteristics** Long term commitment to organisation End user focussed **Role/influences** Long term financial gain Participative approach **Quote** *"We needed to develop a reference site, so PublishCo implementation was an ideal opportunity"* (Salesman3)	**Characteristics** Exploratory, risky development (large or small scale) **Role/Influences** Technology & training provider **Quote** *"At the time we didn't see the effect that using lots of different tools would have, we wanted a reference site and believed using those tools was a means to an end"* (Salesman3)	**Characteristics** Educator **Roles / Influences** Training Emerging technology vision **Quote** *'we expected to use the site as a sales example but the end users have been using all sorts of free resources to cobble together a site that's not as professional looking as we'd hoped for'* (Salesman2

continued on following page

Table 1. continued

External End Users/ Customers	**Characteristics** Innovators **Role/influences** Stimulate technology vision **Quote** *"Online dog breeding forums were starting to emerge but really very few dog breeders were talking on them"* (Customer2)	**Characteristics** Reviewers Testers **Role/influences** External influences	**Characteristics** Explorers **Role / influences** Technology 'path' leaders **Quote** *'the portal effectively gave a voice to the community'* (Customer5)

the presented study the implementation also provided them with a learning opportunity. They were prepared to work with internal, budding developers to provide them with the requisite skills to ensure the systems could evolve with the business. However, they also actively deceived management by presenting illusionary demonstrations and condoned or instigated the use of illegal software. Management's interpretation of the emerging technology was as a distraction and cost overhead. The end users and consultants perceived that they had limited understanding of the presentations and conversations that took place regarding technology, and subsequent interviews confirmed that members of management had sought advice from family members, personal friends, or peer networks rather than question their employees or contracted IT consultants. However, it emerged that they were aware that some covert development was occurring and chose to allow it to remain underground rather than stifle the proactivity that their employees were demonstrating. Customers' interpretation of the technologies has been as communication enabler. The portal facilitates communication in many directions: customer to customer, customer to journalist, customer to management, customer to advertiser. Indeed, it could be argued that the customer has become the final technology path driver: Functionality is added or removed from the portal dependent on customer use, editorial is driven by their discussion forum conversations, whilst the content of the online directories is created and uploaded by the customers themselves. As the portal develops so it is interpreted in different ways, including a sales forum (extensive online directories), an organisational tool (calendar of events), and a dating agency (putting readers with similar requirements in touch for the purposes of pursuing their niche interest).

Reflecting on the findings of the field study, our inquiry illustrates that (a) covert EUD may be driven by a wider stakeholder group than the end user developers themselves, and (b) although EUD activity is traditionally seen as a microlevel activity, the challenge is to harness the developing technical artefacts to achieve maximum business benefit without jeopardizing the stability of the organisation as a whole. We discuss these implications in turn.

The traditional view of EUD is that the activity is small scale and undertaken to satisfy end users' own needs. Yet as end users become more technically sophisticated and as technology itself becomes more homogeneous, user friendly and reconfigurable, so a wider set of stakeholder groups are gaining common or easily transferable technical know-how. This allows traditionally disparate

Figure 1. Technology path

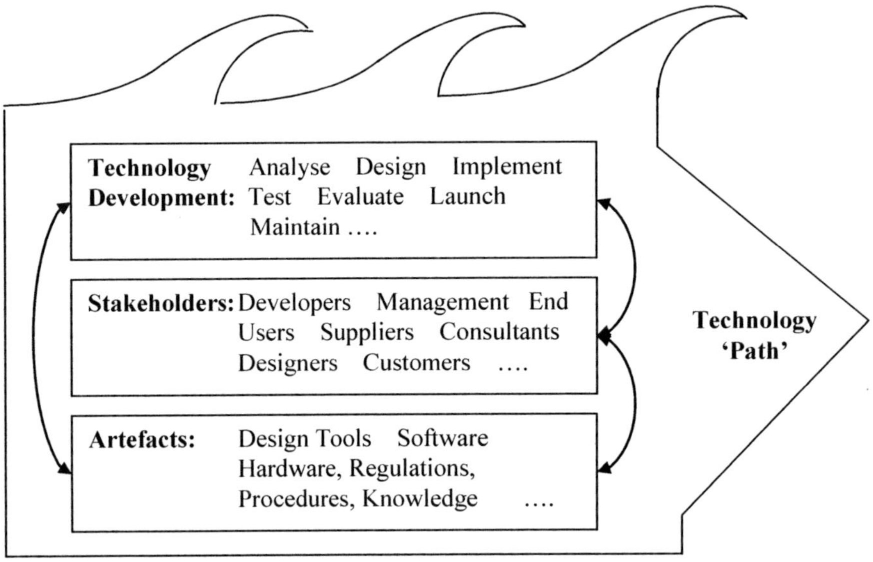

stakeholder groups from both inside and outside organisations to act collectively in covert activity; in the field study, transferred technological know-how enabled the internal end users and the technology vendors to work together to covertly developed information systems solutions. Indeed, whilst the ultimate level of end user involvement has traditionally been seen as the end user taking ownership of the system and becoming the developer himself (Cheney et al., 1986), it is feasible for external stakeholder groups to effectively become the end user developer. In the field study, as the customer stakeholder group has gained technical know-how, it has become able to drive the portal's ever changing design, and it provides the content in the form of uploaded reviews, customer profiles, and commentary via discussion forums. Indeed, the portal could be viewed as an emerging wiki or blog-like environment owned by PublishCo (Leuf & Cunningham, 2001).

REFERENCES

Baker, T., Miner, A. S., & Eesley, D. T. (2003). Improvising firms: Bricolage, account giving and improvisational competencies in the founding process. *Research Policy, 32,* 255-276.

Baskerville, R. (1999). Investigating information systems with action research. *Communications of the Association for Information Systems, 2.*

Baskerville, R., & Wood-Harper, A. T. (1996). A critical perspective on action research as a method for information systems research. *Journal of Information Technology, 11,* 235-246.

Cheney, P. H., Mann, R. I., & Amoroso, D. L. (1986). Organizational factors affecting the success of end user computing. *Journal of Management Information Systems, 3,* 65-80.

Ciborra, C., Braa, K., Cordella, A., Dahlbom, B., Failla, A., Hanseth, O., Hepso, V., Ljungberg, J., Monteiro, E., & Simon, K. A. (2000). *From control to drift.* Oxford: Oxford University Press.

Conti, G. (2005). Why computer scientists should attend hacker conferences. *Communications of the Association for Information Systems, 48,* 23-24.

Dosi, G. (1982). Technological paradigms and technological trajectories. *Research Policy, 11,* 147-162.

Gordon, S. (1996). In *The 6th International Virus Bulletin Conference*Brighton, UK.

Graham, P. (2004). *Hackers and painters: Big ideas from the computer age.* O'Reilly.

Hanseth, O., & Braa, K. (1998). In R. Hirschheim, M. Newmann, & J. I. DeGross (Eds.), *Proceedings of the 19th International Conferenceon Information Systems* (pp. 188-196). Helsinki, Finland.

Hirschheim, R., Klein, H. K., & Lyytinen, K. (1996). Exploring the intellectual structures of information systems development: A social action theory analysis. *Accounting, Management and Information Technologies, 6,* 1-64.

Jawahar, I. M., & Elango, B. (2001). The effect of attitudes, goal setting and self efficacy on end user performance. *Journal of End User Computing, 13,* 40-45.

Kanellis, P., & Paul, R. J. (2005). User behaving badly: Phenomena and paradoxes from an investigation into information systems misfit. *Journal of Organizational and End User Computing, 17,* 64-91.

Kemp, R., Schot, J., & Hoogma, R. (1988). Regime shifts to sustainability through process of niche formation: the approach of strategic niche management. *Technology Analysis & Strategic Management, 10,* 175-195.

Leuf, B., & Cunningham, W. (2001). *The wiki way: Quick collaboration on the Web.* Addison Wesley Longman.

Levi-Strauss, C. (1966). *The savage mind.* Oxford: Oxford University Press.

MacKenzie, D., & Wajcman, J. (1985). *The social shaping of technology.* Open University Press.

McGill, T. (2004). The effect of end user development on end user success. *Journal of Organizational and End User Computing, 16,* 41.

McLean, E. R., Kapperlman, L. A., & Thompson, J. P. (1993). Converging end user and corporate computing. *Communications of the Association for Information Systems, 36,* 76-91.

Mintzberg, H. (1994). *The rise and fall of strategic planning.* New York: Free Press.

Moorman, C., & Miner, A. S. (1998). Organizational improvisation and organizational memory. *Academy of Management Review, 23,* 698-723.

Nelson, R. R., & Todd, P. (1999). Strategies for managing EUC on the Web. *Journal of End User Computing, 11,* 24-31.

Orlikowski, W. J. (1996). Improvising organizational transformation over time: A situated change perspective. *Information Systems Research, 7,* 63-92.

Ouellette, T. (July 26, 1999). Giving users the keys to their Web accounts. *Computerworld*, 66-67.

Rolland, K. H., & Monteiro, E. (2002). Balancing the local and the global in infrastructural information systems. *The Information Society, 18,* 87-100.

Sauer, C., & Burn, M. J. (1997). The Pathology of Alignment. In C. Sauer & P. Yetton (Eds.), *Steps to the Future.* San Francisco: Jossey Bass.

Verton, D. (2001). Hacker conferences highlight security threats. *PC World.*

Webopedia (2005). Retrieved September 14, 2006, from www.webopedia.com

Whyte, W. F., Greenwood, D. J., & Lazes, P. (1991). In W. F. Whyte (Ed.), *Participatory action research* (pp. 19-55). Newbury Park, CA: Sage.

ENDNOTES

[1] Alternative terms in the literature are "regimes" (Kemp et al, 1998), or "trajectories" (Dosi, 1982)

[2] A Wiki is a collaborative Web site comprised of the perpetual collective work of many authors that anyone is allowed to edit, delete, or modify. A blog is similar in structure but does not allow visitors to change the original posted material, only to add comments to the original content (Webopedia, 2005).

This work was previously published in End-User Computing: Concepts, Methodologies, Tools, and Applications, edited by S. Clarke, copyright 2008 by Information Science Reference, formerly known as Idea Group Reference (an imprint of IGI Global).

Chapter XIII
When Technology Does Not Support Learning:
Conflicts Between Epistemological Beliefs and Technology Support in Virtual Learning Environments

Steven Hornik
University of Central Florida, USA

Richard D. Johnson
University of South Florida, USA

Yu Wu
University of Central Florida, USA

ABSTRACT

Central to the design of successful virtual learning initiatives is the matching of technology to the needs of the training environment. The difficulty is that while the technology may be designed to complement and support the learning process, not all users of these systems find the technology supportive. Instead, some users' conceptions of learning, or epistemological beliefs may be in conflict with their perceptions of what the technology supports. Using data from 307 individuals, this research study investigated the process and outcome losses that occur when friction exists between individuals' epistemological beliefs and their perceptions of how the technology supports learning. Specifically, the results indicated that when there was friction between the technology support of learning and an individual's epistemological beliefs, course communication, course satisfaction, and course performance were reduced. Implications for design of virtual learning environments and future research are discussed.

INTRODUCTION

Advances in information technology have enabled organizations and educational institutions to deliver training and learning initiatives free from time and/or place constraints, creating virtual learning environments (VLEs).[1] These environments are becoming central to the design and development of both corporate training programs and university curricula. While there are multiple ways to design these environments, common characteristics of virtual learning environments include the mediation of course interactions and materials through information and communication technologies (Alavi & Leidner, 2001) and greater control over the learning environment (Piccoli, Ahmad, & Ives, 2001).

The market for this type of training is substantial, with recent estimates suggesting that the industry will generate nearly $25 billion by 2006 (IDC, 2003) and grow annually at approximately 37% (Mayor, 2001). Universities are also undertaking distance initiatives, with estimates suggesting that nearly 90% of public universities offer distance education courses, over three million students participate in these courses, and these numbers are projected to grow (Wirt & Livingston, 2004). The major push behind these initiatives has been both convenience and cost. These initiatives have both potential and pitfalls as can be seen through the findings of two recent studies. Although the potential for cost savings is large, with some large companies finding cost savings of between $30-$400 million dollars per year and reductions in training costs of nearly 50% (Salas, DeRouin & Littrell, 2005), another study has suggested that as many as 80% of employees drop out of these programs before they are complete (Flood, 2000).

Thus, it is important to understand the factors that affect the successful implementation of VLE initiatives. Previous research has suggested that instructor characteristics, pedagogical approach or learning models, learner/user characteristics, and the technology each play a key role in creating successful outcomes (Alavi & Leidner, 2001; Piccoli et al., 2001; Webster & Hackley, 1997). Recently it has also been argued that a key to the successful implementation of these environments is the convergence between the technology used in the learning environment and the implemented learning model (cf. Benbunan-Fich, 2002; Leidner & Jarvenpaa, 1995; Robson, 2000).

However, when the technology used to support learning is designed to support a specific learning model, this can often lead to a compulsory learning process that users must follow to reach the course objectives (Vermunt, 1998). For some users, the learning approach supported by the technology can be in direct conflict with their beliefs about how learning should occur (i.e., their epistemological beliefs) (Bakx, Vermetten, & Van der Sanden, 2003; Schommer-Aikins, 2004). Relatively little is known regarding the implications of the conflict between an individual's epistemological beliefs (EBs) and the learning environment supported by the technology, but given the centrality of technology to the learning process in VLEs and the central role of EBs in how individuals approach learning and how they learn (Marton, Dall'Alba, & Beaty, 1993; Marton & Säljö, 1976; Perry, 1968; Vermunt, 1996), the relationship between the two is likely to be important. Thus this research represents the beginning of a systematic examination of the role of EBs in VLEs.

Drawing from research on EB, evidence suggests that when users do not perceive that the technology supports their optimal learning approach (i.e., there is friction between the individual's EBs and the learning approach supported by the technology), there will be both process and outcome losses. If negative expectations regarding the ability of the technology to adequately support a learning environment consistent with the user's EB emerge it can be difficult for the user to accept this novel way of course delivery (Vermunt & Verloop, 1999, 2000). We argue that when users perceive a mismatch between their EBs and the learning model that the technology supports, learning processes and outcomes will be impacted. Thus, the following research question was investigated:

Are learning processes and outcomes negatively affected when there is friction between a user's perceptions of what learning model the technology supports and his or her personal epistemological beliefs?

The remainder of this chapter is organized as follows. First, the chapter briefly introduces the virtual learning environment context. Second the chapter discusses EBs and how these beliefs influence individual learning processes and outcomes. Next, the chapter further builds the argument that friction between individual EBs and beliefs about the technological support of learning models can affect learning processes and outcomes. Fourth, the research context and methods are then discussed. Finally, the results are presented, along with a discussion of the findings, implications, and directions for future research.

VIRTUAL LEARNING ENVIRONMENTS

While training has traditionally taken place in a face-to-face setting, technology has enabled new forms of learning, unconstrained by time or place. In these virtual learning environments, learning processes, communications, shared social context and learning community are mediated through information technology, creating a novel learning environment for users. Specifically, VLEs are characterized by high levels of learner control, computer mediation of communication, and the flexibility for learners to restructure learning in nontraditional ways (Piccoli et al., 2001). As with traditional environments, researchers have focused on how effective VLEs are at producing effective outcomes such as learning, performance, and affective reactions to the training setting. Previous research has found that VLEs can be as effective as face-to-face environments in supporting both learning and affective reactions to the learning environment (cf. Hiltz & Wellman, 1997; Piccoli et al., 2001).

In the development of these environments, the design will reflect some pedagogical approach, or learning model (Leidner & Jarvenpaa, 1995). As such, the learning environment will reflect the instructor's beliefs about what the best way to transfer knowledge is and how the technology will be designed to support this pedagogical approach. However, it is important to note that it is not the methods implemented, but rather student *perceptions* of these methods that most strongly affect student learning most directly (Entwistle, 1998 a, b; Entwistle, McCune, & Hounsell, 2002). Although, there are many learning models that an instructor can choose from this study focuses on three that are among the more widely accepted and which have been of interest to information technology researchers – the objectivist model, constructivist model and collaborative model (cf. Alavi, 1994; Alavi, Marakas, & Yoo, 2002; Liedner & Jarvenpaa, 1995). In the objectivist model, learning is seen as a process of transferring objective knowledge of an expert to the novice. To facilitate this transfer, VLEs will typically provide capabilities such as online presentation of syllabi, lectures, lecture notes, and so forth, in a non-interactive format. In constructivist models, learning is seen as a process where individuals discover knowledge through active participation in the learning process. Constructivist learning best occurs as individuals actively pursue new knowledge. The collaborative model extends the constructivist model by suggesting that learning occurs as individuals work together to create a shared understanding based upon the contributions of multiple individuals. Typically, these latter learning models employ interactive capabilities designed into the VLE including asynchronous communication capabilities such as e-mail, discussion, or chat. Whatever underlying pedagogical approach is desired, we argue that it will be most effective when there is a fit between the technology design and the learning model implemented by the instructor.

Research has also begun investigating the processes through which effective VLEs are developed, finding that effective VLEs are not simply created by the technologies used, but instead are enabled through information and communication technologies (ICTs) as students create a shared social context and feel part of a learning community (Rovai, 2002). It has been argued that for this to occur, learners must communicate and perceive the social presence of others, or "the degree of salience of the other person in the interaction and the consequent salience of the interpersonal relationships" (Short, Williams, & Christie, 1976, p. 65). While communication can facilitate the exchange and sharing of information, social presence enables the connections between learners that create course community and improve learning and satisfaction (Tu & McIsaac, 2002; Tu, 2000).

With ICTs mediating VLE processes, it is also important to understand the influence of user's perceptions about whether or not the technology supports the learning environment and if this perception affects their learning processes and outcomes. To do this, we first focus on an individual's beliefs about learning, or their EBs.

Epistemological Beliefs

Beyond the technical and pedagogical considerations that go into the design of effective VLEs, instructors and designers should also consider student perceptions of how learning best occurs. Just as instructors design the environment around a particular learning model, the users of the system will also have specific beliefs about how learning best occurs. Van der Sanden, Terwel and Vosniadou (2000) describe these beliefs as Individual Learning Theories or internalized frameworks of instruction and learning, which influence the approach individuals take when encountering new learning situations. These EBs are beliefs individuals have regarding knowledge and knowing, including their beliefs about what knowledge is and how one acquires knowledge (Schommer-Aikins, 2004; Schommer, 1994). Schommer also states that individual's EBs affect learning outcomes and suggests that these beliefs are a system of independent dimensions with the following anchors: (1) Certainty—knowledge ranges from absolute to tentative; (2) Structure—knowledge is considered to be either organized as distinct bits or as highly interwoven concepts; (3) Source—knowledge is handed down by authority or derived by reason; (4) Control—An individuals cognitive ability is fixed at birth or their ability can be changed; and (5) Speed —knowledge is either acquired quickly or gradually.[2]

Elen and Lowyck (2000) also identify another aspect of EBs, learning conceptions. Learning conceptions are learner perceptions about what is the most effective way of learning. In this study, we chose to focus on an individual's learning conceptions for two reasons. First, learning conceptions have been shown to affect the learning process and outcomes (Entwistle, 1991; Marton et al., 1993; Vermunt, 1996) including the extent to which learners utilize the capabilities of the environment (Elen & Lowyck, 1998). Second, learning conceptions should be thought of as the mirror of the pedagogical approaches implemented by the instructors. Just as instructors have a specific learning pedagogy in mind when implementing the course, so also learners have specific EBs about the best approach to learning.

These EBs are triggered as individuals engage in the learning process (Hofer, 2004) and are thought to affect how individuals approach learning (cf. Marton et al., 1993; Vermunt, 1996), how individuals learn (Entwistle, 1991), how they devise learning plans and strategies, their self-assessment and monitoring of comprehension (Schommer, Crouse, & Rhodes, 1992), their preferred learning situations (Bakx et al.,

2003), and the extent to which they leverage the environment to their advantage (Elen & Lowyck, 1998, 1999).

EBs are argued to be even more important to learning than the learning model chosen by the instructor because these beliefs filter the learning models implemented by the instructor (Bakx et al., 2003). For example, Bakx et al. (2003) examined the relationship between self-perceived competence, EBs and preferred learning situations, finding that an individual's EBs about learning were related to their preferred learning situations. When individuals believed that learning best occurred as part of an interactive, constructive process they were more inclined to prefer situations that encouraged interaction and more active shared learning. Thus, differences in learning outcomes can be attributed to individual differences in beliefs about the process of learning (Marton & Säljö, 1976). One manner in which users' EBs can act as a filter is through the congruence or friction that is created between the users' EBs and the technological support of learning in the VLE.

Technology and Beliefs: Congruence or Friction

Discrepancy theory states that individuals hold a set of expectations about their environment and also perceptions about how well their expectations of the environment are met (Locke & Latham, 1990). In turn these expectations affect how individuals interpret and interact in their environment. When expectations are met, individuals are positively disposed to the environment, and are more satisfied with the environment than when they are not met. The findings of this stream are also consistent with research on expectation-confirmation theory (ECT), which suggests that when individuals expectations about a product or technology are not met (i.e., they are disconfirmed) they are less satisfied and less likely to continue to use the product or technology (Anderson & Sullivan, 1993; Bhattacherjee, 2001; Oliver, 1980).

In VLEs, although the instructor may design the technological support around a particular pedagogical approach (learning model), users will have their own expectations regarding the effectiveness of the chosen learning approach (Vermunt & Verloop, 1999) and how technology can best be leveraged. As suggested by discrepancy theory, when a match exists between an individual's EBs and the technologically supported learning model, the two are considered to be in congruence. When there is a discrepancy between these beliefs and the technologically supported learning model, friction occurs. In turn, perceptions of congruence or friction affect the learner's expectations of how positive or negative their learning outcomes will be. These expectations often become self-fulfilling, especially in novel learning settings (O'Mara, Allen, Long, & Judd, 1996). As such, congruence and friction are expected to have an impact on both course processes such as course communication, and perceptions of social presence, as well as course outcomes such as performance and satisfaction (Figure 1).

Friction can affect both the user's approach to learning within the environment as well as causing an underutilization of the tools available to them (Lowyck & Elen, 1994). This can occur because individuals do not see the value of the technology in the setting for supporting their learning. The technology can be seen as a barrier to learning, something placed between the user and the learning outcomes that obfuscate the necessary conditions for learning to occur (Fiore, Salas, Cuevas, & Bowers, 2003). In other words, for these users, cognitive focus moves away from learning processes and towards use of the technology; making communication and participation in the environment more difficult, as well as decreasing the value the technology brings to the course. The effect of focusing on the technology as opposed to the training material may lead users to disengage from the course, or to

Figure 1. Epistemological beliefs, technology support and VLE outcomes

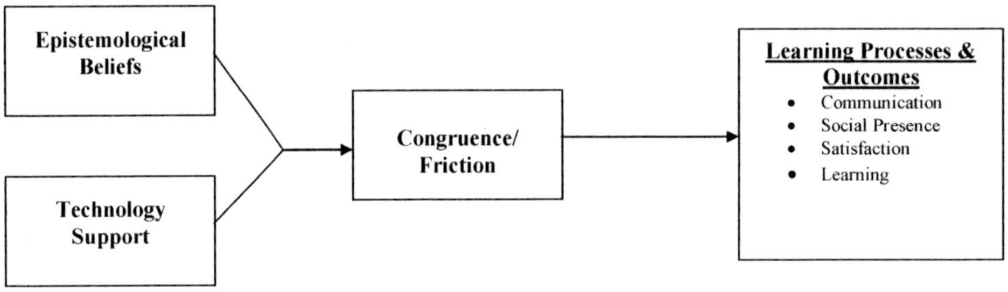

not engage it at all. As individuals disengage, they will interact and communicate less. Conversely, those who perceive congruence will perceive the technology as an important tool in support of learning and be more likely to use the technology capabilities, thus communicating and interacting within the VLE.

In the case of friction, as communication is reduced, individuals will find fewer opportunities to ease isolation and develop perceptions of social presence (Burke & Chidambaram, 1999; Gunawardena, 1995; Walther, 1995). Conversely, those whose EBs are congruent with the technology supporting the learning process, will have greater opportunities to create connections using the technology and therefore, be more likely to feel the connections of the other learners in the environment and feel like they are part of a learning community with increased perceptions of social presence. Thus the following hypotheses were investigated:

H1a:

When there is congruence between a user's epistemological beliefs and beliefs about the learning model supported by the technology, the user will communicate to a greater extent than when there is friction.

H1b:

When there is congruence between a user's epistemological beliefs and beliefs about the learning model supported by the technology, the user will perceive greater social presence than when there is friction.

We also expect friction to have an impact on learners' affective reactions to the VLE. As learning in VLEs can still be novel experiences for learners, users are utilizing technology to communicate in ways in which they are potentially unfamiliar and uncomfortable. Evidence has shown that when faced with a conflict between technology functionality and the way the users wish to use the system, users will attempt to match their approach to the approach designed into the system (cf. Todd & Benbasat, 1991). Vermetten, Vermunt, and Lodewijks (2002) also suggest that learners tend to make best use of the elements in the learning environment that fit their preferred way of learning and ignore or underutilize those that do not. In VLEs, this suggests that while users may feel compelled to adopt a learning strategy supported by the technology, they may not effectively utilize the tool. Friction between the user's EBs and the learning model supported by the technology can exacerbate negative feelings about the environment, because of the user's need to adapt their learning conceptions even though they may not see it as appropriate. As discussed above, the technology can be seen as a barrier to the learning process, increasing the user's frustration with learning in the VLE. Together, these outcomes would be expected to lead to lower levels of satisfaction with the VLE.

Finally, we believe that friction will also lead to a reduction in learning outcomes (Vermunt & Verloop, 1999). An individual's prior educational experiences are reflected in their EBs, which in turn play a role in forming the learner's perceptions of the instructional measures. The match between those perceptions and the types of teaching-learning environment affects the quality of learning achieved (Entwistle et al., 2002). Among other things, for learning to be most successful, connections between individuals are needed (Vygotsky, 1978; Feuerstein, Rand, Hoffman, & Miller, 1980). When friction occurs, learners are more likely to disengage from both the learning process and from their peers, engaging in fewer behaviors that can lead to successful learning outcomes. With other studies also arguing for the importance of ongoing observational learning process (Bandura, 1986; Yi, & Davis, 2003) where individuals learn new behavior and skills through attending and processing the behavior of others, any reduction in connections can lead to reduced attention to the behaviors, ideas and contributions of others. When this occurs, system users are likely to be exposed to less information, process less information, and therefore have reduced learning. In turn, the likelihood of successful learning outcomes will be reduced (Geiger & Cooper, 1996; Harrell, Caldwell, & Doty, 1985). Thus, the following hypotheses were investigated:

H2a:

When there is congruence between a user's epistemological beliefs and beliefs about the learning model supported by the technology, the user will be more satisfied than when there is friction.

H2b:

When there is congruence between a user's epistemological beliefs and beliefs about the learning model supported by the technology, the user will perform better than when there is friction.

METHOD

Research Setting

The study was conducted in an MIS fundamentals course at a large university in the United States. This course was a required course for all business majors and was taught exclusively online using WebCT. The course was taught over 15 weeks and was divided into 6 modules. Each module focused on different topical areas, such as the strategic use of information and technology, e-commerce, decision support, and so forth, with each module lasting approximately two weeks. Students were assigned to groups of approximately 30 and were asked to both post comments to case questions and respond to the comments of others in their group. Student assessment occurred at the end of each module through the use of individual multiple-choice tests. The course was managed by one instructor and two graduate assistants (GAs), who communicated exclusively online (all were not physically on campus during the semester), and one GA held office hours both online and in person.[3] At the end of the fifth module of the course, the survey was made available for one week using WebCT.

The course was designed based on principles from the constructivist and collaborative learning models. To encourage active discovery and knowledge construction (i.e., constructivist learning), students were required to gather information from a variety of sources including text, video, and cases and to integrate these into their understanding of the topic. In support of the collaborative model of learning students wrote, read and posted responses to case questions associated with each module.

Research Participants

A total of 332 students participated in the study, of which usable data was obtained from 324. The sample consisted of 156 males and 152 females.[4]

The average age was 25.9 (SD = 6.5), with a range of 19-54. All of those participating in the course indicated that they were currently employed and had previous computer and Internet experience with over 50% indicating that they had high levels of experience in both.

Measures

Satisfaction

Satisfaction was measured with an 8-item Likert-type scale developed by Biner (1993). The scale used a 7-point strongly disagree to strongly agree response format. The coefficient alpha reliability estimate for this scale was 0.87.

Learning

Two types of learning outcomes, considered to be important by training researchers, were assessed in this study: cognitive/knowledge based outcomes and skill based outcomes (Kraiger, Ford, & Salas, 1993). Each of these outcomes was measured in this study. The form of cognitive knowledge assessed in this study was declarative knowledge. Declarative knowledge was assessed using a 50-point end of module exam score from the next to last course module (Module 5). This exam was chosen because, by the time the participants got to the end of this module, they had enough exposure to and interaction with WebCT for the effect of technology mediation to be manifest. The correlation between Module 5 exam score and participants' overall course grade is 0.41 ($p < .001$). Thus, it is indicative of the participants' overall performance of declarative knowledge. Skill development was measured using the 6-item perceived skill development scale developed by Alavi (1994). This scale used a 7-point strongly disagree to strongly agree response format. The coefficient alpha reliability estimate for this scale was 0.86.

Social Presence

Social presence was measured with a 5-item scale developed by Short et al. (1976). For each question, respondents evaluated the characteristics of the environment using a 5-point, Likert-type scale with anchors such as "unsociable-sociable" and "impersonal-personal." The coefficient alpha reliability estimate for this scale was 0.80. A complete list of all scale items used in the study is found in Appendix A.

Communication

Communication was measured using three types of course communication: the number of discussion postings read, the number of original discussion postings, and the number of follow-up discussion posts. Each of these was standardized and then an aggregate measure was created to represent communication.

Friction

Friction was measured as a gap between what the system users believed was the best way to learn and what they felt the technology supported. Participants were first given descriptions of objectivist, constructivist, and collaborative approaches to learning that an instructor might use. These are shown below:

1. In the objectivist model, learning takes place as the student absorbs the knowledge of the instructor. Therefore, it is the efficiency by which the instructor can transmit his knowledge that will improve a student's ability to learn.
2. In the constructivist model, learning only takes place as the students construct knowledge for themselves. The learners do this through active discovery supported by the instructor.

3. In the collaborative model, students create learning by interacting (discussing and sharing information) with other students.

Next, participants selected the learning approach that would be the most effective way for them to learn. Following this, participants selected the learning approach that they felt WebCT supported. Each scale was scored as follows: 1-objectivist, 2-constructivist, 3-collaborative. Using these two scales, if there was no difference between the learning method they learn best in and the one they feel that WebCT supports, it was coded as a 0. If there was a difference in the approach selected, it was coded as a 1. Overall, 142 people felt that that there was congruence between the technology support of learning and their EBs, while 182 perceived friction to exist. Thus, over 55% of the individuals felt friction between the technology support of learning and their EBs.

Preliminary Analysis

As a manipulation check, we assessed whether or not individuals correctly identified the learning model supported by the technology. Recall that the instructor designed the course to include elements of both constructivist and collaborative learning models. Of those participating, 95% identified WebCT as supporting either the constructivist or collaborative learning models (Table 1). Those who were unable to identify the supported environment were dropped from further analysis, leaving a sample of 307. Of these, 140 individuals felt that there was congruence (45.6%) and 167 individuals perceived friction (54.4%).

To confirm no significant demographic differences between the groups, the groups were compared on a number of variables, including age, gender, GPA, computer experience, Internet experience, confidence in using computers, and previous VLE course experience. Evidence from this analysis suggested that the groups were not different.

Table 1. Learning model identified as desired by learner or supported by WebCT

Learning Model	Frequency	
	Desired	Technology Supported
Objectivist	80	17
Constructivist	98	118
Collaborativist	146	189

Results

Table 2 shows the means and standard deviations of the learning process and outcomes based upon whether the learner perceived congruence or friction and Table 3 shows the correlations among the dependent variables. Learning, communication, social presence, and course satisfaction were all correlated ($p < .001$). An initial multivariate analysis of variance (MANOVA) was run. The overall test was significant (Wilks' lambda $F(4,301) = 4.01$, $p < .01$), which allowed for an individual ANOVA to be performed for each process and outcome variable.

As shown in Table 4, the results of the ANOVA on communication were significant, providing support for H1. Users who perceived congruence between technological support and EBs communicated more ($M = .12$) than when friction was perceived ($M = -.08$), $F(1,305) = 4.98$, $p < 0.05$. H2 predicted that when users perceived congruence, they would experience enhanced social presence ($M = 3.15$) than when they perceived friction ($M = 2.97$); support was not found for this hypothesis ($F(1,305) = 3.08$, $p = .08$). Supporting H3, users who perceived congruence between their EBs and technology support of learning were more satisfied ($M = 5.25$) with the learning environment than those who perceived friction ($M = 4.76$) F

Table 2. Means and standard deviations of processes and outcomes

		Variable									
		Comm.[a]		Social Presence		Satisfaction		Declarative Knowledge		Skill Development	
Group	n	M	SD	M	SD	M	SD	M	SD	M	SD
Congruence	140	0.12	.76	3.15	0.86	5.25	1.10	41.66	8.22	5.52	0.97
Friction	167	-0.08	.77	2.97	0.85	4.76	1.31	38.83	13.43	5.13	1.15
Total	307	0.01	.77	3.05	0.86	4.99	1.24	40.12	11.42	5.31	1.09

[a] *The values listed represent standardized scores.*

Table 3. Correlation of study dependent variables

Construct	1.	2.	3.	4.	5.
1. Communication	—				
2. Social Presence	.22***	—			
3. Satisfaction	.24***	.50***	—		
4. Declarative Knowledge	.24***	.06	.17**	—	
5. Skill Development	.09	.26***	.35***	.16**	—

$*p < .05$, $**, p < .01$, $***p < .001$

Table 4. Results of analysis of variance on course process and outcomes

Source of Variation	SS	df	MS	F
Communication				
Group	2.91	1	2.91	4.98*
Residual	178.44	305	0.59	
Total	181.35	306		
Social Presence				
Group	2.26	1	2.26	3.08[a]
Residual	223.78	305	.73	
Total	226.04	306		
Satisfaction				
Group	17.88	1	17.88	11.97***
Residual	455.57	305	1.49	
Total	473.45	306		
Declarative Knowledge				
Group	613.36	1	613.36	4.76*
Residual	39309.19	305	128.88	
Total	39922.55	306		
Skill Development				
Group	11.61	1	11.61	10.06**
Residual	351.74	305		
Total	363.35	306		

[a] $p < .10$, $*p < .05$, $**p < .01$, $***p < .001$

(1,305) = 11.97, $p < 0.001$. Finally, support was found for H4, with users who perceived congruence between EBs and technology learned more than when friction was perceived. This was true both for declarative knowledge (i.e., score on the skills test) (M = 41.66 vs. 38.83, $F(1,305) = 4.76$, $p < .05$) and for skill development (M = 5.52 vs. 5.13, $F(1,305) = 10.06$, $p < .01$).

DISCUSSION

Summary of Findings

The results of this study provide evidence that congruence between the technology support of learning and an individual's conceptions of the best way to learn can create more effective VLE processes and outcomes than when friction occurs. Specifically, when users feel congruence, they communicate more, learn more, and are more satisfied with the learning experience than when there is friction. While previous research was justified in calling for a fit between the technology used to support learning and the learning model implemented by the instructor, these calls may not go far enough. The shortcoming is that they do not take into account the relationship between an individual's EBs and the technology support of learning. When friction occurs, it becomes difficult for individuals to leverage and appropriate the technology to support their optimal learning strategy or to adapt their learning to match the model supported by the technology. Instead, the findings suggest that the learners will disengage from the course by communicating less, perform less effectively, and become less satisfied with the learning environment.

Implications

One of the greatest advantages from using VLE's is the inherent flexibility of VLEs. Thus, the primary implication of this research is the need to flexibly design VLEs such that the technology support of learning is flexible and adaptable to match users' learning conceptions.

Beyond matching the technological support of an instructor's chosen learning model to the instructional design of the technology, the technology should be flexible enough to adjust to multiple user's beliefs, as some users may not be able to adapt their preferred learning model to the one supported by the technology (Vermunt & Verloop, 2000).

Associated with this, while the technology exists to deliver training via multiple methods, doing so does not currently seem to be the predominant model for delivering VLE initiatives. It is our experience that the vast majority of VLE training and educational settings use the same model despite the realization that people have different preferred learning environments. One reason for the simplification of design is that organizations and instructors choose a specific learning model to implement and then design the technology to support that environment. Various constraints, such as time, technical expertise, and effort to leverage technology for ways in which it was not designed, can make it difficult for the designers of VLEs to take into account these multiple approaches to learning pursued by those using the system. Using a development system based on the convergence of instructional design and Web design, Janicki and Steinberg (2003) found that increased learning occurred when flexible learning content was delivered via multiple methods such as narratives, examples and hands-on exercises. The same type of system could be developed for delivery of learning content based on various learning conceptions and matched with a user's preferred learning approach.

The problem is that when static approaches to technological support occur, some learners are put at a learning disadvantage over their peers. Unlike face-to-face settings, where the instructor can better gauge a persons' learning conceptions, in a VLE they can remain hidden. Thus, these findings suggest that a process for discovering the EBs of those in VLEs needs to be developed. If a learner's beliefs are found to be in conflict with the model used in the VLE, adaptive VLE's can provide the learning content in a manner best suited for the user.

Alternatively in situations where users' will be engaging in multiple VLE experiences over time, those experiencing friction could be provided with training or social behavior modification to change there EBs, as these beliefs have been found to be malleable as one proceeds through various levels of education (Perry, 1968). Echoing the developmental nature of learning conceptions, Vermunt and Verloop (1999) suggest that friction can have positive consequences by catalyzing users to develop new learning strategies. Future research needs to investigate the effectiveness of the various approaches for aligning a VLE technological support of learning with users' learning conceptions. Finally, although this research has focused on the negative consequences of friction, more needs to be done to investigate the potential positive effects that friction might create by changing a user's learning conceptions, specifically what circumstances are needed within the VLE context for this to occur.

Managerial Implications

This research also has implications for managers seeking to maximize returns on their investments of VLEs. This study reinforces the importance of matching user beliefs about learning to the technological support of learning. When this congruence occurs, learning processes and outcomes are enhanced. This benefits managers in two ways. First, congruence leads to better learning, which should translate to improved employee performance. Second, less satisfied individuals are less likely to choose to engage in behaviors that they view negatively, such as enrolling or participating in future VLE initiatives (Ajzen, 1988; Bhattacherjee, 2001). The importance of this cannot be underestimated in environments such as the current one, where over 55% of those participating felt that the technology did not support the way they learned best and because one of the biggest threats to VLEs is the large discontinuance rate (Flood, 2000).

Finally, managers should consider developing their VLE architectures so that they can be flexible enough to tailor course offerings to the preferences of the users. By adopting flexible approaches, managers can provide learning environments that provide the greatest potential for learning. As an example of flexibility in training approaches, Hewlett Packard has developed their corporate training initiatives to allow regional managers to tailor a mix of traditional, blended, and Web-based initiatives to meet the preferences of the region (O'Leonard, 2004). Given these options, users have been able to self-choose the learning approach that they are most comfortable with. This flexibility has led to improved employee performance and improved customer service (O'Leonard, 2004). Future research should investigate how this flexibility translates to improved learning and transfer.

Limitations

As with any study, there are several potential limitations that pertain to the generalizability of these findings. First, the results from this study represent a specific technology implementation in a single course. While there was no evidence to suggest that the results found in this study would be different in different settings, we cannot generalize to other settings. Future research should replicate and extend the findings of this study with different technologies and different contexts. Additionally, participants were required to choose a single learning model between objectivist, constructivist, and collaborative learning models, which did not allow us to investigate the blending of multiple learning models as part of this study. Finally, EBs are more than an individual's learning conceptions but rather a system of interconnected beliefs about knowledge and how knowledge is accumulated (Hofer, 2004; Schommer-Aikins, 2004; Schommer, 1994); as such a more systematic investigation of the aspects of a user's EBs would allow for a deeper understanding of the beliefs on VLE outcomes.

CONCLUSION

This study was motivated by the desire to better understand the implications of friction between the technology support of learning and the EBs of those engaged in the learning process. Results indicated that friction between a user's belief about their learning approach and that provided by a VLE can lead to reduced participation, peer connections, performance and satisfaction. With technology central to the learning processes and outcomes in a VLE, it is important for those designing VLE initiatives to understand that successful VLEs depend not only on the matching of instructor learning models with technology support, but also allowing the technology to be flexible enough for those with differing EBs to have the opportunity to leverage the technology to support their desired learning approach. Without considering the fit between EBs and technology support of learning, the potential exists for organizations to waste large amounts of resources in their investments in their distributed initiatives and for employees participating in these initiatives to learn less, participate less, and ultimately have reduced skills and knowledge than if fit were considered.

REFERENCES

Ajzen, I. (1988). *Attitudes, personality, and behavior*. Chicago: Dorsey.

Alavi, M. (1994). Computer-mediated collaborative learning: An empirical evaluation. *MIS Quarterly, 18*(2), 159-174.

Alavi, M., Marakas, G.M., & Yoo, Y. (2002) Acomparative study of distributed learning environments on learning outcomes. *Information Systems Research, 13*(4), 404-415.

Alavi, M., & Leidner, D.E. (2001). Technology mediated learning: A call for greater depth and breadth of research. *Information Systems Research, 12*(1), 1-10.

Anderson, E.W., & Sullivan, M.W. (1993). The antecedents and consequences of customer satisfaction for firms. *Marketing Science, 12*(2), 125-143.

Bakx, A.W.E.A., Vermetten, Y.J.M., & Van der Sanden, J.M.M. (2003). Self-perceived competence, learning conceptions and preferred learning situations in the domain of communication. *British Journal of Educational Psychology, 73*(2), 223-245.

Bandura, A. (1986). *Social foundations of thought and action: A social cognitive theory*. Englewood Cliffs, NJ: Prentice Hall.

Benbunan-Fich, R. (2002). Improving education and training with IT. *Communications of the ACM, 45*(6), 94-99.

Bhattacherjee, A. (2001). Understanding information systems continuance: An expectation-confirmation model. *MIS Quarterly, 25*(3), 351-370.

Biner, P.M. (1993). The development of an instrument to measure student attitudes toward televised courses. *The American Journal of Distance Education, 7*(1), 63-73.

Burke, K., & Chidambaram, L. (1999). How much bandwidth is enough? A longitudinal examination of media characteristics and group outcomes. *MIS Quarterly, 23*(4), 557-579.

Elen, J., & Lowyck, J. (1998). Students' views on the efficiency of instruction: An exploratory survey of the instructional metacognitive knowledge of university freshmen. *Higher Education, 36*(2), 231-252.

Elen, J., & Lowyck, J. (1999). Metacognitive instructional knowledge: Cognitive mediation and instructional design. *Journal of Structural Learning & Intelligent Systems, 13*(3-4), 145-169.

Elen, J., & Lowyck, J. (2000). Instructional metacognitive knowledge: A qualitative study on conceptions of freshmen about instruction. *Journal of Curriculum Studies, 32*(3), 421-444.

Entwistle, N. J. (1991). Approaches to learning and perceptions of the learning environment. *Higher Education, 22*, 201-204.

Entwistle, N.J. (1998a). Approaches to learning and forms of understanding. In B. Dart & G. Boulton-Lewis (Eds.), *Teaching and learning in higher education* (pp. 72–101). Melbourne: Australian Council for Educational Research.

Entwistle, N.J. (1998b). Improving teaching through research on student learning. In J.J.F. Forest (Ed.), *University teaching: International perspectives* (pp. 73–112). New York: Garland.

Entwistle, N., McCune, V., & Hounsell, J. (2002). *Approaches to studying and perceptions of university teaching-learning environments: Concepts, measures and preliminary findings.* ETL Project Occasional Report.

Feurestein, R., Rand, Y., Hoffman, M., & Miller, R. (1980). *Instrumental enrichment.* Baltimore, MD: University Park Press.

Fiore, S.M., Salas, E., Cuevas, H.M., & Bowers, C.A. (2003). Distributed coordination space: Toward a theory of distributed team process and performance. *Theoretical Issues in Ergonomics Science, 4*(3-4), 340-364.

Flood, J. (2000). Read all about it: Online learning facing 80% attrition rates. *Turkish Online Journal of Distance Education, 3.*

Geiger, M.A., & Cooper, E.A. (1996). Using expectancy to address student motivation. *Issues in Accounting Education, 11*(1), 113-129.

Gunawardena, C.N. (1995). Social presence theory and implications for interaction and collaborative learning in computer conferences. *International Journal of Educational Telecommunications, 1*(2/3), 147-166.

Harrell, A., Caldwell, C., & Doty, E. (1985, October). Expectancy theory predictions of accounting students' academic success motivation. *The Accounting Review,* pp. 724-735.

Hiltz, S.R., & Wellman, B. (1997). Asynchronous learning networks as a virtual classroom. *Communications of the ACM, 40*(9), 44-49.

Hofer, B.K. (2004). Epistemological understanding as a metacognitive process: Thinking aloud during online searching. *Educational Psychologist, 39*(1), 43-55.

IDC. (2003). Press release. Retrieved November 30, 2006, from *http://www.idc.com/getdoc.jhtml?containerId=pr2003_01_14_145111*

Janicki, T., & Steinberg, G. (2003). Evaluation of a computer-supported learning system. *Decision Sciences Journal of Innovative Education, 1*(2), 203-223.

Kraiger, K., Ford, J.K., & Salas, E. (1993). Application of cognitive, skill-based, and affective theories of learning outcomes to new methods of training evaluation. *Journal of Applied Psychology, 78*(2), 311-328.

Leidner, D.E., & Jarvenpaa, S. L. (1995). The use of information technology to enhance management school education: A theoretical view. *MIS Quarterly, 19*(3), 265-291.

Locke, E., & Latham, G. (1990). *A theory of goal setting and task performance*. Englewood Cliffs, NJ: Prentice Hall.

Lowyck, J., & Elen, J. (1994). *Student's instructional metacognition in learning environments (SIMILE).* Paper presented at the Leuven: K.U. Leuven, C.I.P. & T.

Marton, F., Dall'Alba, G. & Beaty, E. (1993). Conceptions of learning. *International Journal of Educational Research, 19*, 277-300.

Marton, F. & Säljö, R. (1976). On qualitative differences in learning: 1 - Outcome and process. *British Journal of Educational Psychology, 46*, 4-11.

Mayor, T. (2001). E-learning: Does it make the grade. *CIO Magazine. 14*(7), 132-144.

O'Leonard, K. (2004). *HP case study: Flexible solutions for multi-cultural learners*. Oakland, CA: Bersin & Associates.

O'Mara, J., Allen, J.L., Long, K.M. & Judd, B. (1996). Communication apprehension, nonverbal immediacy, and negative expectations for learning. *Communication Research Reports, 13*, 109-128.

Oliver, R.L. (1980). A cognitive model for the antecedents and consequences of satisfaction. *Journal of Marketing Research, 17*, 460-469.

Perry, W. (1968). *Patterns of development in thought and values of students in a liberal arts college: A validation of a scheme* (No. 5-0825, Contract No. SAE-8973). Washington, DC: Department of Health, Education and Welfare.

Piccoli, G., Ahmad, R. & Ives, B. (2001). Web-based virtual learning environments: A research framework and a preliminary assessment of effectiveness in basic IT skills training. *MIS Quarterly, 25*(4), 401-426.

Robson, J. (2000). Evaluating on-line teaching. *Open Learning, 15*(2), 151-172.

Rovai, A.P. (2002). A preliminary look at the structural differences of higher education classroom communities in traditional and ALN courses. *Journal of Asynchronous Learning Networks, 6*(1), 41-56.

Salas, E., DeRouin, R., & Littrell, L. (2005). Research-based guidelines for designing distance learning: What we know so far. In H.G. Geutal & D.L. Stone (Eds.), *The brave new world of e-HR: human resources management in the digital age*. San Francisco: Jossey-Bass.

Schommer-Aikins, M. (2004). Explaining the epistemological belief system: Introducing the embedded systemic model and coordinated research approach. *Educational Psychologist, 39*(1), 19-29.

Schommer, M. (1994). An emerging conceptualization of epistemological beliefs and their role in learning. In R. Garner & P.A. Alexander (Eds.), *Beliefs about text and instruction with text* (pp. 25-40). Hillsdale, NJ: Lawrence Erlbaum.

Schommer, M., Crouse, A. & Rhodes, N. (1992). Epistemological beliefs and mathematical text comprehension: Believing it is simple does not make it so. *Journal of Educational Psychology, 84*(4), 435-443.

Short, J., Williams, E., & Christie, B. (1976). *The social psychology of telecommunciations*. New York: John Wiley & Sons.

Todd, P. & I. Benbasat (1991). An experimental investigation of the impact of computer based decision aids on decision making strategies. *Information Systems Research, 2*(2), 87-115.

Tu, C.-H. & McIsaac, M. (2002). The relationship of social presence and interaction in online classes. *American Journal of Distance Education, 16*(3), 131-150.

Tu, C.H. (2000). On-line learning migration: From social learning theory to social presence theory in a CMC environment. *Journal of Network and Computer Applications, 23*(1), 27-37.

Van der Sanden, J.M.M., Terwel, J. & Vosniadou, S. (2000). New learning in science and technology: A competency perspective. In P.R.J. Simons, S.J.L. van der Linden & T.M. Duffy (Eds.), *New learning*. Dordrecht, The Netherlands: Kluwer Academic Publishers.

Vermetten, Y.J., Vermunt, J.D., & Lodewijks, H.G. (2002). Powerful learning environments? How university students differ in their response to instructional measures. *Learning & Instruction, 12*(3), 263-284.

Vermunt, J.D. (1996). Metacognitive, cognitive and affective aspects of learning styles and strategies: Aphenomenographic analysis. *Higher Education, 31*, 25-50.

Vermunt, J.D. (1998). The regulation of constructive learning processes. *British Journal of Educational Psychology, 68*, 149-171.

Vermunt, J.D. & Verloop, N. (1999). Congruence and friction between learning and teaching. *Learning & Instruction, 9*, 257-280.

Vermunt, J.D. & Verloop, N. (2000). Dissonance in students'regulation of learning processes. *European Journal of Psychology of Education, 15*(1), 75-89.

Vygotsky, L.S. (1978). *Mind in society.* Cambridge, MA: Harvard University Press.

Walther, J.B. (1995). Relational aspects of computer-mediated communication: Experimental observations over time. *Organization Science, 6*(2), 186-203.

Webster, J. & Hackley, P. (1997). Teaching effectiveness in technology-mediated distance learning. *Academy of Management Journal, 40*(6), 1282-1309.

Wirt, J. & Livingston, A. (2004). *Condition of education 2002 in brief* (No. NCES 2002011): National Center for Education Statistics.

Yi, M.Y. & Davis, F.D. (2003). Developing and validating an observational learning model of computer software training and skill acquisition. *Information Systems Research, 14*(2), 146-149.

ENDNOTES

[1] While multiple terms have been used to describe these environments, such as virtual learning environments (VLE), distributed training, distance learning, e-learning, and technology-mediated learning (TML), we use the term Virtual Learning Environment in this study.

[2] Individual differences in learning styles have also been suggested as an important criterion in understanding individual learning. Unlike EB, however, learning styles —which vary in their measurement —pertain more to personality traits often having to deal with how an individual views the world and how that view impacts their learning processes. In contrast EB, as defined and measured in this study, are an individual's conceptions about the best way in which learning material are best delivered as either an objectivist, constructivist, or collaborative approach.

[3] Discussions with the last GA found that other than first week assistance in setting up WebCT by a few students (<20), students chose to communicate via course e-mail for assistance.

[4] Sixteen individuals did not indicate their gender.

APPENDIX A. SCALE ITEMS

Social Presence

Higher numbers represent more presence and lower numbers represent less presence.

Impersonal...Personal
Unsociable...Sociable
Insensitive...Sensitive
Cold...Warm
Passive...Active

Satisfaction

I am satisfied with the clarity with which the class assignments were communicated

I am satisfied with the timeliness with which papers, tests, and written assignments were graded and returned.

I am satisfied with the degree to which the types of instructional techniques that were used to teach the class helped me gain a better understanding of the class material.

I am satisfied with the extent to which the instructor made the students feel that they were part of the class and "belonged".

I am satisfied with the instructor's communication skills.

I am satisfied with the accessibility of the instructor outside of class.

I am satisfied with the present means of material exchange between myself and the course instructor.

I am satisfied with the accessibility of the graduate assistants.

Perceived Skill Development

I feel more confident in expressing ideas related to Information Technology.
I improved my ability to critically think about Information Technology.
I improved my ability to integrate facts and develop generalizations from the course material.
I increased my ability to critically analyze issues.
I learned to interrelate the important issues in the course material.
I learned to value other points of view

This work was previously published in End-User Computing: Concepts, Methodologies, Tools, and Applications, edited by S. Clarke, copyright 2008 by Information Science Reference, formerly known as Idea Group Reference (an imprint of IGI Global).

Chapter XIV
A Theoretical Model and Framework for Understanding Knowledge Management System Implementation

Tom Butler
University College Cork, Ireland

Ciara Heavin
University College Cork, Ireland

Finbarra O'Donovan
University College Cork, Ireland

ABSTRACT

The study's objective is to arrive at a theoretical model and framework to guide research into the implementation of KMS, while also seeking to inform practice. In order to achieve this, the chapter applies the critical success factors (CSF) method in a field study of successful KMS implementations across 12 large multinational organisations operating in a range of sectors. The chapter first generates a 'collective set' of CSFs from extant research to construct an a priori model and framework: this is then empirically validated and extended using the field study findings to arrive at a 'collective set' of CSFs for all 12 organisations. These are then employed to refine and extend the theoretical model using insights from the literature on capability theory. It is hoped that the model and framework will aid theory building and future empirical research on this highly important and relevant topic.

INTRODUCTION

KM initiatives fail more often than they succeed (McDermott & O'Dell, 2001). Massey, Montoya-Weiss, and O'Driscoll (2002) argue "that there has been very little research on how to successfully develop and implement KM solutions to enhance performance, particularly in core business processes" (p. 271). The dearth of such research gave rise to calls by practitioners for practical guidelines on how to build and implement KMS, and how to facilitate organizational change to promote knowledge sharing (Alavi & Leidner, 2002; cf. Moffett, McAdam, & Parkinson, 2003). Accordingly, Wong (2005) argues that there is a "need for a more systematic and deliberate study on the critical success factors (CSFs) for implementing KM... [as] Organisations need to be cognizant and aware of the factors that will influence the success of a KM initiative" (p. 261): This study seeks to address such concerns.

It is with these points in mind that this study seeks to arrive at a theoretical model and framework of critical success factors to guide research into the implementation of KMS. It also aims to inform practice, as practitioners in organisations remain unsure as to how to go about planning and deploying KMS (Moffett et al., 2003). In order to achieve its objective, the chapter adopts a qualitative research approach and applies Rockart's (1976) CSF method in a field study of KMS implementations across 12 large multinational organisations operating in a range of sectors. Drawing on Rockart (1979), CSFs may be defined for KM as "the few key areas where "things must go right" for the [KMS implementation] to flourish. If the results in these areas are not adequate, the organisation's efforts [at KM] will be less than desired" (p. 217). In order to attain its stated objective, this study first identifies a collective set of CSFs from the KM literature, which are used to construct a theoretical model and associated framework. Both the framework and the CSFs that constitute it are then empirically validated in the organisations studied; practitioners in these organisations also helped identify additional factors as being of importance. The outcome of this endeavour is a refined and extended model and framework for KMS implementation. In order to undertake the study with the required degree of rigour, the concepts of IS implementation and KMSs, as applied in this study, are first delineated.

IS Implementation Defined

In an early article on IS implementation, Zmud and Cox (1979) argued that "MIS implementation is commonly viewed as involving a series of related activities" (p. 35). Inter alia, these stages are defined by Zmud and Cox as the initiation, strategic design, technical design, development, conversion, and evaluation stages. However, researchers subsequently adopted the convention of referring to the "conversion" stage as the implementation stage and using the term IS development to refer to planning, analysis, design, design, implementation, and use. In essence, IS implementation takes place when the technology dimension is integrated with the people and process dimensions (within particular organisational and institutional contexts and environments) in order to arrive at an organisational IS—furthermore, it overlaps and is intertwined with the "use" phase, as well as the operation and maintenance activities (Iivari, 1990; Iivari & Ervasti, 1994). Thus, when exploring the phenomenon of IS implementation so defined, researchers will attempt to investigate preceding related factors, processes, or activities in order to explain or understand how success in IS implementation is achieved. This is the approach adopted in the present study.

Knowledge Management Systems and Knowledge Sharing

Alavi and Leidner (2001) posit that "Knowledge management systems (KMS) refer to a class of information systems applied to managing organizational knowledge. That is, they are IT-based systems developed to support and enhance the organizational processes of knowledge creation, storage/retrieval, transfer, and application" (p. 114). Drawing on Alavi and Leidner (1999, 2001), Table 1 provides examples of technologies that, researchers argue, help organisations manage their knowledge resources. Given a multiplicity of KM processes (i.e., knowledge creation, storage, etc.) and related IT artefacts, practitioners and researchers decided to simplify matters by focusing on IT for knowledge sharing (Benbya, 2006; Butler & Murphy, 2007). Jennex and Olfman (2004, 2006), for example, posit that KMS, and the knowledge sharing technologies they employ, focus either on processes/tasks or are generic and are infrastructure based. Thus, IT helps organisations share knowledge on processes, tasks, or projects in order to improve their effectiveness; with the infrastructural approach, non-task specific knowledge, or general organisational knowledge is the object of knowledge sharing activities. It is clear from Jennex and Olfman (2004, 2006), however, that a KMS might apply IT to share both task-specific and non-task-specific knowledge in certain organisations. The trend towards focusing on knowledge sharing is also underlined by Benbya (2006), who categorises effective knowledge sharing technologies as being both integrative, highly accessible, and searchable, because "[i]ntegration is a strong predictor of KMS effectiveness, the ability of a system to integrate knowledge from a variety of sources and present it in a manner that enables easy access and reuse is associated with both knowledge quality and knowledge usage" (p. 4). Benbya's conceptualisation is therefore applied in concert with the task/process and generic/infrastructure classification proposed by Jennex and Olfman (2004, 2006) in the present study to help compare the KMS in the organisations studied.

The remainder of this chapter is structured as follows: The second section describes a range of CSFs identified in the literature that are associated with the successful implementation of KM strategies and KMS. This section concludes by presenting a KMS implementation model and research framework for empirical validation in the field prior to comprehensive testing in future research. The third section outlines this study's qualitative research approach. The fourth section then describes and analyses the findings of the field study of 12 organisations. The fifth section presents a refined theoretical model and outlines a path to full theory development. Finally, a number of conclusions are offered.

TOWARDS A KMS IMPLEMENTATION MODEL

There have been several studies on the success factors for KM and KMS—see, for examples,

Table 1. Knowledge management processes and IT artefacts

KM Processes	IT Artefacts	IT Platforms
Knowledge creation	Data mining and learning tools	
Knowledge storage and retrieval	Electronic bulletin boards, nowledge repositories, Databases	Groupware and communication technologies
Knowledge transfer	Electronic bulletin boards, Discussion forums, Knowledge directories (e.g. "Yellow Pages" of subject matter experts)	
Knowledge application	Expert systems, Workflow systems	Intranets

Skyrme and Amidon (1997), Davenport, De Long, & Beers (1998), Holsapple and Joshi (2000); Hasanali (2002); similar factors were also reported in more recent meta-analyses of KM/KMS success factors by Jennex and Olfman (2004, 2006) and Lam and Chua (2005). The challenge for this study will be to build on this body of research to arrive at a set of collective critical success factors that are representative of the key obstacles facing practitioners in implementing KMS.

Zack (1999a) argues that the most important consideration for guiding a knowledge management initiative in an organisation is its strategy. It seems logical therefore to gather together "collective" CSFs under this heading: support for this position is found in Massey et al. (2002). IT-related factors form a second factor grouping; for example, Chua (2004) indicates that "[w]hen used in tandem with an appropriate KM strategy, technology is a powerful enabler of organisational success" (p. 96). The third factor grouping is identified by Alavi and Leidner (1999), who conclude that the "effective resolution of cultural and organizational issues was identified as a major concern in the deployment of KMS. This result is consistent with the IT management literature, which advocates organizational and behavioural change management as critical success factors in the implementation of information systems" (p. 21); thus organisational factors form the final grouping. These three factor groupings—strategy, IT, and organisation—will help the articulation of a parsimonious model of KMS implementation that possesses, what Markus and Robey (1988) term, an "empirical fidelity" with the phenomenon under investigation—the implementation of KMS.

KM Strategy CSFs[1]

While knowledge is recognized as a critical resource for sustained competitive advantages, successful KM remains a key challenge to organisations (Davenport & Prusak, 1998; Lam & Chua, 2005; Wong, 2005). Table 2 illustrates the strategy-based CSFs for KM. According to Hansen et al. (1999) "a company's knowledge management strategy should reflect its competitive strategy" (p. 109); thus, Table 2 indicates that KM strategy must be closely aligned to business strategy (Lam & Chua, 2005). It also indicates that an effective KM strategy should ensure senior management support for, and commitment to, the initiative (Hasanali, 2002). A KMS strategy should also articulate an organisation's knowledge sharing objectives, so that they may be conveyed to all

Table 2. Strategy-based CSFs for KM

Critical Success Factor	Source
Having a close alignment of KM strategy with corporate strategy	Chua (2004); Davenport and Prusak (1998); Hansen, Nohria, and Tierney (1999); Lam and Chua (2005); Sunassee and Sewry (2002), Wong (2005), Zack (1999a, 1999b)
Possessing a comprehensive definition of and communicating KM objectives	Hackett (2000); Jennex and Olfman (2006); Mason and Pauleen (2003)
Ensuring top management commitment	Davenport et al. (1998); Hasanali (2002); Holsapple and Joshi (2000); Jennex and Olfman (2006); Lam and Chua (2005); McDermott and O'Dell (2001); Sunassee and Sewry (2002); Wong (2005)
Developing new roles and responsibilities around KM	Butler, Feller, Pope, Murphy, and Emerson (2006); Butler and Murphy (2007); Davenport and Prusak (1998); Davenport et al. (1998); Hasanali (2002); Roth (2003)

members of staff, not only senior managers and project members (Mason & Pauleen, 2003); it must also provide a clear and unambiguous definition of knowledge (Jennex & Olfman, 2006). The research cited in Table 2 also illustrates that the implementation of KM also requires the establishment of new roles and responsibilities for KM within an organisation (Butler & Murphy, 2007; Davenport et al., 1998).

Information Technology-Related CSFs

The emphasis on implementing IT artefacts for knowledge creation and sharing has several implications for potential success factors, as is indicated in Table 3. Gray and Durcikova (2006) report, for example, that "[a] key limitation on the potential effectiveness of any IT-based system is its ease of use...it follows that one reason why analysts may not source knowledge from a repository is that the technology is not sufficiently easy to use—that is, it may be awkward, slow, or difficult enough to use that analysts may believe that the benefits do not outweigh the costs" (p. 184). Accordingly, Damodaran and Olphert (2000) found that speed and response times of the system are crucial to system success. Thus, KM tools must seamlessly integrate into the day-to-day routine and activities of employees; if it is difficult to use and takes them away from their core activities, they will not see the advantages of using the system (Alavi & Leidner, 1999).

Stenmark (2002) argues that Web-based intranets offer an excellent IT platform for knowledge sharing. Lam and Chua's (2005) empirical findings provide support for this perspective, as do Butler et al. (2006) who illustrate that Web-based technologies form the key components of a core IT artefact for knowledge sharing.

Gold et al. (2001) argue that trust and openness are at the core of knowledge sharing behaviours; however, as knowledge is a valuable firm-specific resource, security is also an important consideration (Alavi & Leidner, 1999; Jennex & Olfman, 2006). In this context, security is viewed as being a technological issue, while openness associated with interpersonal or cultural dimensions (Gold et al., 2001). In their action research study on KMS design, however, Butler et al. (2006) clearly focus on "openness" over security when it comes to developing IT artefacts for knowledge sharing. Indeed, security is low in the hierarchy of success factors, 12th in fact, for KMS, as reported by

Table 3. IT-related CSF

Critical Success Factor	Source
The KMS must be designed so as to be easy to use	Butler and Murphy (2007); Butler et al. (2006); Damodaran and Olphert (2000); Gray and Durcikova (2006); Hasanali (2002); Lam and Chua (2005); Mason and Pauleen (2003)
Build the KMS with Web Technologies	Alavi and Leidner (1999); Butler et al. (2006); Davenport and Prusak (1998); Lam and Chua (2005); Stenmark (2002)
Ensure the KMS presents accurate and appropriate results	Benbya (2006); Damodaran and Olphert (2000); Lam and Chua (2005)
Ensuring that security concerns are balanced with the need for openness	Alavi and Leidner (1999), Butler et al. (2006), Gold, Malhotra, and Segars (2001); Jennex and Olfman (2006)
Having a high degree of IT participation and involvement	Alavi and Leidner (2001); Davenport and Prusak (1998); Malhotra and Galletta (2003).
Having a high degree of user participation and involvement throughout the project	Damodaran and Olphert (2000); Lam and Chua (2005); Malhotra and Galletta (2003); Mason and Pauleen (2003)

Jennex and Olfman (2006). Thus in designing a KMS, the issues of security need to be balanced with openness in KMS design and use.

The IT/IS function in an organisation plays a key supporting role in KMS design, development, and implementation (Davenport & Prusak, 1998): However, the development of such an infrastructure should be business-oriented, as researchers maintain that the development of the KMS should be user-driven and based on the business objectives of an organisation (Damodaran & Olphert, 2000; Mason & Pauleen, 2003). For example, Lam and Chua (2005) report that one KMS project failed due to a dearth of technical and business knowledge required to sustain the programme, the implication here is that it would have been a success had there been a high level of IT and user/business participation throughout.

Organisational CSFs

KM researchers highlight the important influence that organisational actors have in relation to KMS (Moffett et al., 2003). It is hardly surprising then that Bhatt (2001) reports that 56% of executives believe that changing people factors such as behaviour are the most critical elements in KMS implementations (cf. Hackett, 2000). Hislop (2003), for example, states that "personnel issues are now arguably regarded a THE key factor most likely to effect the outcome of knowledge management initiatives" (p. 3).

Alavi and Leidner (1999) argue that culture-based teamwork is a required KM capability; more recently, Wong (2005) emphasises the importance of teamwork at various levels in an organisation, both in the KM implementation team and KMS users. In their study of KM practice, Alavi and Leidner (1999) also note the cross-functional nature of KM teams, with members of relevant business units and the IS function; however, in a general context, practitioners in Hackett's (2000) study illustrate that the "teaming" of knowledge workers and the existence of a culture of teamwork played a critical role in KM success—this has been a recurrent theme in the literature, as Table 4 indicates.

Another major cultural factor is that of trust: Chua and Lam (2005) observe in one organisation, for example, that "[s]taff did not share knowledge within the organisation due to reasons such as the lack of trust and knowledge-hoarding mentality"

Table 4. Organisational CSFs

Critical Success Factor	Source
Focusing on people factors	Bhatt (2001); Butler et al. (2006); Davenport and Prusak 1998; Hackett (2000); Hansen et al. (1999); Hislop (2003); Malhotra and Galletta (2003); McDermott and O' Dell (2001)
Developing a team-oriented culture	Alavi and Leidner (1999); Chua and Lam (2005); Hackett (2000); Davenport et al. (1998); Roth (2003); Wong (2005)
Engendering trust among knowledge workers	Davenport and Prusak (1998); Hansen et al. (1999); Hislop (2003); McDermott and O'Dell (2001)
Ensuring comprehensive user training	Damodaran and Olphert (2000); Hasanali (2002); Storey and Barnett (2000); Malhotra and Galletta (2003), Wong (2005)
Introducing monetary and/or non-monetary incentives and rewards	Davenport et al. (1998); Hislop (2003); Jennex and Olfman (2004, 2006); McDermott and O'Dell (2001); Wong (2005)
Changing organisational structures and processes	Alavi and Leidner (1999); Damodaran and Olphert (2000); Gold et al. (2001); Hackett (2000); Malhotra and Galletta (2003); McDermott and O' Dell (2001); Roth (2003)

(p. 12). Similarly, according to Wong (2005), the development of trust relationships among staff members is essential in order to enable knowledge sharing, this in turn means overcoming the scepticism surrounding the intentions and behaviours of others.

The importance of user training is emphasized across a number of studies (see Table 4); in their analysis of CSFs for KMS, Jennex and Olfman (2004, 2006), for example, include training in two of the CSFs cited. However, even if training is provided, Hasanali (2002) suggests that after the deployment of a KMS, the central KM group should spend most of its time teaching, guiding, and coaching users on how to use the KMS.

Davenport et al. (1998) underline the need for motivational incentives for KM users. There is broad agreement in the literature on the need for incentives in the implementation of KMS; indeed Jennex and Olfman (2004, 2006) underline the need for motivated users who are committed to KMS use—the provision of incentives and training are important factors in achieving this. Accordingly, Wong (2005) points out that "one of the important factors is to establish the right incentives, rewards or motivational aids to encourage people to share and apply knowledge. Giving incentives to employees helps to stimulate and reinforce the positive behaviours and culture needed for effective KM" (p. 271). Malhotra and Galletta (2003) report, however, that in some organisations where formal incentives existed, knowledge sharing was not stimulated. The views of practitioners reported in Hackett (2000) reflect this point, and while monetary incentives are associated with centrally led and driven KM initiatives, non-monetary incentives and intrinsic rewards are linked with "skunk works" type projects. Thus, it may be concluded that the application of incentives, formal or informal, monetary and non-monetary, is contingent on the context of the KMS implementation.

Organisational structures are intended to rationalise and make efficient individual functions or units within an organisation; however, rigid structures and processes encourage individualistic behaviour in which locations, divisions, and functions are rewarded for "hoarding" information and inhibiting successful KM across the organisation (McDermott & O' Dell, 2001). In addition, certain types of organisational structures and processes place limits on communications and can create intentional or unintentional obstacles (Malhotra & Galletta, 2003). Gold et al. (2001) state that a modular organisational design can diminish the costs of coordination and adaptation, thereby increasing flexibility; hence Gold et al. maintain that a non-hierarchical, self-organising organisational structure is the most effective for knowledge sharing. Alavi and Leidner (1999) report that managers worry about managing change around the shift from existing processes to ones that included knowledge sharing: Indeed the change management around structures and process were listed as "key concerns" in their study. Following this line of reasoning it is clear that changing structures and processes, and the management of that change, is important for the successful implementation of KMS.

A Model and Framework of Knowledge Management System Implementation

Based on forgoing arguments, a theoretical model (Figure 1) is proposed to guide the conduct of the present study. Both it, and its associated framework (which is constituted by the CSFs in Tables 2-4 that describe each of the model's high-level constructs) are based on observations drawn from extant research on KM and KMS. The model captures the manner in which KMS implementation success may be directed and effected by: (1) strategic factors, (2) IT factors, and (3) organisational factors. The interaction of these groups of factors is argued to determine KMS success. Benbya (2006) indicates that KMS effectiveness (i.e., the success construct) is

Figure 1. A factors model of knowledge management system implementation

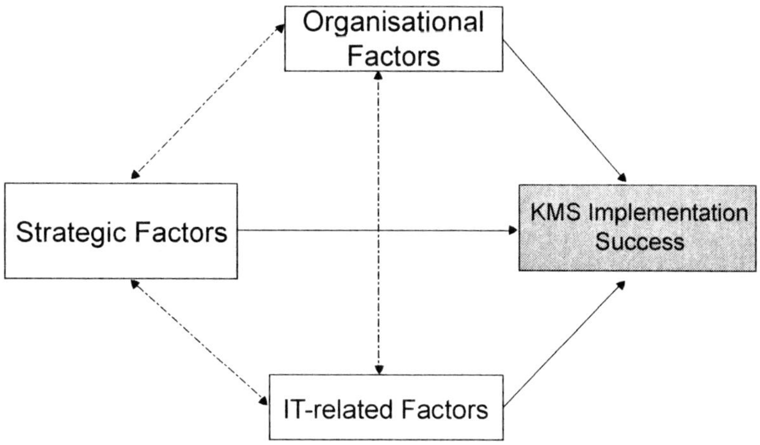

indicated by knowledge quality, usage, and perceived benefits; similar measures are proposed by Jennex and Olfman (2006) viz. perceived benefit and use/user satisfaction leading to net benefits. The primary objective of this study, therefore, is to validate the three groups of CSFs that affect the successful implementation of KMS and the strategic change surrounding the introduction and use of such systems.

RESEARCH APPROACH

In order to examine the factors that affect the implementation of KMSs in several organisations, an interpretive field study approach was adopted (Walsham, 1995). The application of this approach was informed by the CFSs concept and method (Butler & Fitzgerald, 1999; Rockart, 1979).

Twelve organisations that had successfully implemented KMSs were purposively selected to participate in this interpretive field study and application of the CSFs method: these included EMC², Deloitte, Motorola, KPMG, Siemens Corp., Pfizer Corp., IBM, Hewlett Packard, Schering-Plough, Analog Devices Inc., Accenture, and two world-renowned consultancy/professional services organisations. It must be noted, however, that some of these organisations achieved less in the way of success in terms of subsequent use of their KMS. Important selection criteria were that each of these organisations are recognised leaders in KM within their respective industry sectors; furthermore, all had successfully implemented intranet-based KMS based on Web technologies more than one year preceding the study. A recent study by Benbya (2006) adopted similar selection criteria in purposively selecting organisations for study. Purposive sampling was also applied in each organisation to choose the most knowledgeable subject-matter experts (Patton, 1990). Thus 15 interviews were conducted with KM practitioners, with interviewees being purposively chosen using the key informant approach (Patton, 1990)—see Table 5. While organisational anonymity was a requirement for some of the organisations participating in this research, the researchers adopted an approach to effectively anonymize all—Table 5 lists the organisation code employed, while also indicating the sector in which the organisations operate. In addition the table provides a brief analysis of the characteristics of each organisational KMS using criteria adapted from Benbya (2006) and Jennex and Olfman (2006).

Table 5. Organisation code, key informant roles sector and KMS characteristics

Organisation Code	Key Informant Roles	Industry Sector	KMS Characteristics (see legend below)
A	E-service and KM Co-Coordinator	Information Management and Storage (IMS)	II, III, IV
B	IT helpdesk Manager and Local KM Manager	Mobile Technology (MT)	II, III, IV
C	Learning and Leadership Manager	Mobile Technology (MT)	II, III, IV
D	IT Development Manager	Professional Services (PS)	II, III, IV
E	Knowledge and Information Manager Assistant Information Manager	Professional Services (PS)	I, III, IV
F	KM Group Manager	Professional Services (PS)	I, III, IV
G	Development Manager Automation Manager	Pharmaceutics (P)	I, II, III, IV
H	Knowledge Management Supervisor	Pharmaceutics (P)	II, III, IV
I	Knowledge Management Consulting Community Leader Communications Manager for Learning and Knowledge	Global Consulting and Outsourcing (CGO)	I, II, III, IV
J	Senior Partner	Global Consulting and Outsourcing (CGO)	I, II, III, IV
K	Knowledge Management Program Manager	Manufacturing Sector (M)	I, II, III, IV
L	Section Manager and manager of KM initiatives in the Product Development Department	Manufacturing Sector (M)	I, II, III, IV
KMS Characteristics Legend (Adapted from Benbya, 2006; Jennex & Olfman, 2006)	I. Highly accessible Intranet-based KMS that integrates knowledge among general communities of practice II. Highly accessible Intranet-based KMS that integrates knowledge among specific communities of practice III. Knowledge creation and sharing using task/process IV. Knowledge creation and sharing generic/infrastructure approaches		

Given the exploratory and interpretive nature of the study, and the use of the CSF method, each interview was semi-structured, with structure being provided by the application, as an interview guide, of the research framework of "collective" CSFs presented in Tables 2-4. As the KM practitioners interviewed were generally familiar with the CSF concept, or similar approaches such as key performance indicators and so forth, its use permitted a common ground to be established between researchers and researched (Butler & Fitzgerald, 1999). It is consistent with interpretive field research to have social actors narrate their own perspectives of the phenomenon of interest (Walsham, 1995). Researchers therefore encouraged KM practitioners to identify additional CSFs or modify those in the framework. Each interview was taped and up to two hours in duration.

This study's theoretical model and CSF research framework also guided the data analysis, with CSFs acting as "seed categories" to analyse the "content" of each interview transcript and all documentation: This permitted the CSFs for each organisation to be identified in context. Indeed, having interviewees directly validate the a-priori "collective set" of CSFs for KMS implementation, while also nominating additional organisation-specific CSFs, greatly aided the data analysis phase: Hence, CSF-related themes were readily identified in the data. The subsequent comparative analysis of interview transcripts and company documentation confirmed a collective set of CSFs for the organisations studied (cf. Butler & Fitzgerald, 1999; Patton, 1990).

FIELD STUDY FINDINGS

As indicated, the 12 organisations participating in this study had all successfully implemented KMS, but some had subsequent problems with KMS use, as the following sections indicate. That said, the KMS could not be described as failures. Table 6 provides an analytic matrix listing the collective CSFs for all 12 organisations (entries A-L), which emerged from the research data. The factors are grouped under the related high-level headings of strategy, IT, and organisation. The organisational sectors are also identified to help comparison (the legend for each is presented in Table 5). An X signifies whether the CSF was manifested during the KMS implementation process in the organisations (A-L) studied. The following sections provide a descriptive analysis of these CSFs and the influence they exerted on KMS deployment and use in each of the organisations.

Some 23 collective CSFs are presented in Table 6—hence, an additional 7 CSFs were identified in addition to those cited in the literature and appearing in Tables 2-4. The difference arises from claims/observations made by KM practitioners on the existence of additional CSFs (four strategic CSFs) and the need to refine and elaborate on particular CSFs (three additional CSFs emerged from the analysis on IT participation and involvement in IT-related factors and incentives and rewards in the organisational factors). This approach is wholly consistent with the application of an interpretive research approach involving the CSFs method (see Butler & Fitzgerald, 1999).

Strategic CSFs

Practitioners in all but one of the organisations studied (Company L) indicated that it was vital to have KM strategies aligned with corporate business strategies; the reason why Company L differed is due to the application of the KMS to operational processes. The practices of defining, aligning, and communicating KM benefits and goals were present in each of the organisations studied, except Company L. In the majority of firms, KM objectives were formally linked to corporate goals: for example, innovation, attaining competitive advantage, and so on. In Company D, for example, the main objective of KM (capturing solutions to reoccurring problems) was linked to the corporate goal of preventing the "reinvention of the wheel." Organisations adopted similar approaches (e.g., meetings, coffee mornings, workshops, user involvement, and establishing KM slogans) to actively communicate the goals and benefits of KM to the target groups. The e-service and KM Co-Coordinator of Company A stated, for example, that "you must have clear objectives and goals before you implement the system or else it will not work. Employees must be able to see the clear goals and benefits of a KMS." Company A scheduled team meetings and coffee room sessions to communicate KM goals, while also advertising KM on their intranet and making users actively involved in the KM process. The Information Manager of Company E echoed this view and stated: "There has to be a vision, a goal, and you have to see the benefits that you can get out it. If we do x, y, z, and implement it this way then

Table 6. Collective CSFs found to influence KMS implementation in the organisations studiedField Study Findings

Collective CSFs/Companies	A	B	C	D	E	F	G	H	I	J	K	L	Total
Sectors	**IMS**	**MT**	**MT**	**PS**	**PS**	**PS**	**P**	**P**	**GCO**	**GCO**	**M**	**M**	**6**
Strategic Factors													
Having a close alignment of KM Strategy with Corporate Strategy	X	X	X	X	X	X	X	X	X	X		X	11
Possessing a comprehensive definition of and communicating KM Objectives	X		X		X	X	X	X	X	X		X	9
Having a diverse, cross-functional KM Team	X	X	X	X		X	X	X		X	X	X	10
Adopting a suitable Taxonomy of Knowledge				X	X		X	X		X		X	6
Having an Adequate KM budget							X			X			3
Having the project driven by Top/Middle Management	X					X		X					3
Top Management Commitment	X	X	X	X	X	X	X	X	X	X	X	X	12
New Roles & Responsibilities			X	X	X			X				X	5
IT-related Factors													
The KMS must be designed so as to be easy to use	X	X	X	X		X	X	X	X	X	X	X	12
Build the KMS with Web Technologies						X	X	X					3
Ensure the KMS presents accurate and appropriate results				X		X	X					X	4
Security concerns must be balanced with the need for openness			X				X		X				3
Having a high degree of IT Participation and Involvement	X								X		X		3
Having an evolving level of IT Participation and Involvement				X		X	X	X		X		X	6
Having a minimal degree of IT Participation and Involvement		X	X		X								3
Having a high degree of user Participation and Involvement		X	X	X	X	X	X	X	X	X	X	X	11
Organisational Factors													
Focusing on People Factors											X	X	2
Developing a Team-oriented Culture	X	X							X		X	X	5
Engendering trust among knowledge workers				X				X	X				3
Ensuring comprehensive user training		X	X		X	X	X	X			X	X	8
Introducing monetary incentives and rewards	X		X					X		X			4
Introducing non-monetary incentives and rewards		X		X	X	X	X		X		X	X	8
Changing organisational structures and processes			X	X		X	X						4
Total CSFs per Org	9	10	12	12	10	14	15	14	10	10	9	14	

we will get a, b, c out of it." However, in the KM practitioner in Company L viewed that their KMS implementation was not aligned to any corporate goal and stated that as a result KM became largely decentralised with many divisions undertaking their own KM initiatives. This decentralisation resulted in each division setting their own goals for KM and following their own guidelines; he explained, "the local initiatives for KM did not centrally co-ordinate for the maximum benefit across the organisation. Each division went about making their own provision and meeting their own needs in terms of KM, as a result on a global level KM has yet to take off. Currently, it is like ten small companies working in one company." In 9 of the 12 organisations the objectives of the KMS implementation were explicitly defined, whereas in the cases where there was poor communication of benefits (Companies B and D, for example) practitioners recommended increased awareness to improve system use and success.

Five of the organisations established new roles and responsibilities to monitor and support KMS content. Practitioners considered these roles as a "must have" for KM success. The new roles created within the organisations studied varied little, mainly in titles assigned to key personnel (e.g., Knowledge Manager, Knowledge Champion, etc.). In addition, the responsibility assigned (e.g., maintenance, support, and so on) to these roles seldom varied between the organisations studied. In addition, 10 of the 12 organisations established a cross functional KM team. The make-up and responsibilities of the KM teams varied across the organisations studied. In Company H, for example, the KM team was responsible for establishing user needs, prioritizing such needs, implementing the technology, and supporting the users. The team actively sought user feedback on the system and was in constant communication with the IT department when changes were required.

The use of appropriate knowledge taxonomies was identified by six of the KM practitioners as also being key to the success of a KMS. As the Communications Manager for Learning and Knowledge in Company I explained, "creating a taxonomy makes it easier for users to find and submit knowledge." Company I uses a combination of human interaction and technical tools in their KMS (e.g., Lotus Notes) to implement their taxonomy. Company D classifies its organisational knowledge according to the business functions (tax, finance, and consulting) within the organisation and it has designed the KMS to model this structure. Company J, on the other hand, created a detailed level of classifications to store their knowledge. These knowledge categories are further broken down to the time phases of different projects and different processes, for example, sales forecast, project planning, project delivery, and so forth. This KM Practitioner stated that this approach was identified in the user requirements phase to help users navigate to the knowledge captured in the KMS. Other organisations studied went about this by identifying what knowledge they wanted to capture and also the knowledge gaps within the organisation. However, all 15 KM practitioners identified a need for a process to cleanse and categorise captured knowledge. In each of the organisations, this process was assigned to the relevant KM roles (e.g., knowledge champions/managers).

In the majority of the organisations, KM initiatives were implemented as organisational-wide programs requiring input from all levels and functions of the organisation. Organisations achieved this through the establishment of diverse (i.e., in terms of level), cross-functional KM teams that drove the implementation of KM strategies. A distinct overlap arose between establishing heterogeneous, multi-level KM teams and the involvement of top, middle, and lower-level management. KM practitioners in these organisations involved different management levels into the KM teams. They agreed that a successful KM team relied heavily on users who were positioned to have good contact with the different levels within their respective function or community of practice. In essence, members of the KM team represented their function levels (top to lower management).

All KM practitioners emphasised the importance of top management commitment and support. The interviewee from Company I put it thus: "People respond to what their immediate manager asks them to do. If managers are a part of KM and are committed to KM, this will be passed to lower-level management and employees." KM practitioners strongly linked top management to driving required cultural and systems changes. Top management also emerged as having some bearing on budget allocation and employee acceptance of the system. In Companies F, G, and J it was reported that where top management were committed to the KM project, budget did not arise as a barrier (only three organisations identified KM budget as an inhibitor to the project). However, where the KM practitioners questioned the level of top management commitment, they also felt insufficient budget was allocated.

The Role of Information Technology in KM

KMS ease of use was, in the opinion of KM practitioners, the sine qua non for KMS success. All 12 organisations identified that ease of use (e.g., user interface navigation, flexibility, user-friendliness, usability, and speed) was crucial to the success (or effectiveness) of their KMS. The term ease of use, as employed by practitioners, extended to all stages of the knowledge lifecycle from submitting, reviewing, distributing, and searching/locating relevant knowledge. Ease of use was generally established through approaches that incorporated simulated test environments, user involvement, deploying Web technologies, and returning appropriate and accurate results. In Company E, for example, the design phase involved users testing for ease of use in simulated test systems. A number of the systems also replicated their organisational structures to provide categorisation for the knowledge repository. In addition, organisations developed KM roles to monitor data input and categorisation. The importance of this activity was commented upon by a KM practitioner in Company G, who stated that "the knowledge returned must be precise, current and accurate to be of any use to employees"—thereby ensuring accurate and appropriate results.

The dual requirements of security and openness were also identified by three organisations as important factors in the design of a KMS. Users "must have access to as much knowledge as possible but only access to knowledge that is relevant to their needs" (KM Practitioner, Company C). In the case of Companies A, B, C, D, E, F, G, I, J, and L, KM practitioners stated that access to the knowledge repositories and sub-systems belonging to other functional units or departments was typically achieved by obtaining permissions and access rights from the departmental head though e-mail or telephone.

User participation and involvement in KMS implementation was seen as crucial, with 11 out of the 12 companies highlighting it as a critical factor, both in defining user requirements and in creating awareness among users. Many of the organisations achieved user involvement through the establishment of the cross-functional KM teams and by assigning responsibility to key users to link back feedback and developments to the business. Significantly, it emerged from the findings that the stronger the user participation and involvement was in the analysis, design, and testing of KMS the higher the degree of KM success (cf. Cavaye [1995] for evidence of this in traditional IS). For example, the Communications Manager for Learning and Knowledge in Company I pointed out: "Users were involved in giving input in designing the system. They were involved in testing and prototyping the system. Once the system was running they were involved in giving any feedback on the system." Many of the organisations established user groups or steering groups for their respective KM project.

Company F, for example, established an organisation-wide KM team where employees were rotated on a constant basis through user groups to gain extensive feedback. Company E set up a global team to monitor user feedback and to interface with developers user requirements. The Assistant Information Manager in Company E stated that the system "has to come from the users, it has to be what they like and need." The KM Practitioner from Company A identified the outcomes related to a lack of user involvement: he stated that his firm did not get sufficient users involved in the design and development of its KMS, consequently, key functionality was not added to the system. This practitioner argued that this was a major reason users did not see any benefit from using the system.

The IT function's role varied across the organisations studied: For example, three organisations identified that they had strong IT support throughout the duration of the project; Companies B, C, and E had minimal IT involvement; while the role of the IT function evolved over the course of the project for the remaining six organisations. According to the KM Group Manager in Company F, the IT function was brought in at different stages when required to support the KM decision-making process. The IT function also performed the "taken for granted role" of ensuring that the technological infrastructure was in place to allow efficient sharing and access to knowledge. In contrast, in Company K, the IT function was involved in an early stage matching the technology with users' needs. The role of the IT function included introducing the technological capabilities in terms of managing organisational knowledge, while also limiting the user requirements to a certain degree. According to the Automation Manager in Company G, the IT function was involved from the start and contributed to each stage of the KM design, implementation, and support process: Also, the IT function was actively involved in the decision-making process and had a strong presence on the KM team. Equally at Company H, the IT function played a lead role in the design and development of its KMS. A KM team was set up and was led by a software programmer and a financial manager. The various departments submitted their requirements and both the software programmer and the financial manager had the final say in the design of the system. The Development Manager in Company G supported the case for a strong IT presence. He explained: "If knowledge management was mainly driven by IT then, it would not adequately capture the user requirements. However if IT is not part of Knowledge Management, then you are probably going to see the wrong infrastructure, poor development, and poor roll out."

As indicated, three of the respondents supported the view that the IT function should play no part in the KM decision process (Companies B, C, and E): these KM practitioners stated that the IT function was, and should be, restricted to the delivery of the IT infrastructure and in supporting KMS users. However, even though each of these respondents stated that the IT function played little or no role in the design process, it was reported that the IT function had a representative on the KM team. This would indicate that the IT function, even though not visibly seen in the KM decision process, would have been consulted when required and IT professionals were background contributors to the decision-making processes. It is clear then that the IT function played a supporting role in KM in all 12 organisations, but in the pharmaceutical sector (Companies G and H) IT played an important role in the decision-making processes surrounding KMS implementation. Many of the KM practitioners viewed the IT function as being directed by the KM strategy, while feeding into this strategy with IT architecture plans, technical advances and knowledge of any previous systems implementations. Table 6 indicates that what worked best in the majority of organisations was an evolving, but strong participation, by the IT function, as opposed to little or no participation, or having IT lead the KMS project.

Organisational Factors and Their Impact

Creating a knowledge sharing culture was seen by all KM practitioners as being imperative to embedding knowledge sharing in employees. KM practitioners repeated mantra-like that: "People made it happen: They have the knowledge, and they make the decision to share their knowledge" (KM Practitioner Company F). The researchers found that all organisations were progressing to team-oriented and high-trust cultures prior to the introduction of KM. KM practitioners saw this as a fundamental cultural change and the key to knowledge sharing, regardless of the need to implement a KMS. KM practitioners from companies A and K, however, noted that knowledge sharing appeared to be problematic across and between teams—this issue was linked to the absence of KM-related roles in their organisations. This finding points to the importance of new roles and responsibilities as one of the key drivers of knowledge sharing cultures; it also highlights the importance of the link between KM strategy and organisational dimensions.

User training was highlighted by eight of the organisations as a vital factor in KMS implementation. Several organisations implemented comprehensive programmes and conducted KM workshops, held training courses, provided online tutorials, and formed open discussion groups to deliver user training. The leader of the Knowledge Management Consulting Community of Company I explained: "User training is imperative, it's key. It's got to be comfortable for users and one way of making it comfortable is training. If is doesn't integrate well with people, then you got to have more training." Additionally, the Information Manager in Company E stated since "the system is continuously being improved all the time; employees have to be trained to use the system to gain maximum benefit from the system." The KM practitioner from Company B viewed the lack of success of this company's KMS as being directly related to absence of formal training and indicated that the user training associated with the implementation of this firm's global KMS was minimal. They expected that the users would learn through a trial and error approach, the only user training delivered was a one-day demonstration by a knowledge manager to all employees. He believed the lack of user training has led to users finding it difficult "to do simple tasks such as logging solutions or finding knowledge." He noted that as a result of people not being able to use the system, "they became frustrated with the system and could not see the benefit from using the system." He added it was not uncommon to meet employees saying, "I never knew the system could do this." He explained that as a result the system had functionality which many users were unaware of and did not use.

All 12 organisations offered either monetary or non-monetary rewards for knowledge sharing. In the pharmaceutical organisations, monetary incentives were not formally instituted to promote KMS use; however, knowledge sharing was incorporated into each employee's roles. It is significant that both professional services organisations (Company D and F) were attempting to move away from incentives and establish knowledge sharing as a core element in job descriptions. In contrast, KM practitioners from Companies A, C, I, and K revealed that rewards were offered to employees who actively share knowledge. The Leader of the KM Consulting Community in Company I supported the use of monetary rewards and stated "you will always need rewards. Rewards and incentives will make it a bit more interesting, in what's in it for me, and what they are going to get out of it for participating." It is significant that organisations which had poorly developed knowledge sharing cultures (e.g., Company A, C, I) relied heavily on the use of incentives and rewards. The e-service and KM Co-Coordinator in Company A stated that they have established monetary rewards based on "the usage of knowledge." These companies established a "usage count" within the system, (e.g., metering how often a knowledge item is accessed) and the employees are rewarded based on the usage count of the knowledge they have entered. The pharmaceutical organisations, which had well-established knowledge sharing cultures, did not use monetary incentives and rewards. The Development Manager in Company G stated, "sharing knowledge is part of our organisational culture, there is no need to use rewards or incentives. It now has become part of their daily routine." The Automation Manager in Company G commented that "[knowledge sharing] is part of their day-to-day job like any other role they have to carry out." Also the local KM Supervisor in Company H stated that KM "is part of employee's job description. It is embedded in their role to record and share the knowledge about their experiences."

Change to organisational structures and processes did not arise in this study as a barrier to, or critical factor for, KMS implementation. However several KM practitioners reported that the logical design of their KMS reflected closely the structure of their organisation. The KM Group Manager of Company F explained: "Our Knowledge Management System mirrors where the knowledge is physically stored in the organisation by aligning the layout of the Knowledge Management System to the organisational structure." Also, the knowledge taxonomy of Company F's KMS maps readily to core functions in their organisational structure (e.g., tax, finance, etc.). The IT Development Manager of Company D pointed out that his company designed their KMS around audit, tax, management consulting, and financial advisory consulting, which reflects this company's logical structure and key processes. The Learning and Leadership Manager in Company C stated the organisational structure is mirrored in the design of the system: He explained that "our knowledge management strategy embraces structure by how the knowledge is captured and shared. Different functions have different knowledge needs and this must be represented in the Knowledge Management System." The Automation Manager in Company G commented that designing a KMS on the basis of the organisational structure "gives clarity on where to find knowledge." These observations give support for the use of a knowledge taxonomy that can be mapped onto an organisation's structure.

A REFINED THEORETICAL MODEL AND FRAMEWORK FOR KMS IMPLEMENTATION

It is outside the scope of this chapter to present a fully working theory of KMS implementation. Following Teng and Galletta (1991), it presents a "pre-theory" framework to guide research activities enroute to theory development. As Chervany (1973) argues, empirical investigations of IS-related problems require "a research framework that identifies variables (or propositions) to be examined and provides a structure for correlating and synthesizing independent research studies" (p. 181). The CSFs/capabilities model presented here (see Figure 2) attempts to meet these prescriptions and is now formally proposed.

In reflecting on the findings, it was apparent that the link between CSFs and KMS success was mediated by the abilities of organisations and organisational actors to realize the factors. This is an important observation in terms of the proposed model's (Figure 1) explanatory power. Hence, following Wheeler (2002), this chapter proposes to extend the model presented in Figure 1 by proposing the strategic, IT-related, and organisational factors as indicators of strategic, IT, and organisational dynamic capabilities (see Figure 2). Twenty CSFs are included in Figure 2, down from the 23 presented in Table 6, as 3

Figure 2. A critical success factors and capabilities-based model of KMS implementation

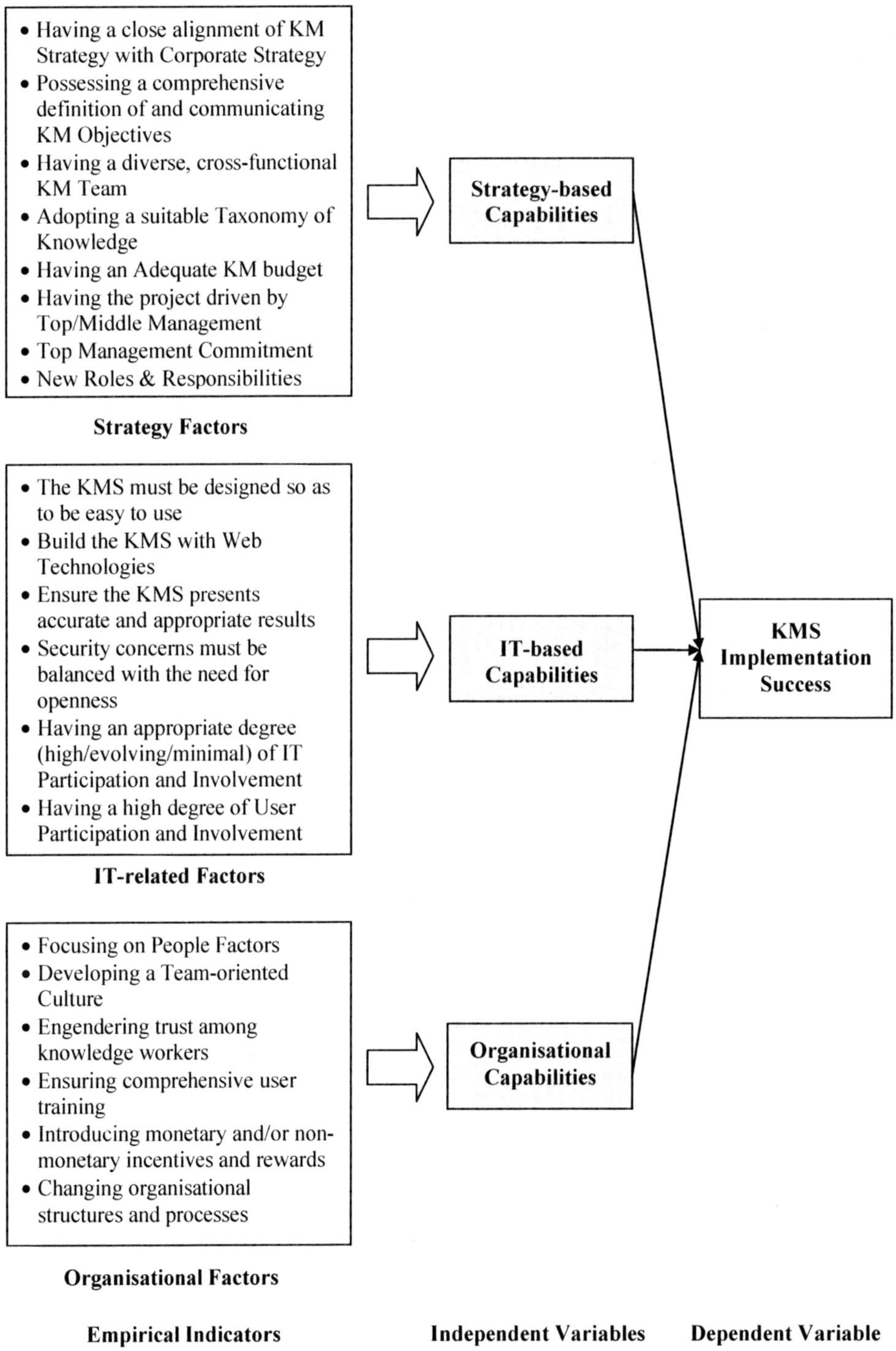

"repeating" CSFs were consolidated (i.e., those dealing with IT participation and involvement and incentives and rewards). In a general context Kangas (1997) argues that organisational "capabilities are developed by combining and using resources with the aid of organizational routines, which are a specific way of doing what the organization has developed and learned" (p. 972). The following broad definition of business and IT capabilities is drawn from Eisenhardt and Martin (2002) conceptualization of dynamic capabilities and Rockart's (1979) CSF concept: business and IT capabilities are the organisational routines that ensure success in the few key areas where "things must go right" for a KMS implementation. The modified model presented in Figure 2 posits that if an organisation is seeking to implement a KMS successfully, then organisational routines (i.e., dynamic capabilities) must be in place to ensure that each of the CSFs are achieved: The failure to succeed in these key areas may result in the failure to implement a KMS, and/or generate user dissatisfaction with the KMS that influences its subsequent use and effectiveness. This constitutes the model's variance theory prediction. The realization of the CSFs are posited as empirical indicators of related strategic, IT, and organisational capabilities (independent variables); the dependent variable of interest, KMS success, may be measured by knowledge quality, usage, and perceived benefits (i.e., KMS effectiveness, Benbya [2006]) or by measures proposed by Jennex and Olfman (2006) viz. perceived benefit and use/user satisfaction leading to net benefits.

CONCLUSION

The evidence provided from KM practitioners participating in this study indicates that the key to the successful deployment of a KMS draws on a range of closely related factors that operate at all levels and functions within an organisation. Nevertheless, there is evidence from the findings that the successful implementation of a KMS does not guarantee ongoing success in the use of the KMS. Indeed, user satisfaction with an implemented KMS may be associated with a lack of success in pre-implementation activities; for example, one of the organisations studied decided not to undertake formal, intensive user training, with poor outcomes for subsequent KMS use.

The findings of this study permitted the theoretical model presented in Figure 1 to be refined and extended to that illustrated in Figure 2. It is significant for the model's validity and the practical relevance of its associated framework (Tables 2-4 and 6) that it was the focus of debate in each of the 15 interviews conducted. KM practitioner feedback helped confirm and identify "collective" CSFs for the successful implementation of KMS. The empirical data suggested the inclusion of additional factors not delineated in the original model; accordingly, these were presented in Table 6 and integrated into the refined model in Figure 2. It is significant that the CSFs identified herein confirm and extend those reported in recent studies (see, for example, Jennex and Olfman, 2004, 2006; Lam and Chua, 2005), while also capturing those reported in reviews of "traditional" IS implementation (see Kwon & Zmud, 1987). The refined model presented in Figure 2 may, therefore, be employed to guide future research (i.e., be tested and confirmed/elaborated) and inform practice (highlight important factors to KM practitioners) on the challenges faced in implementing KMS.

It is accepted within the CSFs literature that not all factors will exert the same influence on related outcomes; some will exert a stronger influence than others, within and across phenomena of interest. In addition, the collective set of CSFs presented in Table 6 warrant further consideration by practitioners and researchers, as the analysis conducted in the Field Study Findings section, along with previous research on CSFs, indicates that relationships exist between CSFs (cf. Butler & Fitzgerald, 1999). In addition, the implication for the model presented in Figure 2 is that there are also relationships between strategic, IT, and organisational capabilities.

In conclusion, this study identified a range of factors deemed to be critical for the implementation of KMS in organisations. The findings on KMS implementation provide further support for the observation that a number of collective CSFs associated with traditional IS development and implementation hold for the implementation of KMS (compare, for example, the factors identified herein with those articulated by Kwon & Zmud, 1987); this observation is congruent with Chua and Lam's (2005) conclusion that "it is meaningful to draw comparisons between KM project abandonment and IS project abandonment" (p. 738) (cf. Davenport et al., 1998). This is to be expected, as Butler (2003) illustrated that "wicked problems" that beset the development of traditional IS also impact Web-based intranet systems. Thus, researchers into KMS implementation should, perhaps, look beyond the KM literature for solutions to enduring problems in business and IS practice; that said, it is also clear that the implementation of a KMS brings its own particular challenges for business and IS practitioners. The challenge for IS researchers will be to progress research into the design, development, implementation, and use of KMS from the foundations provided by the cumulative body of research in the IS field and not fall prey to the temptation to reinvent the wheel in a research context.

REFERENCES

Alavi, M., & Leidner, D. (1999). Knowledge management systems: Issues, challenges and benefits. *Communications of the Associations for Information Systems, 1*(2), 1-36.

Alavi, M., & Leidner, D. E. (2001). Knowledge management and knowledge management systems: Conceptual foundations and research issues. *MIS Quarterly, 25*(1), 107-136.

Alavi, M., & Leidner, D. E. (2002). Knowledge management systems: Issues, challenges and benefits. In S. Barnes (Ed.), Knowledge management systems: Theory and practice. London: Thomson Learning.

Benbya, H. (2006). Mechanisms for knowledge management systems effectiveness: Empirical evidence from the silicon valley. In *Proceedings of the Academy of Management Conference* (pp. 1-6).

Bhatt, G. (2001). Knowledge management in organisations: Examining the interaction between technologies, techniques and people. *Journal of Knowledge Management, 5*(1), 68-75.

Butler, T. (2003). An institutional perspective on the development and implementation of intranet- and Internet-based IS. *Information Systems Journal, 13*(3), 209-232.

Butler, T., Feller, J., Pope, A., Murphy, C., & Emerson, B. (2006, June 12-14). An action research study on the design and development of core it artifacts for knowledge management systems. In *Proceedings of the 14th European Conference on Information Systems*, University of Göteborg, Sweden.

Butler, T., & Fitzgerald, B. (1999). Unpacking the systems development process: An empirical application of the CSF concept in a research context. *Journal of Strategic Information Systems,* 8, 351-371.

Butler, T., & Murphy, C. (2007). Understanding the design of information technologies for knowledge management in organizations: A pragmatic perspective. *Information Systems Journal, 7*(2), 143-163.

Cavaye, A. L. M. (1995). User Participation in System Development Revisited, *Information and Management* (28), 311-323.

Chervany, N. L. (1973). Discussion comments for empirical studies of management information systems. *Data Base, 5*(2-4), 180-181.

Chua, A. (2004). Knowledge management system architecture: A bridge between KM consultants and technologies. *International Journal of Information Management*, 24, 87-98.

Chua, A., & Lam, W. (2005). Why KM projects fail: A multi-case analysis. *Journal of Knowledge Management, 9*(3), 6-17.

Damodaran, L., & Olphert, W. (2000). Barriers and facilitators to the use of knowledge management systems. *Behaviour and information Technology, 19*(6), 405-413.

Davenport, T. H., De Long, D. W., & Beers, M. C. (1998). Successful knowledge management projects. *Sloan Management Review, 39*(2), 43-57.

Davenport, T. H., & Prusak, L. (1998). Working knowledge: How organisations manage what they know. Boston: Harvard Business School Press.

Eisenhardt, K. M., & Martin, J. A. (2000). Dynamic capabilities: What are they? *Strategic Management Journal, 21*(10/11), 1105-1121.

Gold, H., Malhotra, A., & Segars, H. (2001). Knowledge management: An organizational capabilities perspective. *Journal of Management Information Systems, 18*(1), 215-233.

Gray, P. H., & Durcikova, A. (2006). The role of knowledge repositories in technical support environments: Speed versus learning in user performance. *Journal of Management Information Systems, 22*(3), 159-190.

Hackett, B. (2000). *Beyond knowledge management: New ways to work and learn* (Research Rep. No. 1262-00-RR). The Conference Board Inc.

Hansen, M., Nohria, N., & Tierney, T. (1999). What's your strategy for managing knowledge? *Harvard Business Review, 77*(2), 106-116.

Hasanali, F. (2002). *Critical success factors of knowledge management.* Retrieved from http://www.kmadvantage.com/docs/km_articles/Critical_Success_Factors_of_KM.pdf

Hislop, D. (2003). Linking human resource management and knowledge management via commitment: A review and research agenda. *Employee Relations, 25*(2), 182-202.

Holsapple, C., & Joshi, K. (2000). An investigation of factors that influence the management of knowledge in organizations. *Journal of Strategic Information Systems, 9*(2-3), 235-261.

Iivari, J. (1990). Implementability of in-house developed v. application package based information systems. *Data Base*, 1-10.

Iivari, J., & Ervasti, I. (1994). User information satisfaction: IS implementability and effectiveness. *Information & Management*, 27, 205-220.

Jennex, M. E., & Olfman, L. (2004). Assessing knowledge management success/effectiveness models. In *Proceedings of 35th Hawaii International Conference on System, Sciences, HICSS35, IEEE Computer Society* (pp. 1-10).

Jennex, M. E., & Olfman, L. (2006). A model of knowledge management success. *International Journal of Knowledge Management, 2*(3), 51-68.

Kangas, K. (1997). A resource-based analysis of perceived core competencies: Applying the VRIO framework in a case of the Russian trade of a Finnish corporation. In R. Galliers, C. Murphy, S. Carlsson, H. R. Hansen, C. Loebbecke, & R. O'Callaghan (Eds.), *Proceedings of the 5th European Conference on Information Systems*, (Vol. 2, pp. 965-980).

Kwon, T. H., & Zmud, R. W. (1987). Unifying the fragmented models of information systems implementation. In R. J. Boland & R. A. Hirschheim (Eds.), *Critical issues in information systems research.* New York: John Wiley and Sons.

Lam, W., & Chua, A. (2005). Knowledge management project abandonment: An explanatory examination of root causes. *Communications of the Association for Information Systems*, 16, 723-743.

Malhotra, Y., & Galletta, D. F. (2003). The role of commitment and motivation in knowledge management systems: Theory, conceptualisation and measurement of success. In *Proceedings of the 36th Hawaii Conference on Systems Sciences* (pp. 1-10).

Markus, L., & Robey, D. (1988). Information technology and organisational change: Causal structure in theory and research. *Management Science, 34*(5), 583-598.

Mason, D., & Pauleen, D. (2003). Perceptions of knowledge management: A qualitative analysis. *Journal of knowledge Management, 7*(4), 38-48.

Massey, A. P., Montoya-Weiss, M. M., & O'Driscoll, T. M. (2002). Knowledge management in pursuit of performance: Insights from Nortel Networks. *MIS Quarterly, 26*(3), 269-289.

McDermott, R., & O'Dell, C. (2001). Overcoming cultural barriers to sharing knowledge. *Journal of Knowledge Management, 5*(1), 76-85.

Moffett, S., McAdam, R., & Parkinson, S. (2003). An empirical analysis of knowledge management applications. *Journal of Knowledge Management, 7*(3), 6-26.

Patton, M. Q. (1990). *Qualitative evaluation and research methods* (2nd ed.). Newbury Park, CA: Sage.

Rockart, J. F. (1979). Chief executives define their own data needs. *Harvard Business Review, 57*(2), 81-93.

Roth, J. (2003). Enabling knowledge creation: Learning from an R&D organization. *Journal of Knowledge Management, 7*(1), 32-48.

Skyrme, D., & Amidon, D. (1997). The knowledge agenda. *Journal of Knowledge Management, 1*(1), 27-37.

Stenmark, D. (2002). Information vs. knowledge: The role of intranets in knowledge management. In *Proceedings of the 35th Hawaii International Conference on System Sciences* (pp. 1-10).

Storey, J., & Barnett, E. (2000). Knowledge management initiates: Learning from failure. *Journal of Knowledge Management, 4*(2), 145-156.

Sunassee, N. N., & Sewry, D. A. (2002). A theoretical framework for knowledge management implementation. In *Proceedings of SAICSIT 2002* (pp. 235-245).

Teng, J. T. C., & Galletta, D. F. (1991). MIS research directions: A survey of researcher's views. *Data Base*, 53-62.

Walsham, G. (1995). Interpretive case studies in IS research: Nature and method. *European Journal of Information Systems, 4*(2), 74-81.

Wheeler, B. C. (2002). NEBIC: A dynamic capabilities theory for assessing net-enablement. *Information Systems Research, 13*(2), 125-146.

Wong, K. Y. (2005). *Industrial Management and Data Systems, 105*(3), 261-279.

Zack, M. H. (1999a). Developing a knowledge strategy. *California Management Review, 41*(3), 125-145.

Zack, M. H. (1999b). Managing codified knowledge. *Sloan Management Review, 40*(4), 45-58.

Zmud, R. W., & Cox, J. F. (1979). The implementation process: A change approach. *MIS Quarterly, 3*(2), 35-43.

ENDNOTE

[1] The approach adopted in this study to identifying CSFs was to cite those that had been explicitly identified as such in the literature and introduce supportive references where identification was implicit.

This work was previously published in Journal of Organizational and End User Computing, Vol. 19, Issue 4, edited by M. Mahmood, pp. 1-21, copyright 2007 by IGI Publishing, formerly known as Idea Group Publishing (an imprint of IGI Global).

Chapter XV
Exploring the Factors Influencing End Users' Acceptance of Knowledge Management Systems:
Development of a Research Model of Adoption and Continued Use

Jun Xu
Southern Cross University, Australia

Mohammed Quaddus
Curtin University of Technology, Australia

ABSTRACT

This chapter develops a model of adoption and continued use of knowledge management systems (KMSs), which is primarily built on Rogers' (1995) innovation stages model along with two very important social psychology theories—Ajzen and Fishbein's (1980) theory of reasoned action (TRA) and Davis's (1986) technology acceptance model (TAM). It presents various factors and variables in detail. Hypotheses are developed which can be tested via empirical study. The proposed model has both theoretical and practical implications. It can be adapted for application in various organizations in national and international arena.

INTRODUCTION

While the publicity related to organizational learning, intellectual capital, and KM may pass, the need to effectively and systematically manage the knowledge will not diminish. Knowledge is more and more recognized as a key organizational asset for sustaining organizational competitiveness in the market place (Huber, 2001). At present, applying computer-aided knowledge management systems and the aggressive acquisition and retention of knowledge workers are two of the major KM activities (Huber, 2001). A KMS is the infrastructure necessary for the organization to implement its KM processes (Sarvary, 1999) and can be viewed as a "knowledge platform" (Zack, 1999). The objective of the KMS is to support the construction, sharing, and utilization of knowledge in organizations (Alavi & Leidner, 1999).

KMSs have appeared in various formats in different industries. Indeed, there is no single model for a KMS. There is no single role of IT in KM just as there is no single technology comprising KMSs (Thierauf, 1999). Some of the common applications of KMSs are: (1) organizing and sharing/transferring of internal benchmarks/best practices, (2) constructing corporate knowledge directories, such as corporate yellow pages, people information archive, and (3) creating knowledge networks and knowledge maps; among many others (Alavi & Leidner, 2001).

In the past, many ISs, such as management ISs, executive ISs, decision support systems, knowledge-based systems, and so forth focused on codified/explicit knowledge. KMSs provided the opportunity to extend the operating scope of ISs through facilitating the organization's effort in managing both tacit and explicit knowledge (Alavi & Leidner, 2001). Compared to previous systems, such as document management systems, a KMS can provide better help in avoiding the duplication of research effort and assist in a systematic way of capturing people's knowledge and experience (Philips Fox, 1998).

According to Alavi and Leidner (2001) and Junnarkar and Brown (1997), KMSs can play important roles in managing an organization's knowledge. Firstly, KMSs can help connect people to people to share their tacit knowledge and experience without face-to-face meetings. Secondly, KMSs can help make people's tacit knowledge available to others through tools such as groupware, e-mail systems, and online discussion forums, among many others. Thirdly, KMSs can help organize codified knowledge more efficiently through tools such as knowledge repositories and portals, databases, electronic bulletin boards, and intranets. Finally, KMSs can help increase people's own (tacit) knowledge base through learning knowledge in the organization's (codified) knowledge base by applying tools such as data mining and learning tools.

Chait (1999) suggests that KMSs include the key elements of organizations' knowledge capital in many ways, such as information about staff, which improves organizations' ability to identify people with the necessary skills and knowledge; information about customers and clients, which helps organizations to support and serve them better; information about methodologies and tools, which allows organizations to deliver quality and consistent service efficiently and effectively; and information about practices and groups, which keeps everyone in organizations up to date at any time and place. McDermott (1999) emphasizes that the great trap in KM is using tools and concepts of information management to design KMSs. He argues that knowledge sharing requires different concepts and tools from information exchange as a result of the unique characteristics of knowledge. Duffy (1999) states that an effective KMS should contribute in creating an environment in which the organization and its people can be successful through providing systems, tools, and techniques for managing knowledge, without imposing new demands or intruding into day-to-day tasks in the organization.

Although KMSs have been studied widely over the last several years, there is a lack of literature on the end users' adoption and continued use of KMSs. In general, post-adoption behavior of end users has not been studied widely. Some sketchy literature are available on other IT services, for example see Parthasarathy and Bhattacherjee (1998). This chapter thus addresses this gap and attempts to contribute to close this gap. Equipped with a background of high-level factors of adoption and diffusion of generic technologies, this chapter develops a research model to examine and identify the factors affecting the adoption and continued use of KMSs in organizations. Thus two dominant research questions of this study are:

1. What are the factors influencing end users' decision to adopt KMSs?
2. What are the factors influencing end users' decision of continued use of KMSs?

The following section presents relevant background to the study. The research model is then presented and discussed in detail, which is followed by the section on hypotheses development. Finally, future directions and conclusions are presented.

BACKGROUND

There is a deficiency in the literature on the adoption and continued use of KMSs. On the contrary there have been a number of studies of the implementation and the use of new technologies. Reviews and summaries of some of these studies can be found in the literature, such as Swanson (1988); Delone and McLean (1992); Lucas, Schultz, and Ginzberg (1990); Lucas and Spitler (1999), among many others. In this research the adoption and continued use of KMSs is studied using three underlying theories from the literature. They are: (1) theory of diffusion of innovations (Rogers, 1995), (2) TRA (Ajzen & Fishbein, 1980), and (3) TAM (Davis, 1986). These three theories are the primary theoretical foundations for a lot of research projects on IT acceptance and use. While theory of diffusion of innovations focuses on the diffusion process of an innovation, TRA and TAM models attempt to explain the relationship between user attitudes, perceptions, beliefs, and actual use of a technology. The following subsections discuss the three underlying theories as they relate to KMS.

Theory of Diffusion of Innovations

Diffusion is "the process by which an innovation is communicated through certain channels over time among the members of a social system" (Rogers, 1995, p. 5). The four main elements of a diffusion process are: (1) the innovation, (2) communication channels, (3) time, and (4) the social systems (Rogers, 1995). According to Rogers "an innovation is an idea, practice, or object that is perceived as new by an individual" (p. 11). In this research context, the innovation is referred to a KMS. Communication is "the process by which participants create and share information with one another in order to reach a mutual understanding" (Rogers, 1995, p. 5). Communication channels are "the means by which messages get from one individual to another" through the mass media and interpersonal channels (Rogers, 1995, p. 18). The diffusion of KMS occurs through both mass media and interpersonal channels. In this study, the third dimension of time refers to: (1) the time from which an individual first acquires the knowledge about the system through to its adoption (or rejection) to continuing usage, and (2) the relative earliness/lateness with which the system is adopted across the organization (Rogers, 1995). A social system is "a set of interrelated units that is engaged in joint problem-solving to accomplish a common goal. The members of units of a social system may be individuals, informal groups, organizations, and/or subsystems" (Rogers, 1995, p. 23). In this study, the social systems represent sample organizations. At the organizational level, the unit of adoption is the organization where the social system is the organization's external environment. At the individual level, the unit of adoption is the end user, where the primary social system refers to an organization's internal social environment (Brancheau & Wetherbe, 1990).

According to the innovation diffusion research, individuals collect and synthesize information about the innovation; this information process leads to the formation of perceptions about the innovation. The perceptions about the innovation, which are called perceived characteristics in Roger's (1995) theory of innovation diffusion, are relative advantage (how the innovation is seen compared to the one which is currently in place), compatibility (how consistent is the innovation with individual's values and experience), complexity (the difficulty of learning and using the innovation), trialability (the ability to be tested before implementation), and observability (the ability of being able to display the results of using the innovation). In line with these perceptions, a decision to adopt or reject the innovation is made. Some changes of behavior to adapt the innovation are always expected (Hage & Aiken, 1970; Rogers, 1995).

But this initial adoption and use of innovation may not always be enough to fully derive the expected benefits. For the purpose of pursuing the maximum benefits, it is necessary for end users to institutionalize the innovation as part of regular work behaviors. In other words, end users need to infuse, routinize, and implement the innovation into their daily work in a continued-sustained process (Hage & Aiken, 1970; Rogers, 1995; Saga & Zmud, 1994). The institutionalization of a behavior is different from, and possibly more important than, its initial manifestation and adoption. End users may be persuaded to use a new system in the early stage of the diffusion and implementation process, however the benefits of that may never be reached as a result of the lack of continued and sustained usage (Agarwal & Prasad, 1997).

According to Rogers (1995), the "innovation-decision process" is "the process through which an individual (or other decision-making unit) passes from first knowledge of an innovation, to forming an attitude toward the innovation, to a decision to adopt or reject, to implementation of the new idea, and to confirmation of this decision" (p. 161). This process consists of five stages: "knowledge—the individual (or decision-making unit) is exposed to the innovation's existence and gains some understanding of how it functions, (2) persuasion—the individual (or other decision-making unit) forms

Figure 1. A model of stages in innovation process (Rogers, 1995, p.161)

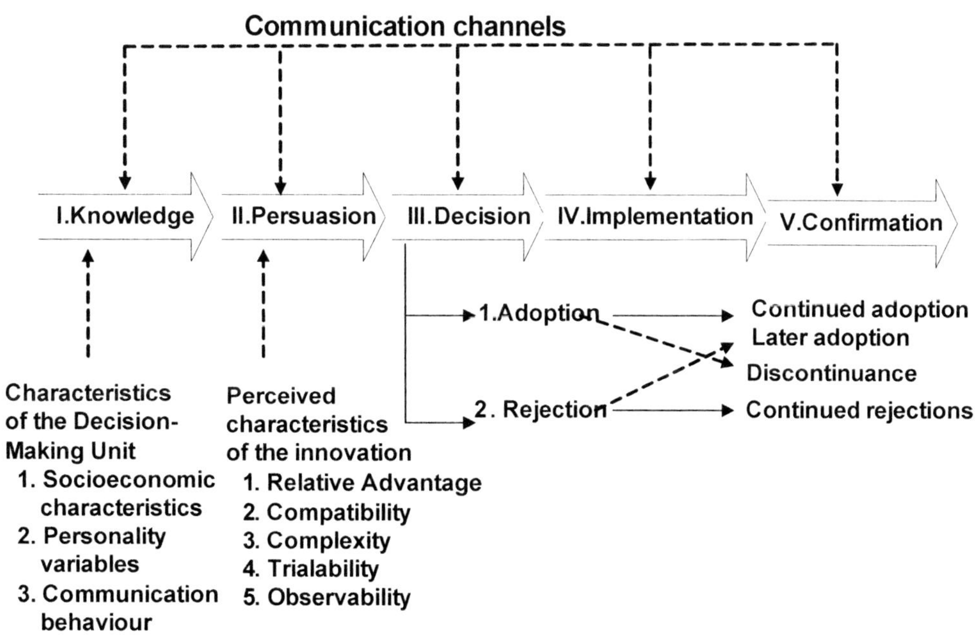

a favourable or unfavourable attitude toward the innovation, (3) decision—the individual (or other decision-making unit) engages in activities that lead to a choice to adopt or reject the innovation, (4) implementation—the individual (or other decision-making unit) puts an innovation into use, and (5) confirmation—the individual (or other decision-making unit) seeks reinforcement for an innovation-decision already made, but may reverse this decision if exposed to conflicting messages about the innovation" (Rogers, 1995, p. 162) (see Figure 1).

The Theory of Reasoned Action

This research focuses on the adoption and continued use of KMSs. Therefore, it is essential to have a good understanding of why people resist adopting a KMS. From this, we can develop practical methods for evaluating KMSs, predict how users will respond to them, and enhance user acceptance by altering the nature of a KMS and the processes by which they are implemented (Davis, Bagozzi, & Warshaw, 1989). Previous researches have suggested the intentional models from social psychology as a potential theoretical foundation for research on the determinants of user behavior (Christie, 1981; Swanson, 1982). This research is thus grounded on two models from social psychology: the TRA (Ajzen & Fishbein, 1980) (see Figure 2) and the theory of TAM (Davis, 1986).

Ajzen and Fishbein's (1980) TRA is "an especially well-researched intention model that has proven successful in predicting and explaining behavior across a wide variety of domains" (Davis et al., 1989). TRA is "designed to explain virtually any human behavior" (Ajzen & Fishbein, 1980). Therefore, it should be an appropriate model for studying the factors affecting the adoption and continued use of KMSs. The TRA has broad applicability in diverse disciplines and has gone through rigorous testing that has proved its robustness in predicting intentions and behavior (Bagozzi, 1981, 1992; Davis et al., 1989; Manstead, Proffitt, & Smart 1983; Sheppard, Hartwick, & Warshaw, 1988).

People consider the implications of their actions before they decide to engage or not to engage in a given behavior. The TRA is built on the assumption that human beings are normally quite rational and make systematic use of the information available to them (Ajzen & Fishbein, 1980). The theory views a person's intention to perform

Figure 2. The theory of reasoned action (Ajzen & Fishbein, 1980)

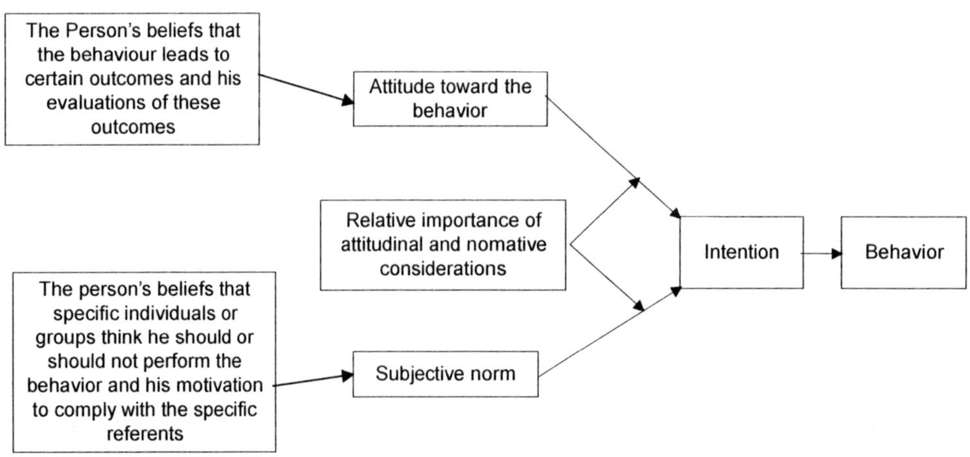

(or not to perform) a behavior as the immediate determinant of the action. In addition, a person's beliefs or perceptions about the characteristics are antecedent to behavior intent to adopt and use the system (Agarwal & Prasad, 1997). Even though it is possible that intention can change with the passage of time, the research has shown that they are good predictors of actual future use (Davis et al., 1989).

Based on the TRA, a person's intention is a function of two basic determinants, one "personal" in nature and the other reflecting "social influence." The personal factor is the individual's positive or negative evaluation of performing the behavior, which is called "attitude toward the behavior" and refers to attitudinal factors. The second determinant of intention is the person's perception of the social pressure put on him/her to perform or not to perform the behavior in question. This factor is termed "subjective norm"—which deals with perceived prescriptions and relates to the normative considerations (Ajzen & Fishbein, 1980). The relative weight of the two determinants of intention is the solution for the situation of conflict between attitude toward the behavior and subjective norm. As a result, it is possible to predict and gain some understanding of a person's intention by measuring his or her attitude toward performing the behavior, his subjective norm, and the relative weights (see Figure 2).

The Technology Acceptance Model

Davis (1986) introduced TAM as an adaptation of TRA (see Figure 3). TAM is one of the widely used models of IT adoption (Gefen & Straub, 2000). TAM has been widely applied, and there are a large number of studies in support of TAM (e.g., Adams, Nelson, & Todd, 1992; Davis 1986, 1989, 1993; Davis et al., 1989; Igbaria, Zinatelli, Cargg, & Cavaye, 1997; Mathiesson, 1991; Straub, Limayem, & Karahanna, 1995; Szjana, 1996; Taylor & Todd, 1995; Venkatesh & Davis, 1996). TAM is not as general as TRA since it is targeted to explain and predict computer usage behavior. However, it has incorporated the accumulated findings of IS research over a decade, and it is especially well-suited for modeling computer acceptance (Davis et al., 1989).

In the TAM, the behavioral intention to use a technology is directly determined by two key beliefs: perceived usefulness (PU) and perceived ease of use (PEOU). PEOU is defined as "the degree of which the prospective user expects the target system to be free of effort" (Davis et al., 1989, p. 985). PU is defined as "the prospective user's subjective probability that using a specific application will increase his or her job performance within an organizational context" (Davis et al., 1989, p. 985). While PU looks at assessment of the extrinsic characteristics of IT, that is, task-oriented outcomes: how IT helps users achieve

Figure 3. Davis' technology acceptance model (Davis et al., 1989)

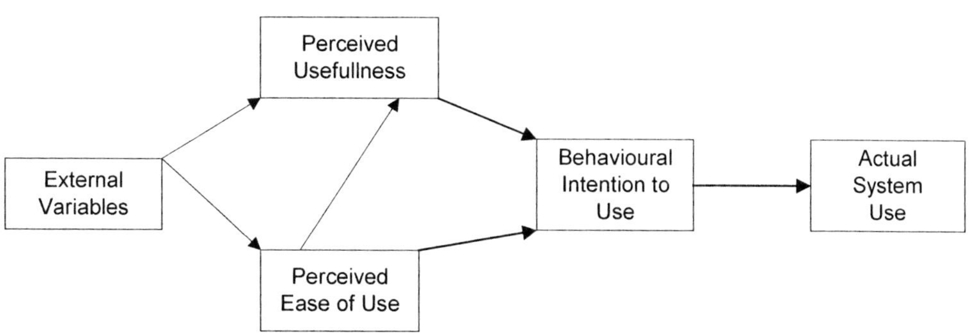

task-related objectives, such as task efficiency and effectiveness, PEOU examines the intrinsic characteristics of IT, such as ease of use, ease of learning, flexibility, and clarity of its interface (Gefen & Straub, 2000).

TAM also argues that PEOU has impact on the PU. The easier a system is to use, the less effort is required to perform a certain task (Davis et al., 1989, 1992). TAM also suggests that PEOU influences the user's decision to adopt an IT primarily through PU, which also posited by various research (e.g., Adam et al., 1992; Gefen, 2000; Keil, Beranek, & Konsynski, 1995; Venkatesh & Davis, 1994).

Many previous studies have documented the role of various external variables on systems usage behavior, including individual factors, management factors, organizational factors, and system features, among many others. Both TRA and TAM also consider that external variables such as the system characteristics, task characteristics, user characteristics, political influences, organizational factors, and the development or implementation process only have indirect influence on behavior by affecting beliefs, attitudes, and intentions (Davis et al., 1989; Szajna, 1996). However, the role of external variables has not been well explored in TAM. Davis (1993) called for: "future research (to) consider the role of additional (external) variables within TAM" (p. 483).

Conceptual Framework

Building on Ajzen and Fishbein's (1980) TRA and Davis's (1986) TAM, this study suggests a research model by combining the theory of innovation diffusion with TRA and TAM. Basically this research suggests that some external factors influence the perceptions about a KMS, which in turn affect the adoption of the innovation and lead to the use of the system, that is, *"External Factors" → "Perceptions" → "Adoption Decision" (for pre-adoption stage)* and *"External Factors" → Realization of Perceptions" → "Continued Use" (for post-adoption stage)*. This simple model is generic in nature and is likely to be applicable, with some adjustments, in various innovation diffusion processes. We use this conceptual framework for the development of the KMS adoption and continued use model, which is discussed in the next section.

THE RESEARCH MODEL

The research model as shown in Figure 4 has two parts: (1) pre-adoption stage and post-adoption stage. The proposed pre- and post-adoption model of KMS is unique and is different from other IS adoption/diffusion models, which only deal with either pre-adoption stage or post-adoption stage.

In the pre-adoption stage, it suggests that the external influences including the individual difference factors, the organizational factors, task complexity factors, and organic factors will influence the adoption of KMSs in an indirect way with their influence being mediated by the perceived benefits (usefulness) and perceived user friendliness (ease of use) of KMSs (Agarwal & Prasad, 1999; Ajzen & Fishbein, 1980; Davis, 1986; Davis et al., 1989; Igbaria, 1994; Igbaria et al., 1997; Igbaria, Guimaraes, & Davis, 1995; Liker & Sindi, 1997; Moore, 1987; Rogers, 1995). The research model also postulates that the perceived factors of perceived benefits, perceived user-friendliness (ease of use), subject norms, and perceived voluntariness have direct effect on the adoption of KMSs. Also as per TAM (Davis 1986, 1989, 1993; Davis et al., 1989), perceived user friendliness affects perceived usefulness (benefits).

In the post-adoption stage, the model suggests that the organizational facilitation factor will influence the continued use of KMSs in an indirect way, with their influence being mediated by the realized benefits (usefulness) and realized user friendliness (ease of use) of KMSs (Agarwal

Figure 4. The proposed research model of adoption and continued use of KMSs

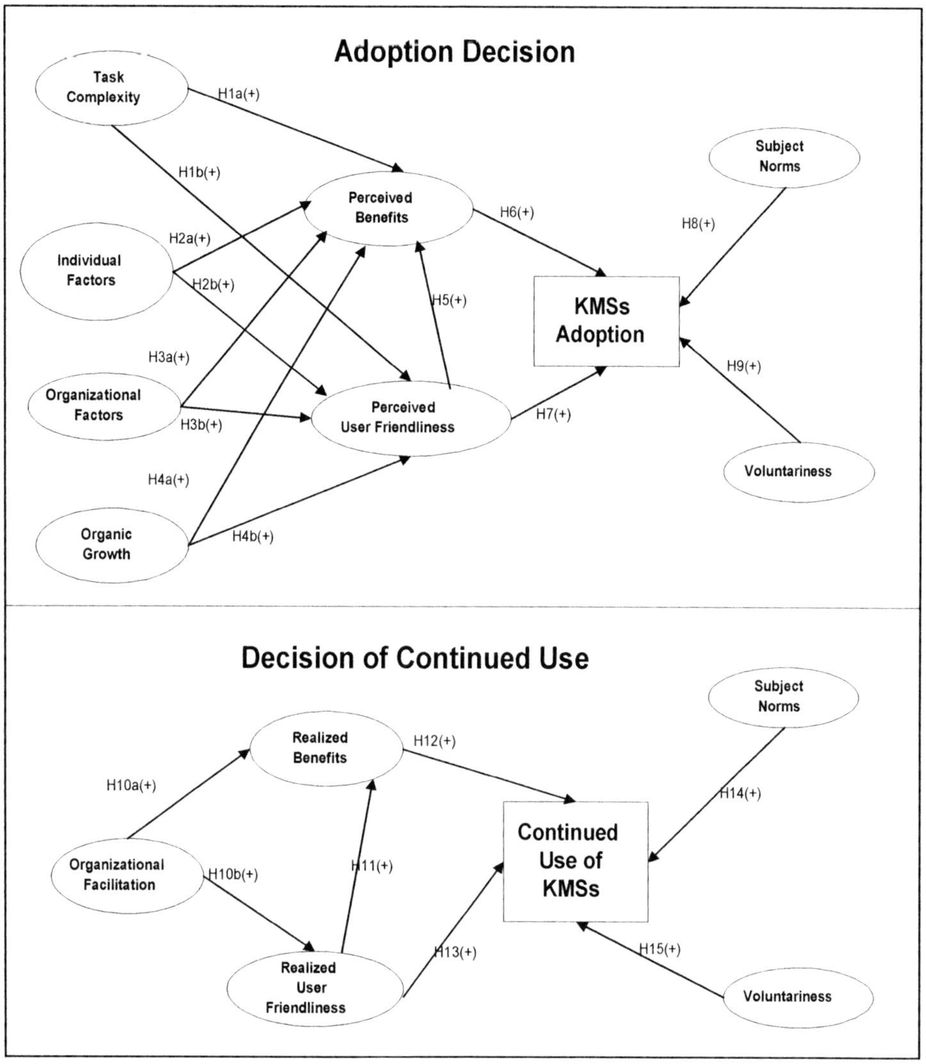

& Prasad, 1999; Ajzen & Fishbein, 1980; Davis, 1986; Davis et al., 1989; Igbaria, 1994; Igbaria et al., 1995; Igbaria et al., 1997; Liker & Sindi, 1997; Moore, 1987; Rogers, 1995). The research model also postulates that the factors of realized benefits, realized user-friendliness (ease of use), subject norms, and voluntariness have direct effect on the continued use of KMSs. Also as per TAM (Davis 1986, 1989, 1993; Davis et al., 1989), realized user friendliness affects realized benefits. Various factors of the model are now discussed in detail.

Individual Differences Factor

In the proposed research model, individual differences factor consists of involvement in KMS, experience/skills, knowledge, roles/responsibilities, personal innovativeness, attitude toward KM, tenure, and position (see Table 1).

Table 1. Individual differences factors (Source: developed for this study)

Individual Differences Factors	**Brief Explanation**
Involvement	End users' involvement in KMSs will encourage the ownership of the systems and lead to their acceptance and use of the systems.
Experience/Skills	End users' experience in using and trying new systems and new technologies will affect their perceptions of KMSs. The more experienced they are, the more likely that they will accept and use the systems.
Knowledge	End users should understand what knowledge their organization has and know how to and where to research and get the knowledge when they need it; so they can reuse other people's experience and knowledge without "reinventing the wheel."
Roles/Responsibilities	End users should have time or be given time off to document what they have done. Roles/responsibilities must be defined to make effort to put knowledge into KMSs.
Personal innovativeness	People in organizations should be encouraged to try new things including KMSs. Personal innovativeness must be encouraged.
Attitude toward KM	End users may have different attitudes toward adopting innovations like KMSs. While there are people who are interested in more effectively and efficiently managing knowledge, some people are protective and will not share knowledge.
Tenure	The longer end users stay in an organization, the more knowledge they may have, and the more they can contribute to the company.
Position	End users who are more senior and in higher positions tend to have more knowledge and experience thus have higher perceived values of KMSs.

Table 2. Organizational factors (Source: developed for this study)

Organizational Factors	**Brief Explanation**
Organizational structure	Gray (2000) implies that organizations with flexible and organic structure are more likely to achieve the perceived benefits of KMS than those organizations that are rigid and bureaucratic.
Organizational culture	When organizations work on building knowledge sharing and pro-KM culture, KM and KMSs begin to take care of themselves. People will manage knowledge automatically and will be more likely to accept and use the systems.
IT infrastructure	IT is an important enabler of KM. Organizations should utilize capacity of IT infrastructure to facilitate their KM efforts.
Business processes	Organizations should have such business processes in place that support and facilitate the use of KMSs.
IT/IS department	IS/IT department, which provides IS/IT services to end users, should support and facilitate the use of KMSs.
Top management support	Top management's support is critical to the acceptance and use of KMSs. Top management should provide resources and leadership for the implementation of the systems.

Organizational Factors

Organizational factors in the proposed research model are represented by organizational structure, organizational culture, IT infrastructure, business processes, IT/IS department, and top management support (see Table 2).

Task Complexity Factors

In the proposed research model, task complexity factors consist of multidisciplinary project, overloaded knowledge, and effective knowledge reuse (see Table 3).

Organic Growth Factors

The concept of organic growth is made of three factors: (1) enticement and education, (2) training, and (3) individual learning (see Table 4).

Perceived Benefits

This construct is represented by six dimensions of potential benefits and driving forces of KMSs. The six dimensions are: (1) effectiveness, (2) creativity, (3) productivity, (4) cost and time reduction, (5) knowledge building, and (6) avoiding same mistakes (see Table 5).

Perceived User Friendliness

The construct of perceived user friendliness reflects the perspectives of end-user focus on the KMSs and is made up of simple to learn and use, cheap to learn and use, speed, accessibility, quality of knowledge, security, complexity, and risk of knowledge (see Table 6).

Perceived Voluntariness

Perceived voluntariness is the degree to which the use of KMSs is perceived as being voluntary, or of free will (Moore & Benbasat, 1991). The con-

Table 3. Task complexity factors (Source: developed for this study)

Task Complexity Factors	Brief Explanation
Multidiscipline project	As a result of the increasing degree of complexity of projects, it is necessary and crucial to manage an organization's knowledge.
Overloaded knowledge	Organizations are facing challenges with overloaded knowledge across boundaries and business lines. Thus, there is a need for KMSs to effectively manage the huge amount of knowledge.
Effective knowledge reuse	Benefits of experience from the past is only realized when it is reused. The huge task of effective reuse of knowledge facilitates the use of KMSs.

Table 4. Organic growth factors (Source: Developed for this study)

Organic Growth Factors	Brief Explanation
Enticement and education	Organizations should promote, educate, persuade, and entice end users to use KMSs rather than forcing them.
Training	Organizations should provide end users with adequate and quality training to facilitate their use of KMSs.
Individual learning	Training is an external facilitating factor. It promotes self-learning, which encourages end users to use the KMS.

Table 5. Dimensions of perceived usefulness (Source: developed for this study)

Perceived Usefulness	Brief Explanation
Creativity	KMSs are able to create new services and innovation, which individuals cannot do on their own.
Productivity	KMSs can enhance productivity (i.e., simplifying workflows, speeding up the projects).
Cost and time reduction	KMSs can save time and money for organizations through avoiding "reinventing the wheel."
Knowledge building	KMSs can help build end users' knowledge base and increase their knowledge, that is, learning from others' experience and knowledge.
Avoiding same mistakes	KMSs can help organizations reduce the chances for mistakes and mediate the risks.
Effectiveness	By having access to appropriate KMSs, individuals can do their job better.

Table 6. Perspectives of user friendliness (Source: developed for this study)

User Friendliness	Brief Explanation
Simple to learn and use	A KMS should be simple to learn and use and should have simple procedures.
Cheap to learn and use	A KMS must be reasonably cheap to learn and use.
Speed	A KMS should be up to speed and must provide knowledge faster.
Accessibility	A KMS must be accessible to everyone in the organization. End users should be able to access past experience, past works, and other people's knowledge and ideas.
Quality of knowledge	There should always be accurate and updated knowledge in a KMS.
Security	Knowledge sharing activities generally involve exchanging and accessing knowledge with both internal and external sources. Correspondingly, there is a need to build appropriate security measures in KMSs.
Complexity of knowledge	Desouza, Awazu, and Wan (2005) identify that users are likely to use simple knowledge products rather than complex ones. Knowledge in a KMS should be properly structured, organized, and stored.
Risk of knowledge	A KMS should have mechanisms in place to mitigate risk and uncertainty of using someone else's knowledge (Desouza et al., 2005).

Table 7. Dimensions of perceived voluntariness (Source: developed for this study)

Perceived Voluntariness	Brief Explanation
Voluntary use	End users will use KMSs on a voluntary basis while forced-use of the systems is not an ideal approach. The end users must feel comfortable to use the systems.
Superior requests	Some organizations believe that rigor, request, and discipline are essential for the successful implementation of KMSs.
Job description	Some organizations may also enforce the use of KMSs by including it in the job description.

struct of perceived voluntariness is reflected in three dimensions: (1) voluntary use, (2) superior request, and (3) job description (see Table 7).

Subject Norm

Subject norms refer to the person's perception that most people who are important to him/her think he/she should or should not use KMSs to perform a task (Ajzen & Fishbein, 1980). End users' use of a KMS can be influenced by others, such as leaders, peers, respected people, superiors, and subordinates.

Table 8. Dimensions of subject norm (Source: developed for this study)

Subject Norm	Brief Explanation
Peer pressure	Peers have influences on end users' use of a KMS. Peers encourage others to use the systems.
Following leaders' lead	Some end users will simply follow leaders or management's lead to use KMSs.
Respected people influence	End users may use KMSs as a result of encouragement from respected people.
Superiors encouragement	Encouragement from superiors may have influence on end users' acceptance and use of KMSs.
Subordinates encouragement	Encouragement from subordinates may have influence on end users' acceptance and use of KMSs.

Table 9. Adoption of KMS (Source: developed for this study)

Adoption	Brief Explanation
Knowledge and expert information search	End users will adopt KMSs to search for required knowledge and information from experts who have required knowledge.
Communication with knowledge holders	End users will adopt KMSs for communication with knowledge holders.
Knowledge sharing	End users will adopt KMSs for sharing knowledge.
Contribution to the system	End users will make contributions to the system by putting in their own knowledge.
Codifying and storing knowledge	End users will adopt KMS to codify and store knowledge.
Knowledge creation	End users will adopt KMSs to create new knowledge.

Adoption of KMS

The construct of adoption of KMS is represented by knowledge and expert information search, communication with knowledge holders, knowledge sharing, contribution to the system, codifying and storing knowledge, and knowledge creation (see Table 9).

Organizational Facilitation

The construct of organizational facilitation is represented by creating imperatives; cutting off old means; having rigor; monitoring system usage; promoting success stories and best practices; satisfying users' needs continuously; making KMS a part of business; making use of KMS a part of an end users' life in the organization; KMS personalization; KMS customization; and KMS invention (see Table 10).

Realized Benefits

This construct is similar to perceived benefits as shown in Table 5. However, variable definitions are different as shown in Table 11, as it represents realized benefits now.

Realized User Friendliness

This construct is similar to perceived user friendliness as shown in Table 6. However, the variable definitions are different now as the construct represents realized user friendliness (see Table 12).

Continued Use of KMSs

The construct of continued use of KMSs is represented by future use of KMSs in the areas of knowledge and expert information search; communication with knowledge holders; knowledge sharing; contribution to the systems; codifying and storing knowledge; knowledge creation; and KMS use habit (see Table 13). Compare and contrast with the variables in Table 9.

Likewise the constructs subject norms and voluntariness must be adapted from pre-adoption to post-adoption of KMS.

Table 10. Organizational facilitation factors (Source: developed for this study)

Organizational Facilitation Factors	**Brief Explanation**
Creating imperatives	Organizations should create imperatives to push end users to ask for help from KMSs and use the systems.
Cutting off old means	Organizations should cut off old means, by which end users used to get knowledge and slowly push people to use the KMS.
Having rigor	Organizations should have rigor and discipline in using KMSs to help ensure end users use the systems.
Monitoring system usage	Organizations should monitor end users' usage of KMSs to enhance the acceptance and use of the systems.
Promoting success stories and best practices	Through promoting success stories and best practices arising from the use of KMSs, more and more end users will be encouraged to get involved and be part of the systems.
Satisfying needs continuously	Organizations should keep end users interests in KMSs by providing what they want.
Making KMS a part of business	Organizations should infuse KMS into their business and have it become a part of the business routine.
Making use of KMS a part of end users' organizational life	Organizations should make use of KMS a part of the end users' life in the organization.
KMS personalization	In the future, predefined user options in KMSs will be modified to meet the needs of individual users (Desouza, Awazu, & Ramaprasad, 2007)
KMS customization	In the future, predefined user options in KMSs will be modified to meet the needs of a collected setting (i.e., a team or organization) (Desouza et al., 2007)
KMS invention	In the future, add-ins or using existing functions for novel purposes will be created in KMSs (Desouza et al., 2007)

Table 11. Dimensions of realized benefits (Source: developed for this study)

Realized Usefulness	**Brief Explanation**
Creativity	KMSs have helped our organization create new services and innovation, which individuals could not do on their own.
Productivity	KMSs have helped our organization enhance productivity (i.e., simplifying workflows, speeding up the projects).
Cost and time reduction	KMSs have helped our organization save time and money for organizations through avoiding "reinventing the wheel."
Knowledge building	KMSs have helped our organization build up end users' knowledge base and increase their knowledge, that is, learning from others' experience and knowledge.
Avoiding same mistakes	KMSs have helped our organization reduce the chances for mistakes and mediate the risks.
Effectiveness	By providing access to appropriate knowledge, KMSs have helped individuals in our organization do their job better.

Table 12. Perspectives of realized user friendliness (Source: developed for this study)

User Friendliness	Brief Explanation
Simple to learn and use	Our KMS is simple to learn and use and has simple procedures.
Cheap to learn and use	Our KMS is reasonably cheap to learn and use.
Speed	Our KMS is up to speed and is quick to get knowledge that end users are after.
Accessibility	Our KMS is available and accessible to everyone in the organization. End users are able to access everything they need and want to know.
Quality of knowledge	There is always accurate and updated knowledge in our KMS.
Security	Our organization has appropriate security measures in our KMS to prevent unauthorized access.
Complexity of knowledge	Knowledge in our KMS is properly structured, organized, and stored. End users can easily retrieve required knowledge from the system (Desouza et al., 2005).
Risk of knowledge	Our KMS has mechanisms in place to mitigate risk and uncertainty of using someone else's knowledge (Desouza et al., 2005).

Table 13. Continued use of KMSs (Source: developed for this study)

Adoption	Brief Explanation
Knowledge and expert information search	In the future, end users will keep on using KMSs to search for required knowledge and information of experts who have required knowledge.
Communication with knowledge holders	In the future, end users will keep on using KMSs for communication with knowledge holders.
Knowledge sharing	In the future, end users will keep on using KMSs for sharing knowledge.
Contribution to the systems	In the future, end users will keep on making contributions to the systems by putting in their own knowledge.
Codifying and storing knowledge	In the future, end users will keep on using KMSs to codify and store knowledge.
Knowledge creation	In the future, end users will keep on using KMSs to create new knowledge.
KMS use habit	In the future, use of KMSs will be become automatic for end users when they are searching for required knowledge (Limayem, Cheung, & Chan, 2003)

HYPOTHESES DEVELOPMENT

This section develops hypotheses as per the proposed research model (see Figure 4). Dishaw and Strong (1999) suggest that task characteristics/requirements have strong effects on system utilization. In addition, Davis's (1986) TAM proposes that external factors, such as task complexity factors, will influence KMS adoption by affecting perceived benefits (usefulness) and user friendliness (ease of use). It is also believed that the increasing demand of knowledge reuse, complexity, and overloaded knowledge will trigger end users' adoption of KMSs.

The preceding discussion results in the following hypotheses:

Hypothesis 1a: *"Task Complexity" factor positively influences the "Perceived Benefits" of KMSs.*
Hypothesis 1b: *"Task Complexity" factor positively influences the "Perceived User Friendliness" of KMSs.*

Individual differences play important roles in determining MIS success and are important determinants of MIS success (Zmud, 1979). Individual differences have also been found to be important in explaining the acceptance and successful implementation of IS (Igbaria et al., 1995). Past research has indicated that the individual/end user characteristics/differences are important factors in explaining/predicting the adoption of innovations (Al-Khaldi & Wallace, 1999; Agarwal & Prasad, 1998, 1999; Brancheau & Wetherbe, 1990; Igbaria et al., 1995; Iivari, 1995; Jackson, Chow, & Leitch, 1997; Larsen & Wetherbe, 1999; Leonard-Barton & Deschams, 1988; Liker & Sindi, 1997; Moore, 1987; Morris & Ventatesh, 2000; Rogers, 1995; Thompson, Higgins, & Howell, 1994; Venkatesh, Morris, & Ackerman, 2000; Zmud, 1979; among many others). People's acceptance of the system is critical to the success of KMSs. An effective way to increase KMS acceptance is involving users in system development (Hahn & Subramani, 2000). If the norm in the organization is to use the KMS to search for the needed knowledge, there is an expectation that every one should use the system. Particular individuals who do not use the system before embarking on the tasks put themselves at odds with the culture in the organization (Brooking, 1999).

Hence, it is hypothesised that:

Hypothesis 2a: *"Individual factors" positively influence the "Perceived Benefits" of KMSs.*
Hypothesis 2b: *"Individual factors" positively influence the "Perceived User Friendliness" of KMSs.*

Gold, Malhotra, and Segars (2001) suggest that knowledge infrastructure capability (technology, structure, and culture) along with knowledge process capability (acquisition, conversion, application, and protection) is an essential precondition for effective KM. The unconditional support of top management and knowledge-sharing culture and reward systems to participate and contribute to the KMS are also important factors for effective KMSs and successful implementation of KMSs (Ma & Hemmje, 2001). Davenport, Long, and Beers (1998) suggest that one of the most important determinants of the successful KM projects is a knowledge-friendly culture, where people have a positive orientation toward knowledge, people are not inhibited in sharing knowledge, and the KM project fits with the existing culture.

Past research finds that organizational factors have significant impact on the adoption of innovations (see Belassi & Fadalla, 1998; Evansiko, 1981; Franz & Robey, 1986; Grover, 1993; Kim & Srivastava, 1998; Kimbley & Lai, 1997; McGowan & Madey, 1998; Rai & Bajwa, 1997; Sarvary, 1999; Sultan & Chan, 2000; Thong, 1999; Thong & Yap, 1995; Yap, Soh, & Raman, 1992; among many others). In addition, Davis's (1986) TAM proposes that external factors, such as organizational factors, will influence KMSs diffusion by affecting perceived benefits and perceived user friendliness.

Therefore the following hypotheses are proposed:

Hypothesis 3a: *"Organizational factors" positively influence the "Perceived Benefits" of KMSs.*
Hypothesis 3b: *"Organizational factors" positively influence the "Perceived User Friendliness" of KMSs.*

Organic growth plays an important role in putting a KMS into an organization since it makes end users interested in the system. Also, it gives them the capability to use the system through sufficient training and self-learning. Organizations should also persuade and educate people to use the system. One of the most difficult tasks in implementing a KMS is to make end users understand that adoption and use of the system will bring benefits not only to the organization but also to themselves. In addition, Davis's (1986) TAM proposes that external factors, such as organic growth, will influence KMSs diffusion by affecting perceived benefits and perceived user friendliness.

The previous discussions lead to the following hypotheses:

Hypothesis 4a: *"Organic growth" positively influences the "Perceived Benefits" of KMSs.*
Hypothesis 4b: *"Organic growth" positively influences the "Perceived User Friendliness" of KMSs.*

TAM (Davis, 1986, 1989, 1993; Davis et al., 1989) and other related studies such as Venkatesh and Davis (2000); Igbaria et al. (1995); Adams et al. (1992); Szajna (1996); and Igbaria et al. (1997) have identified PEOU as an important determinant of system acceptance and use via PU. PEOU has a direct and positive impact on PU.

Kim Sbarcea, the knowledge manager in Phillips Fox, which is the first law firm in Australia to have a KMS with state-of-the-art technology, feels that a KMS must be user friendly: "...Simply put, we needed a means of accessing the knowledge we possess and making it available, simply and easily, to the lawyers and support staff who need it, when they need it" (Phillips Fox, 1998). Davenport and Glaser (2002) suggest that knowledge-sharing programs often fail for the reason that they make it harder, not easier, for people to perform their tasks. Furthermore, Bhattacherjee (2001) finds that users' confirmation of their initial use of ISs has positive impact on their intention of continued use of the systems.

Therefore the following hypotheses are proposed:

Hypothesis 5: *"Perceived User Friendliness" of KMSs positively influences the "Perceived Benefits" of KMSs.*
Hypothesis 11: *"Realized User Friendliness" of KMSs positively influences the "Realized Benefits" of KMSs.*

Bansler and Havn (2002) suggest that expectations/perceptions are key factors in determining an organization's and individual's decision about whether or not to adopt a new KMS. Gray (2000) believes that individuals' perceived value of KMS has direct relationship with their use of KMSs. In addition, the theory of diffusion of innovations, the TRA, and the TAM all propose direct impacts of perceptions on intention to use the system. As a result, it is proposed that perceived benefits and perceived user friendliness have direct impact on KMSs adoption. Bhattacherjee (2001) also finds that users' confirmation of their initial use of ISs has positive impact on their intention of continued use of the systems.

As per the previous discussions, it is hypothesized that:

Hypothesis 6: *"Perceived Benefits" of KMSs positively influence the "Adoption" of KMSs.*
Hypothesis 7: *"Perceived User Friendliness" of KMSs positively influences the "Adoption" of KMSs.*

End users will adopt and use KMSs on a voluntary basis, as forced-use of the systems is not ideal. End users will also use the KMS when they see the value of using the system. Also, when people are forced to use the systems they frequently use them in ways that do not benefit organizations (Bansler & Havn, 2002). In addition, perceived voluntariness focuses on the end users' perceived voluntary adoption and use of KMSs.

The previous discussions lead to the following hypotheses:

Hypothesis 9: *"Voluntary use" of KMSs positively influences the "Adoption" of KMSs.*
Hypothesis 15: *"Voluntary use" of KMSs positively influences the "Continued Use" of KMSs.*

Subject norms reflect the social influence that may affect a person's intention to use KMSs. People often take action based on their perceptions of what others think they should do. Literature (Liker & Sindi, 1997; Lucas & Spitler,

1999; Thompson, Higgins, & Howell 1991) has found that subject norms are positively associated with individual's acceptance of new technology. Huber (2001) suggests that there is considerable ignorance in the literature on the impacts of the social-psychological forces such as the need to adhere to social norms, the need to comply with organizational norms (the right thing to do), the need for recognition, and so forth on knowledge sharing and participation in the KMSs. As a result, there is a great need for future research to explore this area.

The previous discussions results in the following hypotheses:

Hypothesis 8: *Use of KMSs via organizational "norm" positively influences the "Adoption" of KMSs.*
Hypothesis 14: *Use of KMSs via organizational "norm" positively influences the "Continued" use of KMSs.*

Sustained use of KMS by end users is often a challenge faced by many organizations. For the purpose of achieving end users' sustained use of the system, organizations should create imperatives to push them to ask for help from KMSs and use the systems; should cut off old means by which they used to get knowledge and slowly push them to use the systems; and should monitor end users' usage of KMSs to enhance the acceptance and use of the systems. Through promoting success stories and best practices arising from use of KMSs, more and more end users will be encouraged to get involved and be part of the systems. Organizations also should retain end users' interests in KMSs by continuing to provide what they want; infuse KMSs into their business so that they become a part of the business routine; and should make use of KMSs a part of end users' life in organization.

One of the most frequently asked questions about KMSs is "how do you know people will use them"? (Brooking, 1999, p.128). Instead of compensating employees for sharing their knowledge, organizations should make end users understand that the use of KMSs is part of their job and is a part of culture—"the way we do things around here" (Brooking, 1999, p.128). Gray (2000) suggests that increased employee knowledge specialization arising from the use of KMSs will result in the increased use of the systems. In addition, the theory of diffusion of innovations, the TRA, and the TAM all propose direct impacts of perceptions on intention to use the systems. As a result, it is proposed that perceived benefits and perceived user friendliness have direct impact on continued use of KMSs.

The previous discussions result in the following hypotheses:

Hypothesis 10a: *"Organizational Facilitation factors" positively influence the "Realized Benefits" of KMSs.*
Hypothesis 10b: *"Organizational Facilitation factors" positively influence the "Realized User Friendliness" of KMSs.*
Hypothesis 12: *"Realized Benefits" of KMSs positively influence the "Continued use" of KMSs.*
Hypothesis 13: *"Realized User Friendliness" of KMSs positively influence the "Continued Use" of KMSs.*

FUTURE DIRECTIONS AND CONCLUSION

This chapter develops a research model for investigating the factors influencing end users' adoption and continued use of KMSs—a topic that has not been well explored in the literature but represents a primary concern of KMSs. The research model was designed to explore the factors influencing the adoption and continued use of KMSs. Future research could test the entire research model. Parts of the model could also be extracted and investigated in detail. Another interesting future study could be to look at the differentiation among the types of KMSs adopters. According to Rogers' (1995) theory of innovation diffusion, there are five types of adopters. The first to adopt a KMS are the innovators, who adopt it because of its intrinsic values, including perceived user-friendliness/perceived ease of use. Later, the early adopters adopt it since it is able to provide competitive advantage. Only then, the early majority adopt it for pragmatic reasons, such as return-on-investment, cost, and benefit. They are followed by the late adopters and conservatives, who wait until the system is very well established. Future studies could also do longitudinal studies to have a better understanding of the process of adoption and continued use of KMSs, conduct a comparison study between large organizations and small and medium-sized enterprises (SMEs), and investigate the adoption of KMS in different countries.

The model, including both of its main constructs and sub-factors/dimensions, can be taken as they are or fine-tuned to do a comprehensive survey by the researchers of a KM area. Organizations that are embarking on KMSs can use the constructs and factors of the study and do an internal audit to find out how they fare in terms of KMS implementation with respect to these constructs and factors. The proposed model also provides guidelines of KMS adoption and diffusion (continued use) to practitioners and KM consultants. Our immediate plan is to test the proposed research model by surveying end users of KMSs in Australian organizations.

REFERENCES

Adams, D. A., Nelson, R. P., & Todd, P. A. (1992). Perceived usefulness, ease of use, and usage of information technology: A replication. *MIS Quarterly, 16*(2), 227-247.

Agarwal, R., & Prasad, J. (1997). The role of innovation characteristics and perceived voluntariness in the acceptance of information technologies. *Decision Sciences, 28*(3), 557-582.

Agarwal, R., & Prasad, J. (1998). The antecedents and consequents of user perceptions in information technology adoption. *Decision Support Systems, 22*(1), 15-29.

Agarwal, R., & Prasad, J. (1999). Are individual differences germane to the acceptance of new information technologies. *Decision Sciences, 30*(2), 361-391.

Agarwal, R., & Prasad, J. (2000). A field study of the adoption of software process innovations by information systems professionals. *IEEE Transaction on Engineering Management, 47*(3), 295-308.

Ajzen, I., & Fishbein, M. (1980). *Understanding attitudes and predicting social behavior.* Englewood Cliffs, NJ: Prentice Hall.

Al-Khaldi, M. A., & Wallace, R. S. O. (1999). The influence of attitudes on personal computer utilization among knowledge workers: The case of Saudi Arabia. *Information & Management, 36*(4), 185-204.

Alavi, M., & Leidner, D. E. (1999). Knowledge management systems: Issues, challenges, and benefits. *Communications of the Association for Information System, 1*(7). Retrieved May 15, 2001, from http://cais.isworld.org./articels/1-7/article.html

Alavi, M., & Leidner, D. E. (2001). Knowledge management and knowledge management systems: Conceptual foundations and research issues. *MIS Quarterly, 25*(1), 107-146.

Bagozzi, R. P. (1981). Attitudes, intentions, and behavior: The test of some key hypothesis. *Journal of Personality and Social Psychology, 41*(4), 607-627.

Bagozzi, R. P., Baumgartner, H. & Yi, Y. (1992). State versus action orientation and the theory of reasoned action: An application to coupon usage, *Journal of Consumer Research, 18*(4), 505-518.

Bansler, J. P., & Havn, E. (2002, June 6-8). *Exploring the role of networks in IT implementation—The case of knowledge management systems.* Paper presented at the 10th European Conference on Information Systems, Gdansk, Poland.

Belassi, W., & Fadalla, A. (1998). An integrative framework for FMS diffusion. *OMEGA, 26*(6), 699-713.

Bhattacherjee, A. (2001). Understanding information systems continuance: An expectation-confirmation model. *MIS Quarterly, 25*(3), 351-370.

Bowen, T. S. (1999). Making knowledge mean business: Vendors rush to jump on bandwagon. *InfoWorld, 21*(2), 69.

Brancheau, J. C., & Wetherbe, J. C. (1990). The adoption of spreadsheet software: Testing innovation diffusion theory in the context of end-user computing. *Information Systems Research, 1*(2), 123-139.

Brooking, A. (1999). *Corporate memory: Strategies for knowledge management.* London: International Thomas Business Press.

Chait, L. P. (1999). Creating a successful knowledge management system. *Journal of Business Strategy, 20*(2), 23-26.

Christie, B. (1981). Face to file communication: A psychological approach to information systems. New York: John Wiley & Sons.

Davenport, T. H., & Glaser, (2002). Just-in-time delivery comes to knowledge management. *Harvard Business Review,* 107-111.

Davenport, T. H., & Prusak, L. (1998). *Working knowledge: How organizations manage what they know.* Boston: Harvard Business School Press.

Davis, F. D. (1986). *A technology acceptance model for empirically testing new end-user information systems: Theory and results.* Unpublished doctoral dissertation, Massachusetts Institute of Technology, Cambridge.

Davis, F. D. (1989). Perceived usefulness, perceived ease of use, and user acceptance of information technology. *MIS Quarterly, 13*(3), 319-340.

Davis, F.D., Bagozzi, R. P. & Warshaw, P. R. (1992). Extrinsic and intrinsic motivation to use computers in the workplace, *Journal of Applied Social Psychology, 22*(4), 1111-1132.

Davis, F. D. (1993). User acceptance of information technology: System characteristics, user perception and behavioral impacts. *International Journal of Man-Machine Studies, 38,* 475-487.

Davis, F. D., Bagozzi, R. P., & Warshaw, P. R. (1989). User acceptance of computer technology: A comparison of two theoretical models. *Management Science, 35*(8), 982-1002.

Delone, W. H. & McLean, E. R. (1992) Information systems success: the quest for the dependent variable, *Information Systems Research, 3*(1), 60-95.

Desouza, K. C., Awazu, Y., & Ramaprasad, A. (2007). Modifications and innovations to technology artifacts. *Technovation, 27,* 204-220.

Desouza, K. C., Awazu, Y., & Wan, Y. (2005). Factors governing the consumption of explicit knowledge. *Journal of the American Society for Information Science and Technology, 57*(1), 36-43.

Dishaw, M. T., & Strong, D. M. (1999). Extending the technology acceptance model with task-technology fit constructs. *Information & Management, 36*(1), 9-21.

Duffy, J. (1999). Harvesting Experience: Reaping the Benefits of Knowledge, ARMA International, Prairie Village, Kansas.

Franz, C. R., & Robey, D. (1986). Organizational context, user involvement, and the usefulness of information systems. *Decision Sciences, 17*(3), 328-357.

Gefen, D. (2000). Lessons learnt from the successful adoption of an erp: The central role of trust. In S. Zanakis, G. Doukidis, & C. Zopounidis (Eds.), *Recent developments and applications in decision making.* Norwood, MA: Kluwer Academic.

Gefen, D., & Straub, D. W. (1997). Gender differences in the perceptions and use of e-mail: An extension to the technology acceptance model. *MIS Quarterly, 21*(4), 389-400.

Gefen, D., & Straub, D. (2000). The relative importance of perceived ease of use in IS adoption: A study of e-commerce adoption. *Journal of the Association for Information Systems, 1*(8). Retrieved February 14, 2001, from http://jais.aisnet.org/articles/1-8/default.asp?x=95&y=6

Gold, A. H., Malhotra, A., & Segars, A. H. (2001). Knowledge management: An organizational capabilities perspective. *Journal of Management Information Systems, 18*(1), 185-214.

Gottschalk, P. (1999). Use of IT for knowledge management in law firms. *Journal of Information, Law, and Technology.* Retrieved May 15, 2002, from http://elj.warwick.ac.uk/jilt/99~3/gottschalk.html

Gray, P. H. (2000). The effects of knowledge management systems on emergent teams: Towards a research model. *Journal of Strategic Information Systems, 9*(2-3), 175-191.

Grover, V. (1993). An empirically derived model for the adoption of customer-based interorganizational systems. *Decision Sciences, 24*(3), 603-640.

Hage, J., & Aiken, M. (1970). Social change in complex organizations. New York: Random House.

Hahn, J., & Subramani, M. R. (2000). A framework of knowledge management systems: Issues and challenges for theory and practice. In *Proceedings of the Twenty First International Conference on Information Systems,* Brisbane, Queensland, Australia (pp. 302-312). Atlanta, GA: Association for Information Systems.

Huber, G. P. (2001). Transfer of knowledge in knowledge management systems: Unexplored issues and suggested studies. *European Journal of Information Systems, 10*(2), 72-79.

Igbaria, M. (1994). An examination of the factors contributing to microcomputer technology acceptance. *Accounting, Management & Information Technology, 4*(4), 205-224.

Igbaria M., Guimaraes, T., & Davis, G. B. (1995). Testing the determinants of microcomputer usage via a structural equation model. *Journal of Management Information Systems, 11*(4), 87-114.

Igbaria, M., Zinatelli, N., Cargg, P., & Cavaye, A. L. M. (1997). Personal computing acceptance factors in small firms: A structural equation model. *MIS Quarterly, 21*(3), 279-302.

Iivari, J. (1995). Factors affecting perceptions of case effectiveness. *European Journal of Information Systems, 4*(3), 143-158.

Jackson, C. M., Chow, S., & Leitch, R. A. (1997). Toward an understanding of the behavioral intention to use an information system. *Decision Sciences, 28*(2), 357-389.

Junnarkar, B., & Brown, C. V. (1997). Re-assessing the enabling role of information technology in KM. *Journal of Knowledge Management, 1*(2), 142-148.

Karahanna, E., & Straub, D. W. (1999). The psychological origins of perceived usefulness and ease-of-use. *Information & Management, 35*(4), 237-250.

Karahanna, E., Straub, D. W., & Chervany, N. L. (1999). Information technology adoption across time: A cross-sectional comparison of pre-adoption and post-adoption beliefs. *MIS Quarterly, 23*(2), 183-213.

Keil, M., Beranek, P. M., & Konsynski, B. R. (1995). Usefulness and ease of use: Field study evidence regarding task considerations. *Decisions Support Systems, 13*(1), 75-91.

Kim, N., & Srivastava, R. K. (1998). Managing intra-organizational diffusion of technological innovations. *Industrial Marketing Management, 27*(3), 229-246.

Kimberly, J. R., & Evansiko, M. J. (1981). Organizational innovation: The influence of individual, organizational, and contextual factors on hospital adoption of technological and administrative innovation. *Academy of Management Journal, 24*(4), 689-713.

Lai, V. S. (1997). Critical factors of ISDN implementation: An exploratory study. *Information & Management, 33*(2), 87-97.

Larsen, T. J., & Wetherbe, J. C. (1999). An exploratory field study of differences in information technology use between more and less innovative middle manager. *Information & Management, 36*(2), 93-108.

Leonard, D., & Sensiper, S. (1998). The role of tacit knowledge in group innovation. *California Management Review, 40*(3), 112-132.

Liker, J. K., & Sindi, A. A. (1997). User acceptance of expert systems: A test of the theory of reasoned action. *Journal of Engineering and Technology Management, 14*(2), 147-173.

Limayem, M. Cheung, C. M. K., & Chan, G. W. W. (2003, December 14-17). Explaining information systems adoption and post-adoption: Toward an integrative model. In *Proceedings of the Twenty Fourth International Conference on Information Systems,* Seattle, WA (pp. 720-731).Atlanta, GA: Association for Information Systems.

Lucas, H.C., Schultz, R. & Ginzberg, M. (1990). *Information systems implementation: Testing a structural model.* Norwood, NJ, Ablex.

Lucas, H. C., & Spitler, V. K. (1999). Technology use and performance: A field study of broker workstations. *Decision Sciences, 30*(2), 291-311.

Ma, J., & Hemmje, M. (2001, November 8-9). Developing a knowledge management systems step by step. In *Proceedings of the 2nd Europe Conference on Knowledge Management,* Bled, Slovenia (pp. 301-311).

Manstead, A. S., Proffitt, C., & Smart, J. L. (1983). Predicting and understanding mother's infant feeding intentions and behavior: Testing the theory of reasoned action. *Journal of Personality and Social Psychology, 44*(4), 657-671.

Mathieson, K. (1991). Predicting user intentions: Comparing the technology acceptance model with the theory of planned behavior. *Information System Research, 2*(3), 173-191.

Mazzie, M. (2000). Key challenges facing the evolution of knowledge management. In T. K. Srikantaiah & M. E. D. Koenig (Eds.), *Knowledge management for the information professional* (pp. 99-114). Medford, NJ: Information Today, Inc.

McDermott, R. (1999). Why information technology inspired but cannot deliver knowledge management. *California Management Review, 4*(4), 103-117.

McGowan, M. K., & Madey, G. R. (1998). Adoption and implementation of electronic data interchange. In T. J. Larsen & E. McGuire (Eds.), *Information systems innovation and diffusion: Issues and directions* (pp. 116-140). London: Idea Group.

Moore, G. C. (1987). End-user computing and office automation: A diffusion of innovation perspectives. *Information Systems and Operational Research, 25*(3), 214-235.

Moore, G. C., & Benbasat, I. (1991). Development of an instrument to measure the perceptions of adopting an information technology innovation. *Information System Research, 2*(3), 192-222.

Morris, M. G., & Venkatesh, M. (2000). Age differences in technology adoption decisions: Implications for a changing workforce. *Personnel Psychology, 53*(2), 375-403.

Pan, S. L., & Scarbrough, H. (1998). A socio-technical view of knowledge-sharing at Buckman laboratories. *Journal of Knowledge Management, 2*(1), 55-66.

Parthasarathy, M., & Bhattacharjee, A. (1998). Understanding post-adoption behavior in the context of online services. *Information Systems Research, 9*(4), 362-379.

Phillips Fox. (1998). *Leading the way in knowledge management.* Retrieved July 20, 2001, from http://www.phillipsfox.com.au/phillipsfox/publications/pbo98006.htm

Pinker, E. J., & Van Horn, R. L. (2000). Worker incentives to learn in gatekeeper systems: Lessons for the implementation of knowledge management systems. In *Proceedings of the 33rd Hawaii International Conference on System Science.* Retrieved March 21, 2002, from http://computer.org/proceedings/hicss/0493/04933/04933027.pdf

Rogers, E. M. (1995). *Diffusion of innovations* (4th ed.). New York: The Free Press.

Saga, V., & Zmud, R. W. (1994). The nature and determinants of IT acceptance, routinization, and infusion. In L. Levine (Ed.), *Diffusion, transfer, and implementation of information technology* (pp. 67-86). Amsterdam: North-Holland.

Sarvary, M. (1999). Knowledge management and competition in the consulting industry. *California Management Review, 41*(2), 95-107.

Sheppard, B. H., Hartwick, J., & Warshaw, P. R. (1988). The theory of reasoned action: A meta-aanlysis of past research with recommendations for modifications and future research. *Journal of Consumer Research, 15,* 325-343.

Straub, D. W., Limayem, M., & Karahanna, E. (1995). Measuring system usage: Implications for is theory testing. *Management Science, 41*(8), 1328-1342.

Sultan, F., & Chan, L. (2000). The adoption of new technology: The case of object-oriented computing in software companies. *IEEE Transactions on Engineering Management, 47*(1), 106-125.

Swanson, E. B. (1982). Measuring user attitudes in MIS research: A review. *Omega, 10*(2), 157-165.

Swanson, E.B. (1988). *Information systems implementation: Bridging the gap between design and utilization*, IRWIN, Homewood, IL.

Taylor, S., & Todd, P. A. (1995). Understanding information technology usage: A test of competing models. *Information Systems Research, 6*(2), 145-177.

Thierauf, R. J. (1999). *Knowledge management systems for business.* London: Quorum Books.

Thompson, R. L., Higgins, C. A., & Howell, J. M. (1991). Personal computing: Toward a conceptual model of utilization. *MIS Quarterly, 15*(1), 125-143.

Thompson, R. L., Higgins, C. A., & Howell, J. M. (1994). Influence of experience on personal computer utilization: Testing a conceptual model. *Journal of Management Information Systems, 11*(1), 167-187.

Thong, J. Y. L. (1999). An integrated model of information systems adoption in small business. *Journal of Management Information Systems, 15*(4), 187-214.

Thong, J. Y. L., & Yap, C. S. (1995). CEO characteristics, organizational characteristics and information technology adoption in small businesses. *Omega, 23*(4), 429-442.

Venkatesh, V., & Davis, F. D. (1994). Modelling the determinants of perceived ease of use. In J. I. DeGross, S. L. Huff, & M. C. Munro (Eds.), *Proceedings of the Fifteenth International Conference on Information Systems,* Vancouver, British Columbia (pp. 213-227).

Venkatesh, V. & Davis, F. D. (1996). A model of the antecedents of perceived ease of use: development and test, *Decision Sciences, 27*(3), 451-481.

Venkatesh, V., & Davis, F. D. (2000). A theoretical extension of the technology acceptance model: Four longitudinal field studies. *Management Science, 46*(2), 186-204.

Venkatesh, V., & Morris, M. G. (2000). Why don't men ever stop to ask for directions? Gender, social influence, and their role in technology acceptance and usage behavior. *MIS Quarterly, 24*(1), 115-139.

Yap, C. S., Soh, C. P. P, & Raman, K. S. (1992). Information systems success factors in small business. *OMEGA International Journal of Management Science, 20*(5/6), 597-609.

Zack, M. H. (1999). Managing codified knowledge. *Sloan Management Review, 40*(4), 45-58.

Zmud, R. W. (1979). Individual differences and MIS success: A review of the empirical literature. *Management Science, 25*(10), 966-979.

This work was previously published in End-User Computing: Concepts, Methodologies, Tools, and Applications, edited by S. Clarke, copyright 2008 by Information Science Reference, formerly known as Idea Group Reference (an imprint of IGI Global).

Selected Readings

Chapter XVI
Classifying Web Users: A Cultural Value-Based Approach

Wei-Na Lee
University of Texas at Austin, USA

Sejung Marina Choi
University of Texas at Austin, USA

ABSTRACT

In today's global environment, a myriad of communication mechanisms enable cultures around the world to interact with one another and form complex interrelationships. The goal of this chapter is to illustrate an individual-based approach to understanding cultural similarities and differences in the borderless world. Within the context of Web communication, a typology of individual cultural value orientations is proposed. This conceptualization emphasizes the need for making distinctions first at the individual level, before group-level comparisons are meaningful, in order to grasp the complexity of today's global culture. The empirical study reported here further demonstrates the usefulness of this approach by successfully identifying 16 groups among American Web users as postulated in the proposed typology. Future research should follow the implications provided in this chapter in order to broaden our thinking about the role of culture in a world of global communication.

INTRODUCTION

As the adoption of media technology such as the Internet rapidly spreads around the world, communication across cultures increases. Individuals from diverse cultural groups interact with each other regardless of physical distances. On the one hand, such increased communication between cultures might facilitate cultural convergence on the global scale (Kincaid, 1988; Rogers & Kincaid, 1981). On the other hand, online technology's capability to offer individualized communication might further fragment the global culture as people with similar values, outlooks, and interests across the world pursue their personal agendas via the decentralized electronic media (Choi &

Danowski, 2002).

Culture has been a focal issue in global communication. More specifically, cultural similarities and differences have been considered the key to understanding cross-cultural human interactions. Extensive research to date has provided ample evidence of differences between cultures in terms of communication styles and messages. Implicit to this line of research is the assumption that members of a culture are likely to exhibit a pattern of social perception and behavior common within the culture, but different from that of another culture. Given this paradigm of conceptualizing culture, most cross-cultural comparisons are made at the national or cultural level, that is, between nations or cultures, while overlooking potential variations among individuals within a culture.

In today's fast-changing media environment, people are exposed to various cultures through a multitude of channels and formats. While still adhering to the dominant values of the culture in which they belong, people these days rely on multiple frames of cultural reference simultaneously to construct their individual cultural orientations. For these reasons, it would be too simplistic to assume that everyone in the same culture displays the same pattern of thinking and behavior. In fact, individuals' cultural orientations within the same culture could vary widely (Campbell, 2000). Therefore, a thoughtful investigation of today's technology-mediated global culture needs to start from exploring fundamental cultural value orientations at the individual level.

Foremost among the major dimensions of cultural orientations are individualism and collectivism. Generally considered as polar opposites of each other, individualism emphasizes the concept of self, whereas collectivism focuses on other-directedness. Departing from this dichotomous view, recent research has suggested a more in-depth conceptualization of individualism and collectivism where, depending on whether equality (horizontal) or hierarchy (vertical) is underscored, the following four types of orientations can be identified: (1) *horizontal individualism (uniqueness)*, where one can be unique and independent while still maintain status equality with others; (2) *vertical individualism (achievement)*, where one strives to be the best and enjoys privileges that come with it; (3) *horizontal collectivism (cooperativeness)*, where interdependence and equality in status are valued; and (4) *vertical collectivism (dutifulness)*, where people submit to the social hierarchy ascribed by their in-groups (Triandis, 1995, 2001; Triandis & Suh, 2002). Initial empirical evidence supports the viability of this four-way typology in detecting differences across national cultures (Nelson & Shavitt, 2002) and individual differences within a single culture (Lee & Choi, 2005).

Understanding similarities and differences in cultural orientations is the key for successful global online communication. Since the Web has emerged as an ideal medium for tailored communication for people across the world, it is imperative to obtain a baseline understanding of cultural values held by those who are users of the Web. As cultures increasingly interconnect on the Web and national borders gradually vanish, these insights will help prepare us for a future world community that is likely to be dominated by technology-mediated communication. At this juncture, research on cultural values in global communication should focus on the individual, not the nation or culture. Therefore, based on the aforementioned four-way typology, the goal of this chapter is to propose and empirically demonstrate a comprehensive classification framework for assessing cultural orientations at the individual level.

To accomplish the goal, this chapter first explains individualism and collectivism as dimensions of culture, and reviews relevant research developments in this area. Then, a thorough explication of an in-depth typology that encompasses sufficient complexity to reflect cultural differences among individuals is provided. Next, this chapter reports results from an empirical study that classified and compared Web users in the U.S., using the proposed individual level typology. This chapter concludes with implications of the findings and directions for future research.

THE CONCEPTUALIZATION OF INDIVIDUALISM AND COLLECTIVISM

The constructs of individualism and collectivism have been widely regarded as instrumental in helping explain differences between cultures. This can be seen from the vast amount of literature employing these ideas to account for differences in communication patterns and content, business practice, and preferences for communication styles and persuasive message appeals (Cho, Kwan, Gentry, Jun, & Kropp, 1999; de Mooij, 1998; Hall, 1984; Han & Shavitt, 1994; Hofstede, 1983, 1984; Miracle, Chang, & Taylor, 1992). The following sections provide a review of the defining characteristics of individualism and collectivism, and the conceptual advancements in this area.

Individualism and Collectivism

From an analysis of survey data collected from more than 50 countries around the world, Hofstede (1980) identified individualism and collectivism as one of the several fundamental dimensions of culture. He further demonstrated how these constructs can be characterized in people's social perception and behavior. In individualistic cultures, typified as autonomous and independent, people's personal goals are usually valued over the goals of their in-groups. Hence, people's behaviors are primarily based on their own attitudes rather than the norms of their in-groups. In contrast, people in collectivistic cultures are interdependent within their in-groups, and the goals of their in-groups take priority over their own. Collectivists generally behave according to the norms of their in-groups, and place much emphasis on group harmony and hierarchy. In short, collectivists tend to do what is expected of them whereas individualists tend to do what they find personally fulfilling (Triandis, 1995).

Hofstede's (1980, 1984) one-dimensional conceptualization places individualism and collectivism at the opposite ends of a continuum. A majority of the focus in research since then has been placed on explaining national differences using these constructs (Gudykunst & Ting-Toomey, 1988). In applications, comparisons are made by defining nations as residing at one or the other of those two extremes, or between them. Given today's global environment and frequent border-crossing of people and ideas, however, the notion of a homogeneous population within a culture and the nation-based view of those constructs and comparisons may no longer hold true (Singlis & Brown, 1995). Put simply, not every person in an individualistic culture is an individualist. Nor does it mean that people in a collectivistic culture are all collectivists.

Research evidence over the years further suggests that the dichotomous conceptualization of individualism and collectivism in a bipolar manner is limited. There is a great need to expand the conceptualization from uni- to multidimensional in order to capture the complexity of cultural orientations and offer a comprehensive understanding (Singlis, Triandis, Bhawuk, & Gelfand, 1995; Triandis & Gelfand, 1998).

Horizontal / Vertical Individualism / Collectivism

Expanding on the above view, Triandis (1995, 2001) suggested that there are, in fact, different types of individualism and collectivism. Upon careful examination, for example, Korean collectivism is not entirely the same as the collectivism among the Japanese. The individualism in France is different from American individualism. Among the many dimensions that can further distinguish individualism and collectivism is the horizontal-vertical aspect of social relationships. In essence, both individualism and collectivism may be *horizontal* (emphasizing equality) or *vertical* (emphasizing hierarchy). Underneath the horizontal orientation is the assumption that people see themselves as being essentially similar to others in their social relationships. On the contrary, the vertical dimension highlights hierarchy as the key to social relationships where people perceive themselves as being different from others. Adhering largely to Shintoism, the Japanese are highly egalitarian while exhibiting a very strong sense of cooperation. In contrast, social hierarchy is the guiding principle for attitude and behavior in Korea. Korean people have a tendency to value family and group hierarchy, and sacrifice their personal goals for group goals. Research has further shown that some individualistic cultures such as France and Sweden place more emphasis on equality by focusing on "doing one's own things," whereas other individualistic cultures such as the U.S. tend to emphasize hierarchy by embracing superiority.

From this conceptualization, four distinct types of cultures can be identified: (1) *horizontal individualism (HI-uniqueness)*, where people strive to do their own thing and be unique; (2) *vertical individualism (VI-achievement)*, where people strive to be distinguished and the best in competition with others; (3) *horizontal collectivism (HC-cooperativeness)*, where people merge their selves with their in-group and underscore interdependence, harmony, and common goals with others; and (4) *vertical collectivism (VC-dutifulness)*, where people submit to the authorities of the in-group and are willing to sacrifice themselves for their in-group (Triandis, 2001; Triandis & Suh, 2002). Although all individualistic people share the tendency of being independent and giving more priority to personal goals over group goals, in HI, people have little interest in acquiring high status, unlike those in VI. Likewise, in HC, people respect group goals, but do not simply give in to authorities, much different from those in VC.

While this four-way typology was suggested to help identify distinct prototypes of cultures, it also shows promise for an in-depth understanding of within-culture variations (Lee & Choi, 2005). At the individual level, however, the four types are often orthogonal and not mutually exclusive. In other words, people can and may exhibit a number of combinations of these cultural orientations. Given today's global environment where cultures are increasingly fused via a variety of media technology, it is important to recognize the growing need for a more in-depth typology that captures the intricacy of cultural values.

Idiocentrism and Allocentrism

Shifting the focus from the group to the individual, the terms "idiocentrism" and "allocentrism" have been put forth to refer to personal individualism and collectivism, respectively (Yamaguchi, Kuhlman, & Sugimori, 1995). From this conceptual standpoint, a person's cultural orientation is not automatically equated with his or her cultural or national membership. At the individual level of analysis, people's cultural values may no longer coincide with the culture in which they belong. That is, there are idiocentrics (those who possess individualistic characteristics) in collectivistic cultures and allocentrics (those who show collectivistic orientations) in individualistic cultures (Triandis, McCusker, & Hui, 1990). In other words, both idiocentrics and allocentrics can and should be found in any given culture. However, the proportions of the two groups might vary given the dominant values at the national or cultural level (Triandis & Suh, 2002).

Based on the horizontal/vertical individualism/collectivism conceptualization and the need for individual level of understanding, a more thorough classification of cultural orientations is outlined in the following section.

CLASSIFYING INDIVIDUAL CULTURAL ORIENTATIONS

Research evidence to date has demonstrated the utility of Triandis' typology. Studies using the four prototypical patterns (HI, VI, HC, VC) to classify cultures generally assume that one particular type among the four prevails as the cultural orientation. In theory, however, people can have a mixture of attributes as defined by these four types simultaneously. Therefore, the original typology might not effectively describe all possible combinations of the various types of cultural orientations that people exhibit. A more comprehensive multidimensional classification, which identifies those pure types documented in the original conceptualization plus the hybrid types, should serve as a useful starting point to capture differences in individual cultural orientations.

Triandis' distinction between horizontal and vertical individualism and collectivism is primarily developed for group-level comparisons. As an extension, the proposed individual-level typology classifies people based on the extent to which they exhibit patterns of attributes along *horizontal and vertical idiocentrism and allocentrism.* Since a person's orientation can be assessed as either high or low on the four different dimensions, this expanded classification, presented in Table 1, identifies 16 possible individual cultural orientation groups. Below is a description of the major groups under this framework.

Horizontal Idiocentrics and Vertical Idiocentrics

Horizontal and vertical idiocentrics are those individuals who exhibit idiocentric tendencies with either a horizontal or vertical orientation. They score high on horizontal or vertical idi-

Table 1. Individual cultural orientation classification

Group	Name	HI	VI	HA	VA
1	Horizontal Idiocentrics	***High***	Low	Low	Low
2	Vertical Idiocentrics	Low	***High***	Low	Low
3	Horizontal Allocentrics	Low	Low	***High***	Low
4	Vertical Allocentrics	Low	Low	Low	***High***
5	All-Around Idiocentrics	***High***	***High***	Low	Low
6	All-Around Allocentrics	Low	Low	***High***	***High***
7	The Horizontals	***High***	Low	***High***	Low
8	The Verticals	Low	***High***	Low	***High***
9	Bi-Idiocentric Allocentrics	Low	***High***	***High***	Low
10	Bi-Idiocentric Allocentrics	***High***	Low	Low	***High***
11	Tri-Idiocentric Allocentrics	Low	***High***	***High***	***High***
12	Tri-Idiocentric Allocentrics	***High***	Low	***High***	***High***
13	Tri-Idiocentric Allocentrics	***High***	***High***	Low	***High***
14	Tri-Idiocentric Allocentrics	***High***	***High***	***High***	Low
15	Null	*High*	*High*	*High*	*High*
16	Null	Low	Low	Low	Low

ocentrism and low on all of the other dimensions. As idiocentrics, these people value the self and independence. However, *horizontal idiocentrics* focus on self-development and self-reliance, whereas *vertical idiocentrics* are competitive and strive to be better than the rest.

Horizontal Allocentrics and Vertical Allocentrics

People who score high on horizontal or vertical allocentrism and low across all of the other dimensions are referred to as horizontal or vertical allocentrics. *Horizontal allocentrics* work toward group harmony and regard relationships among individuals in the group to be more or less equal. In contrast, *vertical allocentrics* put group goals above individual desires and obey social hierarchy closely.

All-Around Idiocentrics and All-Around Allocentrics

Although distinctions have been made between an orientation toward equality (horizontal) vs. hierarchy (vertical), people could still display both simultaneously. Those who are classified as high on both horizontal and vertical dimensions of idiocentrism, but low on both dimensions under allocentrism, may be regarded as *all-around idiocentrics*. With a focus on the self over groups, their social relations encompass tendencies toward hierarchy as well as equality. In other words, all-around idiocentrics keenly compete with others for status and recognition, as well as strive to be themselves and independent of others. Likewise, those who score high on both horizontal and vertical dimensions of allocentrism, but low on both dimensions of idiocentrism, are referred to as *all-around allocentrics*. Since their social relationships evolve around their in-groups both horizontally and vertically, all-around allocentrics strive for group harmony and hierarchy at the same time. For both all-around idiocentrics and allocentrics, the dominance of either the horizontal or vertical dimension could be a function of other factors.

The Horizontals and the Verticals

People who are predominantly horizontal or vertical in their orientation could behave with a mixture of both idiocentric and allocentric tendencies. They are referred to as *the horizontals* or *the verticals*. The horizontals are classified as high on the horizontal dimension under both idiocentrism and allocentrism, but low on both vertical dimensions. The opposite is true for the verticals. It appears that the horizontals and the verticals value equality and hierarchy, respectively, above and beyond their self or group orientation. With their deep-rooted view of hierarchy-centered social relationships, for example, the verticals respect the authorities in their in-group while endeavor to achieve high social status.

Bi-Idiocentric Allocentrics and Tri-Idiocentric Allocentrics

Bi-idiocentric allocentrics represent a portion of the population who simultaneously display attributes from two diametrically different dimensions. This group includes those who are high on vertical idiocentrism and horizontal allocentrism but low on horizontal idiocentrism and vertical allocentrism. In addition, people who score high on horizontal idiocentrism and vertical allocentrism but low on vertical idiocentrism and horizontal allocentrism also qualify as bi-idiocentric allocentrics. In a similar vein, *tri-idiocentric allocentrics* are people who are high on three out of the four dimensions and low on the remaining one. The level of complexity is significantly higher in this tri-idiocentric allocentric group and, consequently, it could be difficult to disentangle their attitudinal and behavioral differences based solely on cultural orientations.

The Null Group

Those who are equally high or low across all four dimensions are classified as the null group. Theoretically, it is plausible that some individuals simply do not exhibit a strong inclination toward any particular orientation. Likewise, some people might display an equally strong tendency toward all four orientations. Consequently, it might be a rather challenging task to comprehend these null groups' cultural orientations due to their even predisposition along different dimensions. In this situation and other similarly complex situations outlined above, the proposed classification should be used in tandem with other cultural constructs in order to shed light on the multifaceted nature of individual cultural orientations.

CULTURAL ORIENTATIONS AND GLOBAL COMMUNICATION ON THE WEB

With the rapid diffusion of technology, the Web is quickly becoming a significant part of people's daily lives. From information gathering to entertainment, from shopping to personal communication, the Web is omnipresent. As the penetration of the Web reaches the general population, the online population in the U.S. becomes demographically diverse (Schlosser, Shavitt, & Kanfer, 1999). Research evidence shows that this trend is occurring worldwide as well. Web users in different parts of the world become similar to each other in terms of their demographic characteristics and general Web use patterns (Chen, Boase, & Wellman, 2002). As a result, people within the U.S. are less similar to each other than they are to others outside the U.S. In other words, there could be more differences within a country than there are between countries.

The Web is a distinctive medium very much characterized by its ability to offer audience-controlled exposure, selectivity, and interactivity (Wolin, Korgaonkar, & Lund, 2002). Due to the unique nature of the Web, personal characteristics of Web users need to be taken into consideration in order to understand their social behaviors. Among the various demographic and psychological characteristics, Web users' cultural orientations should be of prime importance in this endeavor.

In communications research, individualism and collectivism have served as a useful means to compare similarities and differences in styles and content across cultures (de Mooij, 1998; Hofstede, 1980, 1983). In addition, individualism and collectivism appear to be related to the distinction between low vs. high context. Generally speaking, low-context communication, which is usually straightforward, explicit, and direct, is common in individualistic cultures, whereas communication in collectivistic cultures is highly context dependent, typified by abstract, implicit, and indirect messages (Hall, 1976, 1984).

These constructs of cultural values offer viable means for examining differences in computer-mediated communication between cultures (e.g., Callahan, 2005; Wurtz, 2005). However, as the Web continues to facilitate intercultural flows of ideas and communication between people from all over the world in the forms of message boards, e-mails, newsgroups, and so forth, the need for a thorough investigation of cultural values at the individual level becomes more pertinent (Hewling, 2005). As Scollon and Wong-Scollon (2001) state, "Cultures do not talk to each other; individuals do" (p. 138). Differences in individual cultural orientations might be the most critical factor in understanding Web users from diverse cultures. In this light, the individual-level typology proposed in this chapter should serve as a useful tool for further scrutinizing cultural orientations and classifying Web users.

THE STUDY

A Web survey with a sample drawn from an online panel of consumers in the U.S. was undertaken in October 2003 to provide empirical evidence for the proposed individual-level typology. The method and results of the study are reported in this section.

Method

Sample. Participants were recruited from an established online consumer panel consisting of Web users with diverse demographic characteristics. Those Web users participate in Web-based studies at regular intervals over a period of time for various rewards. The demographics of the online panel track well with the latest online population trend figures—consisting of predominantly female, young, non-Hispanic white, higher education and household income Web users (Pew Internet & American Life Project, 2005). Of the original 1,101 surveys completed, a total of 1,033 were included in the sample after eliminating incomplete surveys.

More than half of the respondents were female (58.5%). The average age of the respondents was 40 years old with a range from 23 to 80 years. Caucasians constituted the majority of the study participants (79.3%), followed by Hispanics (8.2%) and Asians (5.6%). Over half of the respondents were married (54.5%), while one-third (30.5%) were single. The respondents were relatively well-educated, with the vast majority of them having a college degree or higher (98.9%). In terms of economic status, 70.8% of the respondents were employed full time, and about half of the respondents reported an annual household income of $50,000 or higher. Table 2 provides a description of the sample characteristics.

Data collection procedure. An invitation e-mail announcement was sent to members of the online panel. The invitation e-mail contained a brief description of the study and directed panel members to the study site. All those who participated in the study were eligible to win a drawing of a $150 cash prize.

Measures. The questionnaire consisted of four main sections. In the first part of the questionnaire, respondents were asked to indicate how much time on an average weekday they used television, radio, newspapers, magazines, and the Web. Next, respondents' cultural orientations were gauged via the four-way typology measures developed by Triandis (1995). As the scale, originally developed for group-level measurements, was deemed equally suitable for individual-level assessments (Lee & Choi, 2005), the measures were adopted without modifications. Each of the four dimensions was measured via a four-item, seven-point, Likert-type scale: (1) horizontal idiocentrism—HI, (2) vertical idiocentrism—VI, (3) horizontal allocentrism—HA, and (4) vertical allocentrism—VA. Measures tapping the cultural orientations included statements such as "I'd rather depend on myself than others" (HI), "It is important that I do my job better than others" (VI), "If a coworker gets a prize, I would feel proud" (HA), and "Parents and children must stay together as much as possible" (VA).

The third section of the survey examined respondents' experience with the Web. An eight-item, seven-point, Likert-type scale asked for respondents' beliefs about their Web skills and their perceived degree of challenges in their experience using the Web (Novak, Hoffman, & Yung, 2000). Respondents' demographic characteristics such as gender, age, employment, annual household income, ethnicity, highest education level attained, and marital status were obtained toward the end of the survey. The specific items for the major constructs and their respective reliability coefficients are shown in Table 3.

Table 2. Sample characteristics

Characteristic		Frequency	Percent*
Gender	Male	422	58.5%
	Female	604	40.9%
Age	20-29	260	25.2%
	30-39	341	33.0%
	40-49	151	14.6%
	50-59	167	16.2%
	Over 60	104	10.1%
Employment	Full-time	727	70.4%
	Part-time	113	10.9%
	Unemployed	187	18.1%
Ethnicity	Caucasian	814	78.8%
	Hispanic-American	84	8.1%
	Asian-American	57	5.5%
	African-American	14	1.4%
	Native American	3	0.3%
	Multiracial	20	1.9%
	International	22	2.1%
	Other	13	1.3%
Marital Status	Single	313	30.3%
	Married	560	54.2%
	Divorced	59	5.7%
	Living with someone	75	7.3%
	Separated	6	0.6%
	Widowed	11	1.1%
	Other	3	0.3%
Education	Vocational/technical school (2 yrs)	1	0.1%
	Some college	10	1.0%
	College graduate (4 yrs)	579	56.1%
	Master's degree	283	27.4%
	Doctoral degree	46	4.5%
	Professional degree (MD, JD, etc.)	112	10.8%
Household Income	Under $10,000	24	2.3%
	$10,000-$19,999	25	2.4%
	$20,000-$29,999	51	4.9%
	$30,000-$39,999	105	10.2%
	$40,000-$49,999	120	11.6%
	$50,000-$74,999	173	16.7%
	$75,000-$99,999	159	15.4%
	Over $100,000	309	29.9%
	Other	44	4.3%

*Note: * The base of the percentages was the total sample size of 1033.*

Table 3. Specific items for the key measures

Cultural Dimensions
Horizontal Idiocentrism (α = .64)
I'd rather depend on myself than others.
I rely on myself most of the time; I rarely rely on others.
I often do "my own thing."
My personal identity, independent of others, is very important to me.
Vertical Idiocentrism (α = .66)
It is important that I do my job better than others.
Winning is everything.
Competition is the law of nature.
When another person does better than I do, I get tense and aroused.
Horizontal Allocentrism (α = .69)
If a coworker gets a prize, I would feel proud.
The well-being of my coworkers is important to me.
To me, pleasure is spending time with others.*
I feel good when I cooperate with others.
Vertical Allocentrism (α = .64)
Parents and children must stay together as much as possible.
It is my duty to take care of my family, even when I have to sacrifice what I want.
Family members should stick together, no matter what sacrifices are required.
It is important to me that I respect the decisions made by my groups.*

Web Experience
Web Skills (α = .87)
I am extremely skilled at using the Web.
I consider myself knowledgeable about good search techniques on the Web.
I know somewhat less about using the Web than most users. (R)
I know how to find what I am looking for on the Web.
Web Challenges (α = .66)
Using the Web does not challenge me. (R)
Using the Web challenges me to perform to the best of my ability.
Using the Web provides a good test of my skills.
I find that using the Web stretches my capabilities to my limits.

Results

Horizontal/vertical idiocentrism/allocentrism scale validation for individual-level differentiations. Because the group-level measures of horizontal and vertical individualism and collectivism have been previously validated for individual level of analysis (Lee & Choi, 2005), detailed discussion on the horizontal/vertical idiocentrism/allocentrism scale validation is omitted here. In brief, all of the items significantly loaded on the constructs that they were intended to tap into, although two items with a factor loading below 0.4 were dropped from further analysis. Table 4 reports the factor loadings of the indicators for each latent variable and the goodness-of-fit indices of the measurement model. Items for each construct were averaged to form an index score. The descriptive statistics of the four cultural dimensions are shown in Table 5, and the correlations among the constructs are reported in Table 6.

Table 4. Factor loadings of indicators

Factors	Indicators	Unstd.	Std.
Horizontal Idiocentrism	I'd rather depend on myself than others.	1.00	0.72
	I rely on myself most of the time; I rarely rely on others.	1.30	0.72
	I often do "my own thing."	0.74	0.40
	My personal identity, independent of others, is very important to me.	0.54	0.40
Vertical Idiocentrism	It is important that I do my job better than others.	1.00	0.59
	Winning is everything.	1.42	0.68
	Competition is the law of nature.	1.04	0.52
	When another person does better than I do, I get tense and aroused.	1.07	0.53
Horizontal Allocentrism	If a coworker gets a prize, I would feel proud.	1.00	0.65
	The well-being of my coworkers is important to me.	1.03	0.76
	I feel good when I cooperate with others.	0.70	0.58
Vertical Allocentrism	Parents and children must stay together as much as possible.	1.00	0.66
	It is my duty to take care of my family, even when I have to sacrifice what I want.	0.68	0.57
	Family members should stick together, no matter what sacrifices are required.	0.99	0.62

Note: All coefficients are significant ($p < .001$).

Goodness of fit statistics: χ^2 (71) = 561.07, $p < .001$, GFI = .93, AGFI = .89, CFI = .83, RMSEA = .08

Table 5. Descriptive statistics of cultural dimensions

Variables	M	SD
Horizontal Idiocentrism	5.39	0.92
Vertical Idiocentrism	4.03	1.05
Horizontal Allocentrism	5.72	0.84
Vertical Allocentrism	5.32	1.03

Note: All items were measured on a seven-point scale (N=1008).

Table 6. Covariance and correlation matrix of the cultural dimensions

	HI	VI	HC	VC
Horizontal Idiocentrism	1.00	0.33**	0.01	0.03
Vertical Idiocentrism	0.52	1.00	-0.19**	0.09*
Horizontal Allocentrism	0.01	-0.32	1.00	0.29**
Vertical Allocentrism	0.04	0.13	0.39	1.00

*Note: * $p < .01$, ** $p < .001$*

Correlations are in the lower triangle and covariances in the upper triangle.

Table 7. Number of people in cultural orientation groups

Group	Name	HI	VI	HA	VA	No. of Respondents
1	Horizontal Idiocentrics	*High*	Low	Low	Low	39
2	Vertical Idiocentrics	Low	*High*	Low	Low	29
3	Horizontal Allocentrics	Low	Low	*High*	Low	48
4	Vertical Allocentrics	Low	Low	Low	*High*	56
5	All-Around Idiocentrics	*High*	*High*	Low	Low	62
6	All-Around Allocentrics	Low	Low	*High*	*High*	102
7	The Horizontals	*High*	Low	*High*	Low	43
8	The Verticals	Low	*High*	Low	*High*	46
9	Bi-Idiocentric Allocentrics	Low	*High*	*High*	Low	10
10	Bi-Idiocentric Allocentrics	*High*	Low	Low	*High*	37
11	Tri-Idiocentric Allocentrics	Low	*High*	*High*	*High*	30
12	Tri-Idiocentric Allocentrics	*High*	Low	*High*	*High*	72
13	Tri-Idiocentric Allocentrics	*High*	*High*	Low	*High*	69
14	Tri-Idiocentric Allocentrics	*High*	*High*	*High*	Low	25
15	Null	*High*	*High*	*High*	*High*	68
16	Null	Low	Low	Low	Low	78

Note: (N=814 Caucasians)

Horizontal/vertical idiocentrism/allocentrism classification. Given the ethnic diversity in the American culture, differences in cultural orientations among different ethnic groups were examined first. While significant differences between respondents with different ethnic backgrounds were indeed observed, the relatively small sizes of several ethnic groups did not allow for meaningful comparisons across groups in further investigation. Furthermore, an ethnically homogeneous group of Web users was deemed appropriate for the purpose of assessing the utility of the individual-level typology for detecting within-culture variations. For further classification, therefore, the sample included a single majority ethnic group, 814 Caucasian respondents, after eliminating other small ethnic groups. The Web users in the sample were first divided into high vs. low groups on each of the four cultural elements using a median split. The median scores were 5.5 (HI), 4.0 (VI), 5.7 (HA), and 5.3 (VA) respectively. Using the proposed typology, these Web users were then classified into one of the 16 cultural groups depending on their locations along the four cultural dimensions.

Table 7 reports the number of respondents classified as belonging in each of the groups. In support of the viability of the individual-level typology for detecting within-culture variations, all 16 types of cultural orientations were represented and successfully identified among members of the single culture tested here. Surprisingly, among the 16 groups, *all-around allocentrics* appeared to be the largest cultural group with 102 people. The next major groups identified included two *null* and two *tri-idiocentric allocentric* groups, with the number of members ranging from 68 to 78. While 62 *all-around idiocentrics* comprised the sixth largest group, two other types of allocentrics followed as the next largest groups, with 56 people and 48 people classified as *vertical allocentrics* and *horizontal allocentrics*, respectively. Furthermore, fewer people were categorized as *horizontal idiocentrics* (39) or *vertical idiocentrics* (29) than *the verticals* (46) or *the horizontals* (43). Taken together, these results show that, contrary to the literature in which the U.S. is constantly characterized as a predominantly individualistic culture at the national culture level, collectivistic tendencies were commonly observed in the cultural orientations among American Web users at the individual level.

Table 8. Time spent with media on an average weekday

	Television	Radio	Newspapers	Magazines	The Web
Do not use on a daily basis	38 (4.7%)	43 (5.3%)	227 (27.9%)	205 (23.8%)	1 (0.6%)
30 min.	68 (8.4%)	209 (25.7%)	320 (39.3%)	415 (54.7%)	84 (12.2%)
1 hour	87 (10.7%)	167 (20.5%)	153 (18.8%)	109 (8.7%)	120 (15.7%)
1 hr. 30 min.	68 (8.4%)	76 (9.3%)	33 (4.1%)	14 (1.2%)	78 (7.6%)
2 hours	155 (19.0%)	95 (11.7%)	33 (4.1%)	40 (6.4%)	145 (16.3%)
2 hr 30 min.	47 (5.8%)	31 (3.8%)	7 (0.9%)	0 (0.0%)	31 (2.3%)
3 hours	117 (14.4%)	36 (4.4%)	18 (2.2%)	16 (2.3%)	87 (11.6%)
3 hr 30 min.	21 (2.6%)	6 (0.7%)	0 (0.0%)	1 (0.0%)	11 (0.6%)
4 hours	82 (10.1%)	31 (3.8%)	7 (0.9%)	2 (0.6%)	65 (7.6%)
4 hr 30 min.	11 (1.4%)	5 (0.6%)	1 (0.1%)	1 (0.0%)	8 (1.7%)
5 hours	26 (3.2%)	23 (2.8%)	4 (0.5%)	0 (0.0%)	23 (2.3%)
More than 5 hr.	90 (11.1%)	87 (10.7%)	6 (0.7%)	5 (1.2%)	155 (20.3%)

Note: The base of the percentages was the sample size of 814 Caucasians.

Of note was that there appeared to be a large number of people who exhibited both idiocentric and allocentric orientations, belonging in *bi-idiocentric allocentrics* or *tri-idiocentric allocentrics.* Consistent with previous theoretical discussion on the potential influence of the global culture in the Web environment, the Web users examined in this study indeed fused a combination of cultural values across the vertical/horizontal and the idiocentric/allocentric dimensions to guide their social perception and behavior. In summary, within a seemingly homogenous cultural group, the diversity of individuals' cultural orientations was observed using the individual-level typology.

Characteristics across cultural groups. In the next phase of the analysis, media use, Web experience, and demographic characteristics of the people classified into the different cultural groups were examined. When asked to indicate the amount of time spent on an average weekday watching television, reading magazines and newspapers, listening to the radio, and using the Web, a great majority of the respondents reported using all of the media on a daily basis. The media use patterns appeared to be relatively consistent across the groups, and no significant differences between the groups were observed. Not surprisingly, the Web was the most popular media, with about 19% of the respondents using it for more than five hours a day, followed by television, radio, newspapers, and magazines. Table 8 reports respondents' time spent with each of the media. Additionally, respondents' perceived skills and challenges pertaining to the Web were assessed. Respondents' self-judged Web skills were quite

Table 9. Demographic characteristics of cultural orientation groups

No.	Group Name	Gender		Employment			Marital Status		
		Female	*Male*	*Full-time*	*Part-time*	*Unemployed*	*Single*	*Married*	*Others*
1	Horizontal Idiocentrics	24	15	25	3	10	14	14	11
2	Vertical Idiocentrics	13	15	27	1	1	10	15	4
3	Horizontal Allocentrics	31	17	31	9	8	5	34	9
4	Vertical Allocentrics	33	23	38	**10**	8	11	**42**	1
5	All-Around Idiocentrics	41	21	**52**	5	3	**29**	21	12
6	All-Around Allocentrics	**64**	**36**	**61**	**13**	**28**	11	**76**	**15**
7	The Horizontals	30	13	32	3	8	11	21	11
8	The Verticals	21	25	35	4	7	8	37	1
9	Bi-Idiocentric Allocentrics	5	5	6	2	2	3	7	0
10	Bi-Idiocentric Allocentrics	23	14	22	5	10	8	26	3
11	Tri-Idiocentric Allocentrics	12	18	22	3	5	3	20	7
12	Tri-Idiocentric Allocentrics	**48**	24	48	8	**16**	19	36	**17**
13	Tri-Idiocentric Allocentrics	24	**45**	50	4	**14**	**26**	36	6
14	Tri-Idiocentric Allocentrics	20	5	19	4	2	8	10	7
15	Null	37	**30**	46	**10**	12	19	36	12
16	Null	**53**	25	**59**	9	9	**23**	**39**	**16**
	Total	479	331	573	93	143	208	470	133

Note: The three highest numbers per each group are highlighted in bold type.

high with an average rating of 5.85 on a seven-point scale, whereas they perceived the Web as not very challenging, displaying a mean score of 2.90.

As summarized in Table 9, people from the different cultural orientation groups showed similar demographic characteristics as well. Across almost all of the demographic categories, *all-around allocentrics* were identified as the major type of cultural orientation. This is similar to the overall rankings of the groups. One exception was that single people were mostly *all-around idiocentrics,* whereas the two most common cultural orientations among married people were *all-around allocentrics* and *vertical allocentrics.* Of additional interest was that among those who worked full time, *all-around idiocentrics* were identified as the second frequent type, following *all-around allocentrics.*

SUMMARY

Understanding culture is central to global communication because it serves as a meaningful platform that helps articulate communication needs and delivery for people around the world. As the penetration of the Web increases and technology-mediated communication proliferates, cultural gaps between countries and regions are often said to become narrower. Yet, multiplicity, as opposed to uniformity, of cultural values is still commonly observed. The decentralized and individualized nature of today's media technology might have resulted in further fragmentation of people's cultural orientations, since reinforcements from others with similar views and preferences could be just a click away. Perhaps cultural convergence takes place between like-minded people across nations, whereas divergence of cultural values might be witnessed among people within a country. A thoughtful investigation is warranted to unravel the dynamic role of communication technology in culture change.

In today's global environment, geographical perimeters are blurred and people are exposed to many different cultures through various means. Global trends, growing communication between cultures, and shifting frames of cultural reference make the scrutiny of individual cultural orientations a pressing issue. Based on recent conceptual developments of individualism and collectivism, an in-depth typology classifying individual cultural orientations is proposed in this chapter. Altogether, 16 individual groups are delineated. This theoretical classification was tested with a sample of American Web users where all 16 groups were identified. While appearing homogeneous on the surface with the same national membership and ethnicity, as well as similar income and education levels, Web users in the study showed a wide assortment of cultural orientations as postulated by the classification typology. This empirical evidence provides the much-needed impetus for future research using a cultural value-based approach to understanding users of technology-mediated communication around the globe.

Three general theoretical issues are evidenced from the proposed classification and empirical findings. First, the simple fact that the individual-level typology was successfully implemented suggests that, in today's highly interconnected world, it is necessary to start assessing cultural level constructs at the individual level. Other cultural value dimensions such as power distance, uncertainty avoidance, and long-term orientation might also benefit from this approach.

Second, given today's global environment, people with complex cultural orientations are not only common but should be expected. This can be seen from the identification and prevalence of the horizontals, the verticals, and the bi and tri-idiocentric allocentrics in the study. These groups were theoretically derived and empirically verified. This observation further illustrates the importance of an individual-level approach to understanding the role of culture in global communication. Finally, because of the complex mixing of cultural value orientations in some groups, it might be necessary that the proposed classification be further explored in conjunction with other constructs such as communication context dependency and situational variations, to name but a few. Among the multitude of cultural values characterizing the groups, for example, a certain dimension might become particularly pertinent depending on the context. This endeavor will help propel valuable progress toward a comprehensive framework of culture and global communication.

Certainly, the individual cultural value orientation approach does not negate the need for group-level comparisons. In fact, group-level comparisons are of practical concern because operationally they provide effective means for the differentiation and implementation of cross-country communication, both in terms of content and format. After applying the proposed classification to individuals, comparisons of dissimilar groups within and similar groups across countries or cultures are possible. The traditional national or cultural borders are no longer an issue of concern in this conceptualization. Future research should follow the implications provided in this chapter in order to broaden our thinking about the role of culture in a world of borderless communication.

REFERENCES

Callahan, E. (2005). Cultural similarities and differences in the design of university Web sites. *Journal of Computer-Mediated Communication, 11*(1). Retrieved February 15, 2006, from http://jcmc.indiana.edu/vol11/issue1/callahan.html

Campbell, A. (2000). Cultural identity as a social construct. *Intercultural Education, 11*(1), 31-39.

Chen, W., Boase, J., & Wellman, B. (2002). The global villagers: Comparing Internet users and uses around the world. In B. Wellman & C. Haythornthwaite (Eds.), *The Internet in everyday life* (pp. 74-113). Malden, MA: Blackwell.

Cho, B., Kwan, U., Gentry, J. W., Jun, S., & Kropp, F. (1999). Cultural values reflected in theme and execution: A comparative study of U.S. and Korean television commercials. *Journal of Advertising, 28*(4), 59-73.

Choi, J. H., & Danowski, J. (2002). Making a global community on the Net—global village or global metropolis?: A network analysis of Usenet newsgroups. *Journal of Computer-Mediated Communication, 7*(3). Retrieved December 9, 2005, from http://jcmc.indiana.edu/vol7/issue3/choi.html#theories

de Mooij, M. (1998). *Global marketing and advertising: Understanding cultural paradoxes.* Thousand Oaks, CA: Sage.

Gudykunst, W. B., & Ting-Toomey, S. (1988). *Culture and interpersonal communication.* Newbury Park, CA: Sage.

Hall, E. T. (1976). *Beyond culture.* Garden City, NY: Doubleday & Company.

Hall, E. T. (1984). *The dance of life.* Garden City, NY: Doubleday/Anchor.

Han, S.-P., & Shavitt, S. (1994). Persuasion and culture: Advertising appeals in individualistic and collectivistic cultures. *Journal of Experimental Social Psychology, 30,* 326-350.

Hewling, A. (2005). Culture in the online class: Using message analysis to look beyond nationality-based frames of reference. *Journal of Computer-Mediated Communication, 11*(1). Retrieved December 15, 2005, from http://jcmc.indiana.edu/vol11/issue1/hewling.html

Hofstede, G. (1980). *Culture's consequences.* Beverly Hills, CA: Sage.

Hofstede, G. (1983). Dimensions of national cultures in fifty countries and three regions. In

J. Deregowski, S. Dzirawiec, & R. Annis (Eds.), *Explications in cross-cultural psychology* (pp. 389-407). Lisse, The Netherlands: Swets and Zeitlinger.

Hofstede, G. (1984). *Culture's consequences: International differences in work-related values.* Beverly Hills, CA: Sage.

Kincaid, D. L. (1988). The convergence theory and intercultural communication. In Y. Y. Kim & W. B. Gudykunst (Eds.), *Theories in intercultural communication* (pp. 280-298). Newbury Park, CA: Sage.

Lee, W., & Choi, S. M. (2005). The role of horizontal and vertical individualism and collectivism in online consumers' response toward persuasive communication on the Web. *Journal of Computer-Mediated Communication, 11*(1). Retrieved December 15, 2005, from http://jcmc.indiana.edu/vol11/issue1/wnlee.html

Miracle, G. E., Chang, K. Y., & Taylor, C. R. (1992). Culture and advertising executives: A comparison of selected characteristics of Korean and U.S. television commercials. *International Marketing Review, 9*(4), 5-17.

Nelson, M. R., & Shavitt, S. (2002). Horizontal and vertical individualism and achievement values: A multimethod examination of Denmark and the United States. *Journal of Cross-Cultural Psychology, 33*(5), 439-458.

Novak, T. P., Hoffman, D. L., & Yung, Y.-F. (2000). Measuring the customer experience in online environments: A structural modeling approach. *Marketing Science, 19*(1), 22-42.

Pew Internet & American Life Project. (2005). *Who's online: Demographics of Internet users* (February-March 2005 tracking survey). Retrieved July 29, 2005, from http://www.pewinternet.org/trends/User_Demo_05.18.05.htm

Rogers, E. M., & Kincaid, D. L. (1981). *Communication networks: Toward a new paradigm for research.* New York: The Free Press.

Schlosser, A. E., Shavitt, S., & Kanfer, A. (1999). Survey of Internet users' attitudes toward Internet advertising. *Journal of Interactive Marketing, 13*(3), 34-54.

Scollon, R., & Wong-Scollon, S. (2001). *Intercultural communication* (2nd ed.). Oxford: Blackwell.

Singlis, T. M., & Brown, W. J. (1995). Culture, self, and collectivist communication: Linking culture to individual behavior. *Human Communication Research, 21,* 354-389.

Singlis, T. M., Triandis, H. C., Bhawuk, D., & Gelfand, M. J. (1995). Horizontal and vertical dimensions of individualism and collectivism: A theoretical and measurement refinement. *Cross-Cultural Research, 29,* 240-275.

Triandis, H. C. (1995). *Individualism and collectivism.* Boulder, CO: Westview.

Triandis, H. C. (2001). Individualism-collectivism and personality. *Journal of Personality, 69,* 907-924.

Triandis, H. C., & Gelfand, M. J. (1998). Converging measurements of horizontal and vertical individualism and collectivism. *Journal of Personality and Social Psychology, 74,* 118-128.

Triandis, H. C., & Suh, E. M. (2002). Cultural influences on personality. *Annual Review of Psychology, 53,* 133-160.

Triandis, H. C., McCusker, C., & Hui, C. H. (1990). Multimethod probes of individualism and collectivism. *Journal of Personality and Social Psychology, 59*(5), 1006-1020.

Wolin, L. D., Korgaonkar, P., & Lund, D. (2002). Beliefs, attitudes and behavior towards Web advertising. *International Journal of Advertising, 21*(1), 87-113.

Wurtz, E. (2005). A cross-cultural analysis of Websites from high-context cultures and low-context cultures. *Journal of Computer-Mediated Communication, 11*(1). Retrieved December 15, 2005, from http://jcmc.indiana.edu/vol11/issue1/wuertz.html

Yamaguchi, S., Kuhlman, D. M., & Sugimori, S. (1995). Personality correlates of allocentric tendencies in individualist and collectivist cultures. *Journal of Cross-Cultural Psychology, 26*, 658-672.

This work was previously published in Linguistic and Cultural Online Communication Issues in the Global Age, edited by K. St.Amant, pp. 45-62, copyright 2007 by Information Science Reference, formerly known as Idea Group Reference (an imprint of IGI Global).

Chapter XVII
mCity:
User Focused Development of Mobile Services Within the City of Stockholm

Annette Hallin
Royal Institute of Technology (KTH), Sweden

Kristina Lundevall
The City of Stockholm, Sweden

ABSTRACT

This chapter presents the mCity Project, a project owned by the City of Stockholm, aiming at creating user-friendly mobile services in collaboration with businesses. Starting from the end-users' perspective, mCity focuses on how to satisfy existing needs in the community, initiating test pilots within a wide range of areas, from health care and education, to tourism and business. The lesson learned is that user focus creates involvement among end users and leads to the development of sustainable systems that are actually used after they have been implemented. This is naturally vital input not only to municipalities and governments but also for the IT/telecom industry at large. Using the knowledge from mCity, the authors suggest a new, broader definition of "m-government" which focuses on mobile people rather than mobile technology.

INTRODUCTION

All over the world, ICT technologies are used to an increasing extent within the public sector. For cities, ICTs not only provide the possibilities of improving the efficiency among its employees and its service towards tourists, citizens, and companies; it is also an important factor in the development of the city and its region, as ICTs today generally are considered to constitute the driving force of economy and social change (Castells, 1997). It is also argued that ICTs can improve efficiency, enhance transparency, control, networking and innovation (Winden, 2003). Thus, several cities are involved in projects concerning the development, testing, and implementation of ICTs. A few examples include Crossroads Copenhagen in Denmark, Testbed Botnia, and TelecomCity from the cities of Luleå and Karlskrona in Sweden. Within all these projects, triple-helix like organizations are used involving the local municipality or national government, the local university, and the locally-based companies (Jazic & Lundevall, 2003)

Also within the City of Stockholm, there is such a project—the mCity Project. This was launched by the City of Stockholm in January of 2002, with the aim of organizing "the mobile city" through the implementation of relevant ICTs. The mCity Project consists of several small pilot projects, focusing on identifying needs in the community and creating solutions to these. In this chapter, we intend to describe this project, its organization, work processes, and the results. We also discuss the experiences made and how the project can serve as an inspiration towards a broader understanding of "m-government".

BRIEFLY ABOUT THE CITY OF STOCKHOLM

The City of Stockholm is Sweden's largest municipality with about 760,000 inhabitants,[1] but is, compared to other capitals in the world, a small city. Due to the Swedish form of government, Stockholm—as well as all other Swedish cities—has large responsibilities, including child care, primary and secondary education, care of the elderly, fire-fighting, city planning, and maintenance, and so forth. All these responsibilities are financed through income taxes, at levels set by the cities themselves, with no national interference. The operational responsibility lies, in the case of Stockholm, on 18 district councils and on 16 special administrations, depending on the issue. Through 15 different fully-owned or majority-interest, joint-stock and associated companies (hereafter called "municipal companies"), the City of Stockholm also provides water, optical fibre-infrastructure, housing (the City of Stockholm has the largest housing corporation in the country), shipping-facilities (the ports in the Stockholm area), parking, tourist information, the city theatre, the Globe Arena (for sports, concerts and other events) etc. In total, the city has an organization comprising 50,000 employees, and a yearly turn over of 31.5 billion SEK,[2] which is equivalent to about 5 million USD. For the City of Stockholm, it is only natural to engage in ICT projects of different kinds, as this could be expected to have both financial and pedagogical benefits within this large organization—just as it had for other public organizations in Sweden (Grenblad, 2003).

In fact, ICT projects are encouraged by the City of Stockholm through the Stockholm "E-Strategy". This is a visionary and strategic document, issued by the City Council[3] in the beginning of 2001 which—among other things—firmly states the role of the citizen as the central figure for all activities in the city organization; the development of mobile technologies to enhance flexibility, as well as the importance of the city acting to aid Swedish ICT industry (*The City of Stockholm's E-Strategy, 19th of February 2001*). It is the City Executive Board[4] which is responsible for implementing the resolution of the City Council, but the "E-Strategy" document also points to the responsibility of the management of the different district councils, special administrations, and municipal companies for the strategic development of ICTs within each organization. The document also describes the function of "the IT Council", which is to ensure that the e-strategy is implemented in a good way within the municipal organization, that is, not as a separate strategy, but in close contact with the activities for which the organizations are responsible.

BACKGROUND, ORGANIZATION, AND GOALS

The idea of mCity was born in 2000 when the former EU Commissioner Martin Bangeman suggested a cooperation between European cities in order to stimulate the use of the upcoming 3G network and its services. In January 2001, a workshop was held with representatives from a number of major cities, telecom operators, vendors, and investors. A project proposal was submitted by Bangeman, suggesting that a few other European cities—Stockholm, Bremen, and Berlin among others—should start a holding corporation in order to develop and sell 3G services. However, this collaboration project did not become a reality. Instead, the City of Stockholm decided to proceed with a smaller scale project—mCity.

The following goals have been specified for the mCity Project in Stockholm:

- **To improve the working environment for the employees of the City of Stockholm.** By putting people in the center and letting them lead the development of mobile services, they will help develop services that will ease their own work tasks and their everyday lives.
- **To increase the quality of services for citizens.** The mCity Project strives to improve the service of the city to its citizens and visitors by improving the work environment for employees and by introducing citizen-specific solutions.
- **To stimulate the regional business (IT/Telecom).** By developing new solutions in collaboration with industry, new opportunities for the ICT industry within Stockholm, and throughout Sweden are developed, thereby creating a strong home market for companies in Stockholm.
- **To reinforce Stockholm's profile as an IT capital.** By developing new and useful mobile services, Stockholm's reputation as a leading IT capital will be further reinforced.
- **To spread the good example.** By working with small-scale test environments and small-scale tests, the results can be duplicated if successful. By involving the end users closely in the project, sustainability is ensured. An effect of more deeply involved users is that the users themselves become spokespersons for the services and actually help spread the word.

During its first year, the project was located in one of Stockholm's district councils, which meant close contact with the end users. The project manager felt, however, that in order to keep up with the ICT development in other parts of the city, the project would be better off if it could be located more centrally in the organization. Since then, the project has been moved closer to the central administrative organization in the city.

The project organization of mCity is described in Figure 1. The Steering Committee, organized with representatives from different parts of the city, for example the IT Department and the City of Stockholm Executive Office,[5] make strategic decisions about budget issues, what projects to initiate, and so forth. Different heads have chaired the Steering Committee during the course of the project. There are also members from the Stockholm IT Council in the Steering Committee, to ensure that the mCity Project follows Stockholm's E-Strategy. The different pilot projects are initiated together with district councils, special administrations, or municipal companies which undertake the responsibility of local project management in each case. The mCity Project Manager is in charge of initiating and setting up the local projects in collaboration with the local project management and then keeps track of the day-to-day development of the projects. He/She is also responsible for collecting and spreading information about the projects, and for preparing the meetings with the Steering Committee as well

Figure 1. Organization of the mCity Project

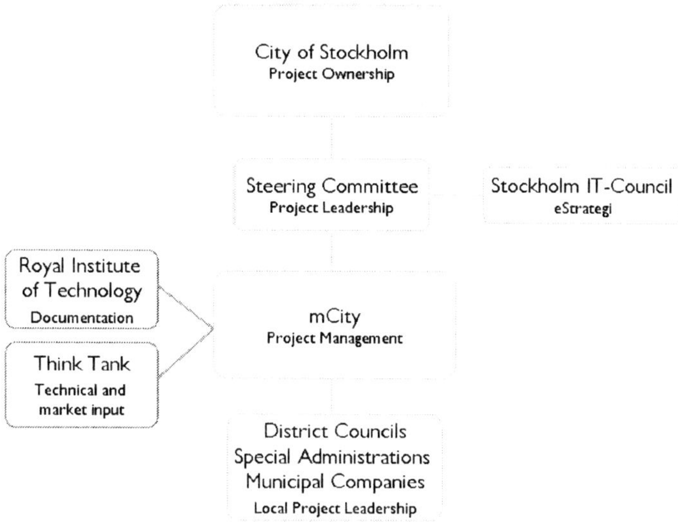

as implementing the decisions of the Steering Committee. In their work, he/she can also use the Think Tank, to which a number of companies within the mobile technology industry belong, to ask for advice concerning technology or market requirements/development. Finally, a researcher from KTH, the Royal Institute of Technology, has been responsible for documenting the project.

WORKING PROCESS

Within the mCity Project, services for both private and public sectors are tested and thereafter developed in a larger scale if proven relevant. The services are operated and tested in "small islands" because it makes it easier to get close to the users and to change the tested services if something needs to be improved. Using this model, mCity has been able to connect groups with specific needs with companies developing mobile services that can satisfy these needs.

End-user needs, that is, the needs of citizens, visitors and employees within the City of Stockholm, form the starting point of every initiative within mCity—see Figure 2. One way of creating situations where users can make their voices heard is by initiating hearings, focus groups, interviews, and so forth. In some cases, the mCity Project Manager has been involved in this first part of the process; in other cases, the local management of the different district councils, special administrations, or municipal companies take the initiative of formulating an application, specifying the need. The exact details of the working process have shifted, depending on the organizational setting of the project.

In the next step, the mCity Project management uses their Competence Network to form a group with technical expertise to which the user's need is presented. The group ponders about the possible technical solutions suitable to solve the problems and in this process, the end-users' knowledge of ICTs, their workload, and the financial/technical situation of the user environment are also taken into account. Depending on the situation, mCity can also contribute financially to the pilot project.

Through mCity, the hope is to accomplish a better every day life for end users. Therefore, the

Figure 2. Working process within mCity

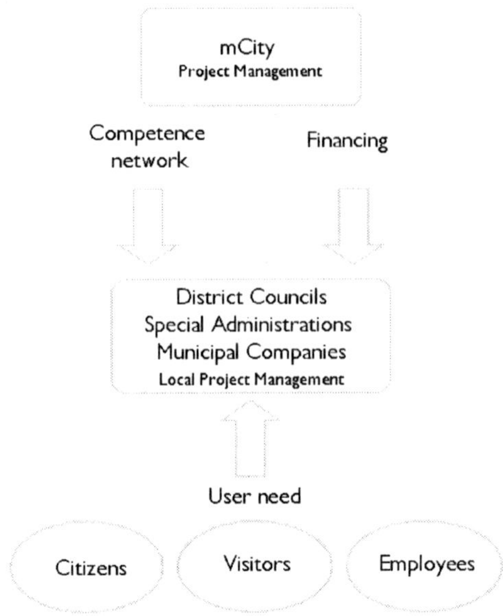

benefits of the services developed in relation to the concrete needs of users, are of high interest and hopefully, it is also possible to measure the added value. End solutions should be easy to use—it should be almost intuitive to understand how to use the provided service. This is one reason for why simple technology is mostly used in mCity Projects—technology is seldom the problem, the focus is rather on what to introduce and how to introduce it. To summarize, the working process can be described in three keywords:

- User-oriented
- Benefit-driven
- Simple

It should be pointed out that mCity primarily does small-scale pilot projects; when these have been launched, it is the responsibility of the district councils, the special administrations, or the municipal companies involved to decide whether to keep running the project, to enlarge it, and also to take the full operational and financial responsibility for the future project.

PILOT PROJECTS

mCity has started and financed several pilot projects since its launch in 2002. Different user groups are in need of different services and the largest segments identified are people who work, live in, and visit the City of Stockholm, as shown earlier in Figure 2. Through the pilot projects, these segments have been further specified, as described in Figure 3: tourists, students, SMEs, commuters, and city employees.

Tourists

The very first project within mCity was carried out in 2002 for tourists, when the official event database owned by Stockholm Visitors' Board[6] was made available via mobile Internet. The city

Figure 3. Focus groups of the mCity Project

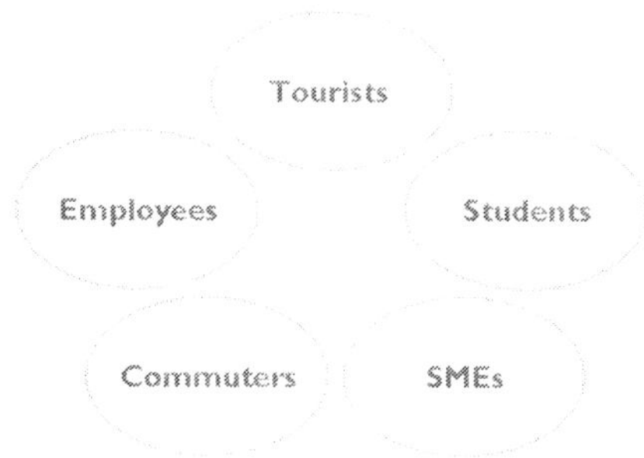

wanted to do this in relation to its 750th anniversary which was to be celebrated that year, and it was decided that something new should be tested, which is why WAP was chosen.

A few years later in 2004, another service targeting tourists was developed by mCity. This time the development process was conducted by a group of talented students taking a project course at the Royal Institute of Technology. This project, *.tourism*, was initiated by the Art Council at the Cultural Administration in order to find new ways of making information about Stockholm available through new technology. The result, a Web site with information on statues, art objects, and buildings of interest is available via mobile or fixed Internet on the address, www.explore.stockholm.se. The server recognizes if the user is accessing the Web site from a PDA, a laptop, or a mobile phone. By using XML functionality, separate interfaces for the different devices are shown, giving the user the best experience possible depending on the device used.

On the Web site, it is possible to search by the name of an object, a location, or a street. It is also possible to list all attractions within a city district. One can also make a guided tour through the Web site, and making this accessible for others to benefit from. Naturally, the personal tours have to be authorized by an administrator in order to filter non-ethic information.

Students

mStudent is a joint project venture between the Federation of Student Unions in Stockholm (SSCO), the Stockholm Academic Forum, and the City of Stockholm within the framework of mCity. The objective is to develop mobile services which are useful to 80,000 students in the Stockholm region. For example, if students can receive an SMS telling them that a lecture has been cancelled, they might not have to come to the university campus at all that day, saving time to be better used for studies or other activities.

During the spring of 2003, 28 students from eight different universities and university-colleges in the Stockholm region participated in a feasibility study to identify a number of services interesting to students. This first phase of the project was carried out together with Telia,[7] Ericsson, and Föreningssparbanken.[8] The objective was to identify mobile services that would be useful to students in their everyday life. In order to really use the most of the students' innovative minds, they were all given one of Ericsson's most modern mobile phones and were allowed to use them without limitations. This made them experts on the available services and also good judges on new services.

Today, mStudent initiates and administrates different forms of tests and evaluations of mobile services in cooperation with businesses in Stockholm. The purpose is to encourage companies and universities to develop and use improved mobile services and thereby increase the quality of service to students as a group. The activities carried out are based on the list of mobile services that the students identified as interesting in the first phase; but apart from this, mStudent has also become a testbed which tests and evaluates all types of mobile services that can be useful for students. The "test pilots" are all students from Stockholm's universities and university colleges, and mStudent gathers the students in focus groups for workshops, evaluations, and other activities. Some companies are already working together with student reference groups in order to gain feedback on their planned services.

SMEs

mCity has been involved in one project aiming toward higher use of mobile services among SMEs.[9] In one of the shopping malls in central Stockholm, Söderhallarna, the stores can use the Internet and mobile technology to communicate, both with customers and the mall administration. The choice of Söderhallarna was not a coincidence. The property is actually owned by the City of Stockholm, and it is of importance to the mall administration to keep up with the technological development to be able to attract stores to the premises.

By working closely with the storeowners and the mall administration, mCity managed not only to improve the internal communication, but also to provide new ways of treating customer relations with the aid of mobile services. For instance, stores can now inform their customers of last-minute offers or arrivals of new products with SMS or e-mail. Also, customers can easier interact with some of the companies. One of the lunch providers receives the orders from their customers via SMS. This increases the probability of preparing the food on time when not having to take orders on the phone. The technology is also used by the Head of Marketing for the mall, in order to create VIP offers to customers, and to communicate with SME owners and other mall staff, such as janitors.

Commuters

Up-to-date traffic information, provided by the City of Stockholm and the Swedish Road Administration among others, is today available on the Internet site, www.trafiken.nu. The information can be reached via WAP and Internet, but more ways of accessing the information have been developed. To make traffic information available regardless of place or time is important since it brings the choice to commute at a given time to the commuter. The commuters can improve their itinerary and choice of transportation based on the information about the current traffic situation.

mCity is involved in several pilot projects within the traffic area all initiated with a pre-study to find out what kind of information commuters are interested in and would benefit from. In one project, mCity has financed the development of the use of dynamic voice to present information available on the Internet site. The synthetic voice starts reading the new information when a commuter calls a special telephone number available from both fixed and mobile telephones. In another project, commuters are able to subscribe to information on specific routes. The commuter submits information about specific time spans during which he/she is interested in knowing about traffic disturbances on a Web page. As soon as something happens on the route of interest on the specific time span, an SMS is sent out with this information.

Employees

mCity has initiated several SMS management systems within the municipal organizations of Stockholm. Even though the technology used often is the most basic one, the impact has been extensive. Three examples of SMS solutions developed within mCity are described in the following sub-sections.

Schools: Absence Management

A few compulsory and upper-secondary schools have been provided with an absence management system. By keying in their social security number and a four-digit code, pupils can report themselves absent into an automated solution provided by the school. The information is then automatically sent as an e-mail or an SMS to the teachers, thereby reducing administrative work. The flow of information between the school and the parents is also improved since parents may receive an SMS when the child skips class or when parents should remember to pack extra clothes for special extracurricular activities.

The Care Sector: Scheduling Services

Within the care sector, scheduling is a time-consuming effort. Now, staff can plan and book time slots through the Internet, and changes can be made by management through SMS. Positive effects with the solution is that staff motivation has increased and the Head of Staff can now work with core activities as the administrative workload is reduced. This solution was tested together with the SMS solution described next.

The Care Sector: Substitute Management

Within the care sector, a group SMS service has been implemented to facilitate substitute management. Instead of trying to reach substitutes through regular phone calls, managers can send SMSs to groups of staff, saving several hours every time. This creates better opportunities for planning, resulting in less stress for care staff and great financial benefits for the City of Stockholm. Also, managers have discovered the possibilities of encouraging staff through group SMS; an occasional "Have a nice weekend!", or the like, is very much appreciated by the staff working in mobile care units, not seeing much of their colleagues and managers when spending much time out in the field.

This SMS system has been so successful that it has now been made available to all employees within the City of Stockholm to use and benefit from. An interesting fact is that as more people are getting the opportunity to use the system, new areas of use are discovered every day by the users themselves.

MCITY EXPERIENCES

Looking at the mCity Project, it is clear that by focusing on and involving users who traditionally are considered underdeveloped within the field of ICTs, mCity reduces the digital divide. Areas like education and the care sector present great potential for municipalities and ICT companies as large savings of time and money can be made when administrative tasks are simplified. Also, by focusing on the areas with largest potential, one can increase average levels of use and knowledge of ICTs in the organization, even if simple technology is used. Thus, even the use of SMS might be an important step toward the use of more advanced mobile services (Williamson & Öst, 2004).

By involving the end user early on, the development process becomes more time consuming. On the other hand, there seems to be a higher chance of successful development and implementation. The involvement of end users in the development of mobile services leads to the appreciation of the users who feel that their experiences are valuable and have real impact. It is important to note that the "end user" is the very person who will use the system in the end, not his or her supervisor or manager. Thus, in small-scale projects, it is often necessary to involve several levels of management, involving the ones who will use the system, the ones who can oversee work processes, as well as the ones who will pay for the system.

It is not always easy to involve people with limited skills and knowledge in technology in projects involving technology. Some people are also more skeptical of changes than others; they may have gone through several organizational changes within a short time span, or might not be interested in revising their working processes at all. This is especially obvious when implementing new technology. Thus, it is important to recognize that technological artefacts are as much social as technological objects, affecting people's way of life as time and space are changing (Brown, 2002; Glimell & Juhlin, 2001; Urry, 2000).

In order to involve the end users, the project must be presented in a way which makes it come across as a project which will lead to obvious changes for the better and not primarily as a technological project. "We're not necessarily positive to technology per se, but we are positive to all new projects and ideas that will improve our work", a manager involved in the SMS project for substitute management stated in an interview (Hallin, 2003).

The information generated through the process also provides the companies involved with valuable input on user behavior and preferences. To engage companies in an m-government project like mCity has been very rewarding for all parties, but even though the pilot projects have been too small to make it necessary to issue invitations to tender, a discussion about the delimitations of working together with the private sector in development projects has taken place within the Steering Committee. This discussion has been similar to the general discussion going on in Sweden, as several public institutions find that the Public Procurement Act makes innovation in the area of public e-services difficult (Grenblad, 2003). In Sweden, there are not many precedents concerning these kinds of simple and quick forms of cooperations between the private and the public sectors. Clear directives as to how and when companies should be involved are needed.

A final lesson from the mCity Project is that simple technology offers great possibilities. mCity has not per se been interested in testing new technology just for the sake of testing new technology; the effects should be real and readily measurable, as described previously. This said, new technological inventions may also be tested and used, as has been the case within the mStudent Project and within the early tourist project. The clue is to always have in mind who is going to use the service. Students are in the forefront when it comes to usage of technology, and tourists also tend to be open minded to use new technology when travelling. Administrators in elderly care or in the school sector might not be as mature in their use of ICTs.

The choice of technology is also often subjected to other types of limitations. When developing new systems based on new technology, you have to be able to answer a lot of questions. One is whether the service should be available for all or just for a small group of people. In the case of mCity, this has been a difficult aspect since all services are tested on a small scale, enlarged when proving relevant. In small-scale environments, technological integration is not really necessary, but when making a service available on a larger scale, it is. In the projects in elderly care and in school administration, this was clearly evident. When making the group SMS project a large-scale implementation, integration to several internal programs was necessary, such as the mail system and the identification portal. This was not impossible, but of course involved more work and thorough consideration.

In a municipality, it is also necessary to consider the cost of implementing new technology. The new services have to deliver lowered cost or some other kind of gain for the city; developing services just for fun or because they are high-tech at the moment is not good enough.

TOWARD A NEW DEFINITION OF M-GOVERNMENT

Is mCity an m-government project? Generally, "m-government" is defined as "a subset of e-government", involving the use of mobile/wireless applications in the public sector, making the public information and services available anytime, anywhere (Lallana, 2004). According to this definition, it could be questioned whether mCity is an m-government project, as there are pilot projects with other goals than the one stated earlier. The mStudent Project, for example, aims at improving the life for students in the Stockholm region by introducing new mobile services from different providers, and in the SME project, small- and middle-sized companies and their customers benefit from the mobile service introduced.

It is clear that the City of Stockholm through the mCity Project takes a broader grip on the task of providing people with the possibility of accessing public information and services, by also taking on a pedagogical role of encouraging people to use ICTs in different areas of city life, and by stimulating the ICT industry to develop new applications as well as rethink old applications. In order to understand this approach we must establish the relationship between the mCity Project and the municipal and national ICT strategies as well as the project's relationship to the vision of Stockholm as an IT capital.

mCity in Relation to Municipal and National Strategy

As described previously, the Stockholm E-Strategy is the policy document according to which the ICT work in the city is done. On its very first page, the document points out that the globalization process inevitably will lead to a new Europe where Stockholm will face tougher competition from other European cities, and that in order to face these challenges, the use of ICTs is an important factor. "IT must help to make Stockholm more attractive by securing the city's long-term goals that Stockholm should continue to be a fine place to live and work in"(*The City of Stockholm's E-Strategy, 19th of February 2001*).

The "E-Strategy" of Stockholm is, on a municipal level, what the "24/7 Agency" is on the national level. The "24/7 Agency" was issued in 2000 by the Swedish government, aiming at extending the public sector's use of ICTs, making services available 24 hours a day, 7 days a week (*The 24/7 Delegation*). The vision entails all parts of the public sector—municipalities, county councils as well as central government—and is the Swedish government's way of trying to cope with expected demographic changes leading to a larger aging population which will demand more of a public administration with fewer employees. At the same time, citizens in general are expected to demand more value for money and a growing internationalization is thought to increase the competitive pressure on public bodies. The development of e-government in Sweden is a way of meeting these challenges (Lund) and the belief of the 24/7 Agency is that the Swedish administrative model, with independently managed central government agencies, is a factor for the success of rapid development of digital applications and e-services (Lundbergh, 2004).

Swedish authorities primarily call for the most appropriate services, not specific technologies. Thus, the name of "the 24/7 Agency" places focus on the time aspects of service-provision—public services should be provided around the clock—not on specific technologies. The question "how" is subordinated, as, "Accessibility, irrespective of time of day and geographical location, may be achieved through a range of established service channels" (Östberg, 2000). Also, the Stockholm E-Strategy is on purpose called the "E-Strategy", and not the "IT-policy", in order to shift "...focus from IT to activities and show [..] how enhanced integration of *electronic services* ('e-services') can develop the municipality's work" (*The City of Stockholm's E-Strategy, 19th of February*

2001). According to this, the E-Strategy does not prescribe certain technologies, but only points at different areas that the city should work with: Internet, information management, mobile technologies (in general), and so forth.

mCity and the Vision of Stockholm as an IT Capital

The mCity Project not only aims at developing technology which make the city available around the clock. It is also a project used to enhance the image of Stockholm as an IT capital; an image based for example on the fact that Ericsson and other major players within the ICT sector have their development offices within the area. According to the Stockholm E-Strategy, IT can play an important role in making Stockholm an attractive city for people to live and work in, and therefore, the city must take an active part in creating business opportunities for ICT companies. One of the goals stated in the E-Strategy is to, "Be one of the most attractive municipalities for relocation, start-up and running of businesses, in competition with the foremost European cities" (*The City of Stockholm's E-Strategy, 19th of February 2001,* p. 14).

Through the mCity Project, the city has given several ICT companies in the Stockholm area the opportunity to test ideas, develop new applications and market themselves in and outside the country—naturally, in compliance with the Public Procurement Act. This strife to encourage local development conveys an entrepreneurial stance which might be perceived as contrasting with the managerial practices of earlier decades which primarily focused on the local provision of services, facilities, and benefits to the population (Harvey, 1989). However, when cities find themselves competing on a global—not only on a national—arena, a new kind of city management develops, involving proactive management of the images of the city as a management tool. Today, city managers are not only active administrators of the traditional areas of responsibility (Czarniawska, 2000) and the "branding" of the city involves much more than producing colorful brochures (Ward, 1998).

mCity and M-Government

As described earlier, the mCity Project aims at creating "the mobile city", as this is thought to be a good place for people to live, work, and spend their holidays in, since mobility means flexibility. But this does not necessarily mean that the project only deals with the development of mobile technologies, which makes information and services of the City of Stockholm available. "Mobile" here does not refer to the technology, but to the people using it, and "the mobile city" is the city where people have the flexibility to do what they want, where and when they choose. The mobile city can be achieved by the city becoming a role model, using mobile technology for its own activities, for example in schools, in homes for the elderly, or through mobile services which give commuters information about traffic, but also by stimulating the use of mobile technology in general, for example by encouraging students in the Stockholm area to ask for and use mobile services.

It is also obvious, that for mCity, the traditional m-government definition is not sufficient, as the city itself is not limited to its municipal organization. As we have showed earlier, the projects within mCity involve cooperation with both national institutions (for example, within the traffic projects), regional institutions (for example, within the mStudent Project) as well as private companies. Thus, rather than focusing on technology or the municipal organization, mCity focuses on people, and to see this project as an m-government project is to broaden the definition of m-goverment from only encompassing the use of mobile/wireless applications in the public sector, making the public information and services available anytime, anywhere. And rather than having the municipal organization as the starting point for its activities, the city, as it is perceived by its citizens, visitors, and employees, is the unit from where the project takes off. Thus, we suggest a new definition of m-government:

A public body which supports the mobility of its people, by providing its services when and where the people need them, and by supporting the development of whatever wireless technologies are needed, and the education of people in these.

THE FUTURE OF M-CITIES

It has been argued that the organizing capacity of a city determines whether the city will be able to develop in a sustainable way, and that the ability to include ICTs is becoming a more important aspect of the organizing capacity of cities (Winden, 2003). This, we believe is true.

Once a small, local initiative, mCity has grown into a project which covers many application areas. Through the project, it has become clear that mobile services can help Stockholm simplify routines, minimize administration, save both time and money, and make life a bit easier for people thus contributing to a better working and living environment by improving the service quality offered by the city. These results further strengthen the notion that the building of m-government is important, probably not only for cities, but for all public bodies. But in order to be successful, a people's perspective has to be adopted and the traditional borders of the public body might have to be challenged. To start with people rather than with technology or with the organization, is an important prerequisite for success.

REFERENCES

The 24/7 Delegation. (No. Dir 2003:81). Retrieved November 19, 2006, from http://www.sou.gov.se/24timmarsdel/PDF/Eng%20version.pdf

Brown, B. (2002). Studying the use of mobile technology. In B. Brown, N. Green, & R. Harper (Eds.), *Wireless world. Social and interactional aspects of the mobile age* (pp. 3-15). London: Springer.

Castells, M. (1997). *The power of identity* (vol. 2). Oxford: Blackwell.

The City of Stockholm's E-Strategy. (2001, February 19). Available through the City of Stockholm +46 (0)8-508 00 000. The Swedish version can be retrieved (last retrieval, November 20, 2006) from http://www.stockholm.se/files/16100-16199/file_16185.pdf

Czarniawska, B. (2000). The European capital of the 2000s: On image construction and modeling. *Corporate Reputation Review, 3,* 202-217.

Glimell, H., & Juhlin, O. (Eds.). (2001). *The social production of technology. On the everyday life with things.* Göteborg: BAS.

Grenblad, D. (2003). *Growth area – E-services in the public sector, analyses of the innovation system in 2003.* Vinnova (The Swedish Agency for Innovation Systems).

Hallin, A. (2003). *Mobile technology and social development – Dialogic spaces in msociety.* EGOS Annual Conference, Copenhagen, July 2-5.

Harvey, D. (1989). From managerialism to entrepreneurialism: The transformation in urban governance in late capitalism. *Geografiska Annaler, 71B*(1), 3-17.

Jazic, A., & Lundevall, K. (2003). *mWatch – A survey on mobile readiness in the Baltic Sea Region.* Presented at the 5th Annual Baltic Development Forum Summit, Riga, Latvia. Retrieved November 20, 2006, from http://www.bdforum.org/download.asp?id=49

Lallana, E. C. (2004). *eGovernment for development, mgovernment definition on and models page.* Retrieved January 13, 2005, from http://

www.e-devexchange.org/eGov/mgovdefn.htm

Lund, G. *The Swedish vision of 24-hour public administration and e-government – Speech by Gunnar Lund, Minister for International Economic Affairs and Financial Markets, held December 9th.* Unpublished manuscript.

Lundbergh, A. (2004). *Infra services – A Swedish way to facilitate public e-services development: MEMO.* The Swedish Agency for Public Management.

Urry, J. (2000). *Sociology beyond societies, mobilities for the twenty-first century.* London & New York: Routledge.

Ward, S. V. (1998). *Selling places. The marketing and promotion of towns and cities 1850-2000.* New York: E & Fn Spon.

Williamson, S., & Öst, F. (2004). *The Swedish telecommunications market 2003.* (No. PTS-ER-2004-24), The Swedish National Post and Telecom Agency.

Winden, W. v. (2003). *Essays on Urban ICT Policies.* Rotterdam: Erasmus University Rotterdam.

Östberg, O. (2000). *The 24/7 agency. Criteria for 24/7 agencies in the networked public administration.* Stockholm: The Swedish Agency for Administrative Development.

RELEVANT WEB SITES

www.stockholm.se/mCity
www.stockholm.se/english/
www.mstudent.se
www.telecomcity.org
www.testplats.com
www.24-timmarsmyndigheten.se
www.pts.se
www.trafiken.nu
www.explore.stockholm.se

ENDNOTES

1 All of Sweden has about nine million inhabitants.

2 2004.

3 The City Council is the supreme decision-making body in the City of Stockholm, consisting of 101 members from the six parties represented in the council, and are elected by the Stockholmers every 4th year.

4 The City Executive Board consists of 13 members, who proportionally represent the parties in the City Council.

5 The Office of the City Executive Board.

6 The municipal company in Stockholm providing service to visitors.

7 The largest telecom operator in Sweden today known as TeliaSonera after a merge with the Finish company Sonera.

8 One of the major bank corporations in Sweden.

9 Small- and middle-sized enterprises.

This work was previously published in End-User Computing: Concepts, Methodologies, Tools, and Applications, edited by S. Clarke, copyright 2008 by Information Science Reference, formerly known as Idea Group Reference (an imprint of IGI Global).

Chapter XVIII
End-User Quality of Experience-Aware Personalized E-Learning

Cristina Hava Muntean
National College of Ireland, Ireland

Gabriel-Miro Muntean
Dublin City University, Ireland

ABSTRACT

Lately, user quality of experience (QoE) during their interaction with a system is a significant factor in the assessment of most systems. However, user QoE is dependent not only on the content served to the users, but also on the performance of the service provided. This chapter describes a novel QoE layer that extends the features of classic adaptive e-learning systems in order to consider delivery performance in the adaptation process and help in providing good user perceived QoE during the learning process. An experimental study compared a classic adaptive e-learning system with one enhanced with the proposed QoE layer. The result analysis compares learner outcome, learning performance, visual quality and usability of the two systems and shows how the QoE layer brings significant benefits to user satisfaction improving the overall learning process.

INTRODUCTION

It is widely acknowledged that e-learners differ in skills, aptitudes and preferences, may have different perceptions of the same factors and some of them may have special needs due to disabilities. People also seek different information when accessing Web-based educational systems and may prefer certain learning styles. Therefore, various adaptive and personalized e-learning systems such as ApeLS (Conlan & Wade, 2004), WINDS (Specht et al., 2002), iClass (O'Keeffe, 2006), INSPIRE (Papanikolaou et al., 2003) and AES-CS (Triantafillou et al., 2002) were proposed in order to capture and analyze these user-related features, and personalize the educational material thus optimizing users' learning experience.

With the latest communication-oriented devices like smart phones, PDAs, laptops and network technologies such as 3G, WiFi, IEEE 802.11 family of standards (IEEE802.11, 1999), WiMax, IEEE 802.16 family (IEEE802.16, 2004), e-learners can access personalized information "anytime and anywhere." However, the network environments allowing this universal access have widely varying performance-related characteristics such as bandwidth, level of congestion, mobility support and cost of transmission.

It is unrealistic to expect that the personalized content delivery quality can be maintained at the same level in this variable environment. Rather an effort must be made to tailor the material served to each person to their operational environment including current network delivery conditions, ensuring high quality of experience (QoE) during the learning process.

QoE focuses on the learner and is considered in (Empirix, 2003) as a collection of all the perception elements of the network and performance relative to users' expectations. The QoE concept applies to any kind of network interaction such as Web navigation, multimedia streaming, voice over IP, etc. Different QoE metrics that assess user experience with the systems in term of responsiveness and availability have been proposed. QoE metrics may involve subjective elements and may be influenced by any sub-system between the service provider and the end-user.

It should be noted that some adaptive e-learning systems have already taken into consideration performance features such as device capabilities, the type of access to the network, download time, etc. in order to improve learning QoE (Chou et al., 2004; Brady et al., 2004; Smyth & Cotter, 2002; Apostolopoulos & Kefala, 2003). However, these account for only a limited range of factors affecting QoE. Also, they were considered separately one from another, unlike the real life situation when there is a simultaneous influence on user interaction with the e-learning systems.

In order to address the effect the complex operational environment has on e-learning, a detailed analysis of the key factors that affect learner QoE was conducted. A ***QoE adaptation layer*** that extends the adaptation features of classic e-learning systems was proposed. It aims to provide high level QoE when users engage in a learning process via network environments with variable connectivity characteristics.

This chapter presents, in details, the proposed QoE layer in the context of a classic architecture for adaptive e-learning systems (AeLS). The most significant AeLS proposed to date are presented in the "Related Work" section that also includes a summarization of the methods most often used in AeLS evaluation. Results of a detailed experimental study that involved a well-known AeLS and a version of the same system enhanced with the proposed QoE layer are then presented. The consequent result analysis compares learner outcome, learning performance, usability and visual quality of the two systems and shows how the QoE layer brings significant benefits to the learning process. The chapter ends with conclusions.

RELATED WORKS

Adaptive E-Learning Systems (AeLS)

Most adaptive e-learning systems are adaptive hypermedia systems (AHS) with applicability in education. In general, AHS aim to help in any application area where the hyperspace is large enough and the system is used by heterogeneous groups of users that have different goals, knowledge, interests, preferences and tasks. Education is one of the major areas of AHS applicability that also includes: online information systems, online help systems, information retrieval, institutional information, and personalized views systems (Brusilovsky, 1996, 2001).

Adaptive e-learning systems (AeLS) in general and mainly Web-based AeLS have attracted considerable interest due to their potential to facilitate personalized learning. They are used by heterogeneous groups of students with different levels of knowledge on a particular subject. The goal of the students is to learn all the material or a reasonable part of it. These systems consider, as the most important feature of the user, the knowledge level of the subject being studied. In order to provide different content to different users and to the same user at different knowledge stages, the system "watches" the students during their learning process.

Before 1996, very few AeLS were developed and mainly in the form of lab systems built to explore some new methods that used adaptivity in an educational context (Brusilovsky, 2001). Examples include a hypertext-based system for teaching the C programming language (Kay & Kummerfeld, 1994), Anatom-Tutor, an intelligent anatomy tutoring system for use at university level (Beaumont, 1994) and ELM-PE, an on-site intelligent learning environment that supports learning of the LISP programming language through examples (Weber & Möllenberg, 1995).

After 1996, with the exponential increase in Internet popularity, the Web started to have an important effect on teaching and learning, mainly in higher education. Many online lecture notes or complex tutoring applications were distributed on the Web. The realization that there is a need to address heterogeneous audience of Web-based courses has led the development of a large number of Web-based AeLS, among which the most important are presented next.

ELM-ART

ELM-ART updated ELM-PE and provided live examples and intelligent diagnoses of problem solutions. Later on, new enhancements were added leading to the ELM-ART II (Weber & Specht, 1997). This system supports online exercises and tests, student-tutor communications via e-mail and student-student discussions via chat rooms. The exercises and tests results allowed the system to assess the student's knowledge more carefully and to infer user's knowledge state. In the next version of the ELM-ART, a multi-layered overlay model was introduced (Weber, 1999). Apart of the knowledge states, now users were able to declare knowledge units as already known. The users could change their associated student model whenever they wanted or switch back to the original state without any loss of information. The system was also extended with two new communication tools: discussion list and user group. The latest version of ELM-ART has been combined with NetCoach, an authoring tool for developing Web-based courses. With NetCoach (Weber et al., 2001), authors can create adaptive Web-based courses that are based on the multi-layered overlay model, that support different types of test items, and include all the communication tools mentioned.

InterBook

InterBook is a system for authoring and delivering adaptive electronic textbooks via the Web (Brusilovsky et al., 1996). It is an environment in which structured textbooks could be presented in a multiply navigable interface. All InterBook-served electronic textbooks have a generated table of content, a glossary, and a search interface. The system uses colored annotations to inform the user about the status of the node referred to by a link.

InterBook stores a domain model of concepts and their structure and an overlay model that helps the system to assess the user's knowledge on different topics and is built based on user-visited pages. These models are used by the system to provide adaptive guidance, adaptive navigation support, and adaptive help. The system also provides different options to the user in the form of direct guidance via the "teach me" button that links the most suitable nodes to be read in the current context. It also includes a glossary index of the concepts.

AHA!

AHA!, developed by the Database and Hypermedia group from Eindhoven University of Technology (De Bra & Calvi, 1998), does not offer support for developing and delivering adaptive courseware only, it is a general-purpose server-side Web-based adaptive system. However AHA! was exemplified and used in education for delivering adaptive university courses at Eindhoven. The first version was developed in 1998, based on the AHAM model (De Bra et al., 1999), and since then the system has undergone several revisions. AHA! includes a domain model, a user-model and an adaptation model. An adaptive engine both performs content and link adaptation and updates information in these models, based on level of user knowledge about concepts. User knowledge is accumulated while the users read pages and take tests. Content adaptation is performed based on the *fragment variants technique.* Unlike InterBook that uses link annotation only, *AHA! link adaptation is performed by using* both link hiding and *link annotation* techniques. The color scheme can be configured by the author and overridden by the user. The user is also allowed to choose between *link annotation* and *link hiding.*

More recently AHA! was enhanced with an authoring tool that implements the principles of LAOS authoring model for adaptive hypermedia systems proposed in (Cristea & de Mooij, 2003).

Since AHA! system is open source, one of the latest introduced and well-known, it was used for the tests presented in his chapter.

INSPIRE

Many researchers are trying to integrate learning styles in the design of their AeLS, along with the classic learner's features such as goals/tasks, knowledge level, background, preferences and interests. INSPIRE is an AeLS that monitors learner's activity and dynamically adapts the generated lessons to accommodate diversity in learner's knowledge state and learning style (Grigoriadou et al., 2001). It emphasizes the fact that learners perceive and process information in very different ways, and integrates ideas from theories of instructional design and learning styles.

With regards to the adaptive dimension of INSPIRE, the selection of the lesson contents and the provided navigation support are both based on the domain model of the system which is represented in three hierarchical levels of knowledge abstraction: learning goals, concepts and educational material (Papanikolaou et al., 2003). The system makes, also, use of a learner model (user model) in order to exploit learners' knowledge level and individual traits (such as its dominant learning style) and to determine the appropriate instructional strategy. This strategy helps in the selection of lessons' contents, the presentation of the educational material, and the annotation of hyperlinks in the domain hyperspace. Several levels of adaptation are supported: from full system-control to full learner-control. It offers learners the option to decide on the level of adaptation of the system by intervening in different stages of the lesson generation process and formulating the lesson contents and presentation.

INSPIRE is used to support a course on Computer Architecture offered by the Informatics and Telecommunications Department, at Athens University, Greece.

JointZone

JointZone (Ng et al., 2002) is a Web-based learning application in Rheumatology for medical students. It combines user modeling, domain modeling and adaptive techniques in order to deliver personalized Web-based learning.

It uses keyword indexing and site layout structure information for domain modeling giving a conceptual and structural representation of the content. This reduces the involvement of a domain expert in organizing and labeling the content.

The content of JointZone exists in the form of an online electronic textbook, which is illustrated with photo images and videos taken on various forms of rheumatic diseases. In an additional section, there are a total of 30 interactive case studies that simulate a variety of rheumatic clinical scenarios where students can actively engage in problem solving rather than being passive recipients of information. The cases are subdivided into three groups designated "Beginner," "Intermediate" and "Advanced." The layout of these cases differs according to the degree of expertise of the user. In JointZone, the user model captures two aspects of the students' differences: individual browsing history and knowledge level in the Rheumatology domain. The model also involves the novel idea of using individual effective reading speed to better identify if a student has read a page. The user's knowledge level is initialized based on his/her first entry registration details. This knowledge value evolves through the user engagement with the application, based on student performance in the case study. The adaptation uses two adaptive techniques: link hiding and link annotation. Based on these techniques and the information from the user model, different personalized features are provided such as: personalized reading room, personalized site map, and personalized topic map.

AeLS Evaluation Methods

The method mostly used in the evaluation of adaptive educational systems adopts a "**with or without adaptation approach**" (Karagiannidis et al., 2001) and considers that the evaluated system can have adaptive and non-adaptive versions. The experiments are conducted between two groups of learners, one working with the adaptive version of the system and the other—with its non-adaptive version. This conventional method of comparing the adaptive and non-adaptive versions of an application is highly debatable as it depends on how the non-adaptive version was obtained. Possibilities may involve an original system to which enhancements were added to obtain the adaptive system, a system resulted from the adaptive system by switching off its enhancements and a system that maintains only some of all adaptive features.

Evaluation Strategy

Looking from **evaluation strategy** point of view, two main directions were taken.

A first approach targets system evaluation **"as a whole"** and is very often used in education (Brusilovsky et al., 2001). The evaluation process focuses mainly on overall learner performance and their satisfaction related to the use of the adaptive system. This user satisfaction can be quantified by selected and measurable criteria. The most used criteria in the evaluation process are: task completion time, learning performance assessed by comparing the results of a pre-test and post-test, number of navigation steps, number of times the subjects revisited "concepts" they were attempting to learn, and user's satisfaction reflected through questionnaires.

Recently, a novel **layered-based** evaluation of adaptive applications was recommended by a number of researchers such as Weibelzahl and Weber (2002) and Elissavet and Economides (2003). Unlike the previous approach, layered evaluation assesses the success of the adaptation by decomposing the system into different layers and evaluating them one by one. The different layers reflect various aspects and stages of the adaptation.

Although the proposed frameworks are described at different levels of granularity, the evaluation process was originally divided in two main phases: evaluation of the interaction assessment phase and evaluation of the adaptation decision-making phase (Karagiannidis et al., 2001):

- **Layer 1: Interaction Assessment Evaluation.** This layer tests if the system detected the learner's goals, knowledge, preferences, interests, and the user's experience with the respect of hyperspace. It also assesses whether the assumption drawn by the system concerning characteristics of the user-computer interaction is valid.
- **Layer 2: Adaptation Making Evaluation.** Layer 2 tests if the selected adaptive technique is appropriate, valid and meaningful for learner's goal or improves interaction for specific learner's interests, knowledge, etc.

The division of the evaluation process into the two layers that also reflect the main phases of the adaptation may help to determine where the fault (if any) of the adaptive system may be and to target the solutions accordingly. For example, it can be the case that adaptation decisions are reasonable but they are based on incorrect system assumptions, or that the system assumptions are correct but the adaptation decision is not meaningful.

A more detailed approach that consists of a four-layer framework for adaptive system evaluation was proposed in Weibelzahl and Weber (2002):

- **Layer 1: Evaluation of the Reliability and Input Data.** This evaluation prevents unreliable input data to result in miss-adaptation.
- **Layer 2: Evaluation of Inference.** This layer evaluation test the inference mechanism in different environments under real world conditions
- **Layer 3: Evaluation of Adaptation Decision**. The idea of the evaluation is that if some user properties have been inferred, several adaptation possibilities exist. (e.g., with/without adaptive guiding, with/without link annotations).
- **Layer 4: Evaluation of Interaction.** In this case human system interaction has to be evaluated to prevent confusion and dissatisfaction of the users. Different objective and subjective measures are taken into account such as: system usability, solution quality, frequency of tasks success, number of required hints, etc.

When assessing AeLS, most often application usability, learner achievement and learning performance are considered. Next, they are discussed in details.

Usability Evaluation Tests

One of the most important features of any software application is its usability. According to ISO 9241 standard, usability represents the effectiveness, efficiency and satisfaction that a software application offers to its users in a given context of use and task. In an educational environment the usability of software application is related to its pedagogical value. Although there is a large amount of knowledge related to educational software usability evaluation strategies, currently there are not well-defined techniques for usability evaluation of e-learning applications (Heines, 2000). This is mainly as e-learning is an area of relatively short history, users of e-learning tools can access them through various computer, network and social contexts and the characteristics of typical users of e-learning services can not be easily predicted.

Some of the most used methods proposed in the literature to be applied during the usability evaluation are: *query techniques* (interviews and questionnaires), *logging of user performance* in laboratory conditions, *timing and keystroke level measurements*, *subjects' observation* through adequate equipment, *heuristic evaluation*, etc. These methods are applied during or after the subjects have interacted with the system when performing one or multiple tasks. Usually the usability is analyzed through five major characteristics: usage efficiency, ease of remembrance, pleasant to use, easy to learn with and number of errors.

Questionnaires and interviews are the most widely used technique since they provide a quantitative measure of usability and they serve as an objective comparison of two systems. This technique offers a concise test of usability, it gets directly the users' viewpoint and attitude and it is suitable for wide range of end-users, especially students. A big advantage is that it does not require the presence of an evaluator. In this context, Preece (2000) suggested a list of guidelines for creating questions for the questionnaires, currently widely used for the usability evaluation of the Web-based systems.

Heuristic evaluation is also a widely accepted method for diagnosing the system's usability due to the fact that it can be completed in a relatively short period of time. This methodology involves an expert that evaluates the system using a set of recognized usability principles, called "heuristics" (Nielsen, 1994).

Learner Achievement Evaluation

Learning process evaluation should include assessment of quality and quantity of learning (learning outcome). Therefore, **learner achievement** (defined as the degree of knowledge accumulation by a person after studying a certain material) continues to be a widely used barometer for determining the utility and value of learning technologies. It is analyzed in the form of *course grades, pre/post-test scores, or standardized test scores.*

A course grade is a certification of competence that should reflect, as accurately as possible, a student's performance in a course. There are multiple methods for assigning grades, such as weighting, distribution gap method, curve, percent grading, relative grading, and absolute standard grading.

Pre/Post test scores are also a viable methodology to assess the extent to which an educational intervention has had an impact on student "learning." Pre-tests and post-tests are used to determine the subject's knowledge prior and after the study, respectively.

Standardized tests scores give a "standard" of measure of students' performance when a large number of students, often geographically distributed, take the same test.

Tests, quizzes or exams are methods used to evaluate students and assess whether they learned what is expected. Jacobs and Chase (1992) made a distinction between the three terms: tests, quizzes and exams, based on the scope of content covered. An examination is the most comprehensive form of testing. A test is more limited in scope, focusing on particular aspects of the course material. A quiz is very limited and usually is administered in fifteen minutes or less.

Among them, tests are most important in the evaluation of AeLS as they offer a feedback in the learning process, helping to optimise it and their results are the most reliable evidence that users have learned.

The evaluation based on tests, quizzes or exams may consist of five different types of test items:

- **Yes-No (True-False):** Users have to answer questions by selecting "Yes" or "No" only.
- **Forced-Choice:** Users have to answer by selecting only one of the alternative answers.
- **Multi-Choice:** Users have to answer by selecting all correct answers provided.
- **Essay (Free-Form)/Short Answer:** Users can type freely an answer to the question. Short answers are one to three paragraphs long.
- **Gap-Filling (Completion):** Users have to type in characters or numbers to complete a word or sentence.

Each type of test item has its relative strengths and weaknesses. Each has also a general value of difficulty and relevance for the tested concept.

Learning Performance Evaluation

Learning performance term refers to how fast a study task (e.g., learning task, searching for a piece of information or memorizing information displayed on the computer screen) takes place. The most used metric for measuring AeLS learning performance is *study session time*. The completion time for a study session is measured from the start of the session, when the subject logs into the system and starts to study, until the subject starts answering the questions from the evaluation test. Other metrics worth mentioning are: *number of navigation steps* performed during a study session, *number of pages re-visited, average time spent per page* for studying the information, and *average access time.*

QoE-AWARE ADAPTIVE E-LEARNING SYSTEM

Figure 1 illustrates the architecture of the proposed **quality of experience-aware adaptive e-learning system** (QoE-AeLS) resulted from the addition of the novel **QoE layer** to the classic AeLS. Apart from the QoE layer—represented in the figure with a different color and presented in more details in the following section, AeLS has four main components: domain model, user model, adaptation model and adaptation engine. These components provide adaptation functionality following the principles presented in the AHAM model (De Bra et al., 1999). Most adaptive e-learning systems based on AHAM include these components.

The **domain model (DM)** organizes the educational material in a hierarchical structure of concepts, among which logical relationships exist. At the lowest level, the concepts correspond to fragments of information. These fragments—stored in a Domain Database—are combined into composite concepts (also called pages) by defining relationships among them. Composite concepts may be further combined using relationships to eventually form more complex units of information. The content is selected from the DM and delivered to the learner based not only on these relationships, but also on learners' characteristics.

The **user model (UM)** maintains and stores in a user database various demographic information

Figure 1. Architecture of the quality of experience-aware adaptive e-learning system

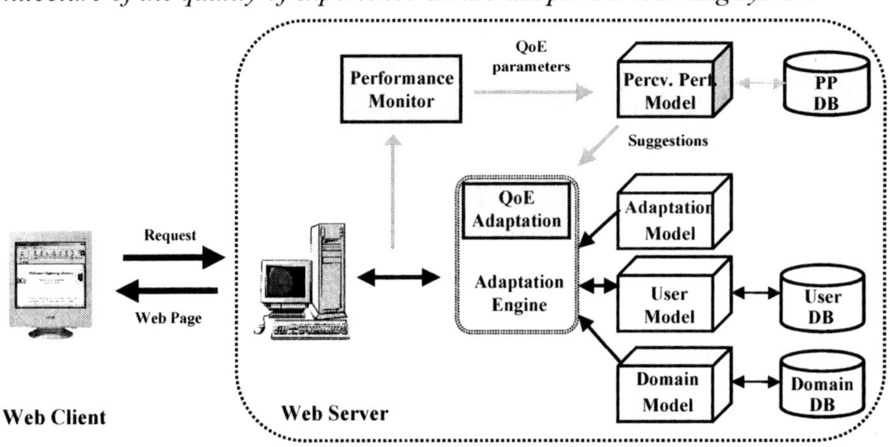

related to the learner (e.g., age, gender, learner's current goal and interest in the educational material), learner's navigational history, etc. Both explicit (via registration) and implicit (through normal navigation and content selection) information is used to generate and update the UM. In order to construct UM, to analyze the user profile and to derive new facts about the user, different user modeling methods have been proposed. The most common ones are the overlay method (De Bra & Calvi, 1998; Pilar da Silva et al., 1998) and the stereotype method (Boyle & Encarnacion, 1994; Murphy & McTear, 1997). Lately, Bayesian networks have become popular for modeling user knowledge and goals and to identify the best actions to be taken under uncertainty (Nejdl & Wolpers, 1999; Conati et al., 1997).

The **adaptation model (AM)** provides the adaptive functionality of the system. The main goal is to define how content adaptation, navigation support adaptation and updates of the UM are performed. Condition-action rules are used to express the adaptation mechanism. These rules combine information from the UM and DM and determine how UM is updated and which information will be delivered to the learner.

The **adaptation engine (AE)** interprets the condition-action rules described in the AM, performs the content selection from the DM, creates the navigational support (links), and delivers a personalized Web page to the learner according to its profile built by the UM.

QoE LAYER FOR QOE-AELS

The proposed *QoE layer* includes the following components: perceived performance model, performance monitor and QoE adaptation unit.

The **performance monitor (PM)** monitors different performance metrics (e.g., download time, round-trip time, throughput, user tolerance for delay) and learner behavior-related actions (e.g., abort and reload requests) in real-time during user navigation and delivers them to the Perceived Performance Model.

The **perceived performance model (PPM)** models this information using a stereotype-based technique, probability and distribution theory, in order to learn about the learner's operational environment characteristics, about changes in the network connectivity and the consequences of these changes on the learner's QoE. The PPM also considers the learner's subjective opinion about their QoE as explicitly expressed by the user. This introduces a degree of subjective assessment, specific to each user. Based on the gathered information, the PPM suggests optimal Web-based educational material characteristics (e.g., the number of embedded objects in the Web page, the dimension of the based-Web page without components and the total dimension of the embedded components) that will provide a satisfactory QoE. The PPM aims to ensure that the download time per delivered page, as perceived by the learner, respects the user tolerance for delay and stays within the user satisfaction zone (Sevcik, 2002; Bhatti et al., 2000; Bouch et al., 2000; Servidge, 1999, Ramsay et al., 1998).

The **QoE adaptation unit (QoEAU)** deploys a QoE adaptation algorithm (Muntean & McManis, 2006a) that uses PPM's content-related suggestions. Its objective is to determine and apply the correct transformations (e.g., modifications in the properties of the embedded images and/or elimination of some of them and placing a link to the image location) to the personalized Web page. This is performed in order to match the PPM suggestions on the Web page characteristics and thus to provide high QoE during learning process.

QoE-AWARE E-LEARNING PROCESS

The main goal of the QoE-AeLS is to provide personalized material that suits both learners' individual characteristics (e.g., goals, knowledge, learning style) and their operational environment in order to ensure high QoE during learning process. Therefore the process of adaptation and personalization of educational material allows for both *user-based* and *QoE-based* adaptations. *User-based* adaptation selects those pieces of information from the DM for inclusion in a learner-tailored document, based on the user profile from the UM. *QoE-based* adaptation is applied when the delivery of the personalized document in given environment characterized by certain connectivity would not provide a satisfactory QoE. Based on the result of PM communication monitoring, PPM makes suggestions and QoEAU applies them in order to increase user QoE.

More details about the proposed QoE-AeLS architecture and QoE-aware adaptation process can be found in Muntean and McManis, 2006a, 2006b, 2006c.

EVALUATION OF THE QoE LAYER

The proposed QoE layer has been assessed through both simulations and qualitative evaluation in the educational area (mainly distance learning), when learners interact with the system in a variable residential-like low bit rate operational environment. Simulation-based testing results that show the benefit of using the QoE layer for delivering content in low-bitrate environments are presented in (Muntean & McManis, 2006a; Muntean et al., 2006). In order to perform qualitative evaluation, the proposed QoE layer was deployed on the open-source AHA! system (AHA, 2006), creating QoEAHA.

The experimental evaluation was performed in the Performance Engineering Laboratory, School of Electronic Engineering, Faculty of Engineering and Computing, at Dublin City University, Ireland. Two sets of task-based scenarios were developed and carried out in laboratory settings. The scenarios were created in order to provide real usage context for participants, as they would interact with the system in real study conditions.

The subjects involved in the tests were randomly divided into two groups. One group used the original AHA! system, whereas the second one used QoEAHA. The subjects were not aware of what version of the system they were using during the experiment. No time limitation was imposed on the execution of the required tasks. None of the students had previously used any of the two versions of the AHA! system and none of them had accessed the test material prior taking the tests. Therefore no previous practice with the environments was assumed for any of them. The material on which the students consisted of the original adaptive tutorial delivered with the AHA! system version 2.0.

Figure 2 graphically presents the network-based setup used for testing. It involved four PC Fujitsu Siemens desktops with single Pentium III (800MHz) processors and 128 MB memory each that acted as clients, an IBM NetFinity 6600 with dual Pentium III (800 MHz) processors and 1 GB memory as Web server and one Fujitsu Siemens desktop computer with Pentium III (800 MHz) processor and 512 MB RAM that acts as a router and has a NISTNET network emulator installed on it. NISTNET that allows for the emulation of various network conditions characterized by certain bandwidth, delay, loss rate and loss pattern was used to create low bit rate residential-like operational environments with bandwidth in the 56 kbps to 128 kbps range. For both groups of subjects, same network conditions were emulated between their computers and the AeLS. These setup conditions offer similar connectivity to that experienced by residential users. These conditions determined performance-related adaptations when QoEAHA version was used.

The goal of the experimental study was to compare the learning outcome and learning performance, system usability, visual quality and user satisfaction when AHA! and QoEAHA systems were used respectively.

Figure 2. Laboratory testing setup

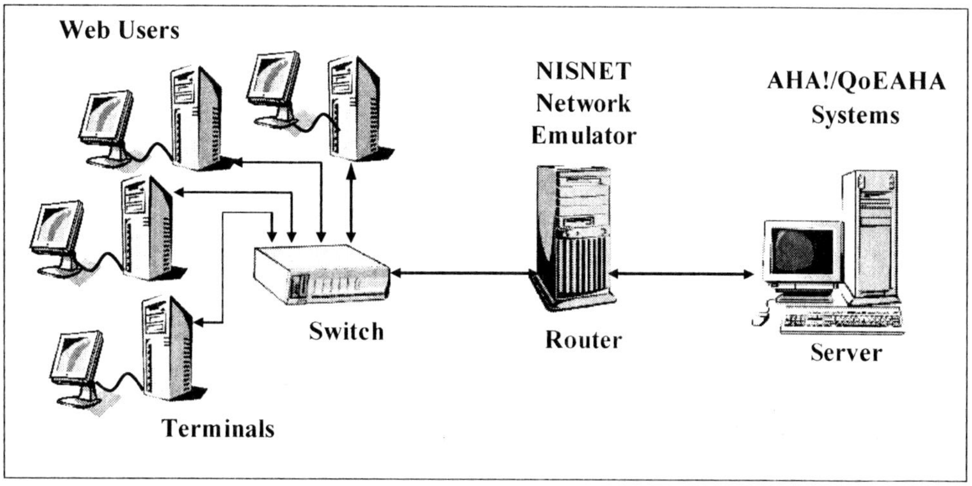

Scenario 1: Interactive Study Session

The first testing scenario covered an interactive study session of one chapter from the adaptive AHA! tutorial and delivered a network with emulated 56 kbps connectivity. This experimental test involved forty-two postgraduate students from the Faculty of Engineering and Computing, at Dublin City University, Ireland as subjects. Before and after the subjects completed the study task, they were asked to take online evaluation tests in order to assess their knowledge levels.

At the start of the study session, the subjects were given a short explanation about the AeLS usage and their required duties. They were asked to perform the following steps:

1. *Complete an online pre-test evaluation* in order to determine subjects' prior knowledge about the studied domain. It consisted of a questionnaire with six questions related to the learning topic.
2. *Log onto the system* and proceed to browse and *study the material.* Back and forward actions through the studied material were permitted.
3. *Complete an online post-test* at the end of the study period in order to determine the subjects' level of knowledge. The post-test consisted of a questionnaire with fifteen questions that tested recollection of facts, terms and concepts from the supplied material. The students were not allowed to return to the studied material.
4. *Answer a usability questionnaire* that assessed system usability and user QoE level. It consisted of ten questions categorized into navigation, accessibility, presentation, perceived performance and subjective feedback.

In order to fully assess the subjects learning outcome, both pre-test and post-test were devised that consisted of a combination of four different types of test-items most commonly used in the educational area: "yes-no," "forced-choice," "multi-choice" and "gap-filling." For time-related reasons, 6 questions were included in the pre-test evaluation as follows: 3 "yes/no," 2 "forced choice" and 1 "multi-choice." The post-test evaluation consisted of 15 questions: 5 "yes/no," 6 "forced choice," 3 "multi choice" and 1 "gap filling." As the test-items have different degrees of difficulty, different corresponding weights in the final score have been assigned for a correct answer as follows: one point for "yes/no" questions, two points for "forced-choice" questions, three points for "multi-choice" questions and four points for "gap-filling" questions. For incorrect answers no points were given.

As the maximum scores were 10 and 30 points for pre-test and post-test respectively, the final scores of both the tests were normalized and were expressed in the 0-10 range.

Scenario 2: Search for Information

The second testing scenario focused on a search for information and involved subjective visual quality assessment in the case with the worse network connection—28kbps. Since the QoE-based adaptation mechanism involves modifications on the properties of the embedded images, the goal was to assess whether the quality of content is good enough to perform the required task. The subjects were asked to look up for two different terms and to answer two questions related to these terms. The terms were described in the embedded images. Objective and subjective visual quality assessments that involved the measurement of the time taken to complete the task and questionnaire-based evaluation techniques were used. The subjective assessment on a five-point quality scale (1-"bad," 2-"poor," 3-"fair," 4-"good," 5-"excellent") ascertained the impact of QoE-based content adaptation on subjects' learning experience.

Scenario 2 involved twenty postgraduate students from the faculty of Engineering and Computing at Dublin City University, Ireland. The goal of these tests was to assess whether the resulted quality of images is good enough for the subjects to be able to perform the required task.

Learning Outcome

Learning outcome was analyzed in terms of pre-test/post-test scores of the two groups after a study session. Table 1 presents both pre-test and post-test scores resulted after the tests were performed according to scenario 1. Pre-test scores (AHA_{mean} = 0.35, $QoEAHA_{mean}$ = 0.30) showed that both groups of students had the same prior knowledge on the studied domain. The mean scores for the post-test were 7.05 for the subjects

Table 1. Scenario 1: Pre-test and post-test results

	Score	Mean	Min	Max	St.Dev
pre-test	AHA!	0.35	0.0	2.0	0.55
	QoEAHA	0.30	0.0	2.0	0.53
post-test	AHA!	6.70	4.30	9.30	1.401
	QoEAHA	7.05	4.60	9.00	1.395

Table 2. Scenario 2: Post-test results (answers found in embedded images)

	Score	Mean	Min	Max	St.Dev
post-test	AHA!	6.40	2.0	10.0	3.25
	QoEAHA	6.30	2.0	10.0	3.15

Table 3. Study time for the learning tasks when AHA! and QoEAHA were used

Study Time (min)	Mean	Min	Max	St.Dev
AHA!	21.23	12.95	31.84	5.90
QoEAHA	17.77	9.37	30.38	5.44

that used QoEAHA and 6.70 for the AHA! group. A two-sample t-test analysis on these values does not indicate a significant difference in the marks received by the two groups of subjects ($\alpha = 0.05$, $t = -0.79$, t-critical = 1.68, p(t) = 0.21).

Since answers for three of questions from the post-test questionnaire have required the subjects to study the images embedded in the Web pages, an analysis of the students' learning outcome on these questions was also performed. After the scores related to these three questions were normalized in the 0 to 10 range, the mean value of the students' scores was 6.3 for the QoEAHA group and 6.4 for AHA! group. More details about these results are presented in Table 2. A two-sample t-test analysis, with equal variance assumed, performed on the two sets of results indicates with a 99% level of confidence that there is no significant difference in the students' learning achievement ($t = -0.08$, t-critical = 2.71, p(t) = 0.93, $\alpha = 0.01$). This result is very important as an adaptive degradation in the image quality (up to 34 % in size) was applied by the QoEAHA.

Therefore, it can be concluded that the addition of the QoE layer does not affect the learning outcome and that QoEAHA offers similar learning capabilities to the classic AHA! system, regardless of the characteristics of the operational environment.

Learning Performance

The impact of the QoE-based content adaptation on the **learning performance** was assessed through the following metrics: *study session time, study time per page* and *number of accesses to a page performed by a person.* These metrics were analyzed and compared for both groups of subjects.

Study Session Time

The distribution of the ***study time*** taken by the students in order to accumulate the information provided during the first scenario using the AHA! and QoEAHA systems respectively is presented in Figure 3. One can notice that on average students that made use of the QoEAHA system (Average Study Time = 17.77 min) have performed better than the ones that used the AHA! (Average Study Time = 21.23 min) (see Table 3). The very large majority of the students that used QoEAHA (71.43%) performed the task in up to 20 minutes with a large number of students (42.87 %) requiring between 15 minutes and 20 minutes of study time. In comparison, when the AHA! system was used, only 42.85% of the students succeeded to finish the learning task in 20 min. The majority of them (71.42%) required up to 25 minutes with the largest number of students (28.57%) in the interval 20-25 minutes.

In Figure 3, one can also notice that 9.5% of the students from group 1 (using QoEAHA) succeeded to learn in less the 10 minutes while none of the students from group 2 (using AHA!) had this performance.

Study Time per Page

In order to assess the results of the comparison between the two AeLS, in terms of ***study time per page,*** two Web pages, out of those stud-

Figure 3. Study time distributions for students involved in a learning task

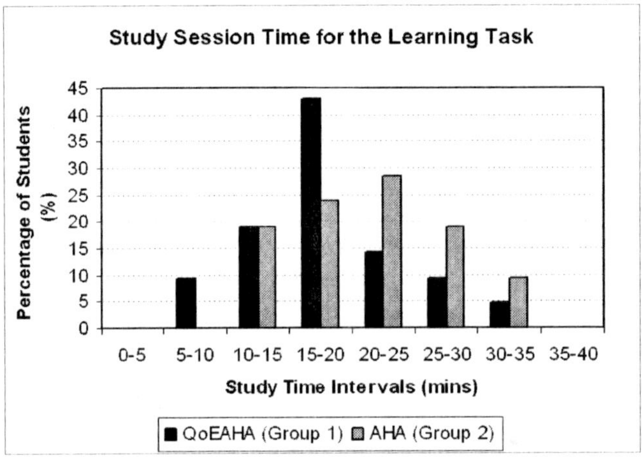

Table 4. Number of accesses per Web page per person during learning session

Number Accesses per Page	Page 1					Page 2				
	Avg	Min	Max	St.dev	σ^2	Avg	Min	Max	St.Dev	σ^2
QoEAHA (Group 1)	1.43	1.0	3.0	0.60	0.4	1.38	1.0	3.0	0.59	0.4
AHA! (Group 2)	1.76	1.0	4.0	0.83	0.7	1.71	1.0	4.0	0.90	0.8

ied by the students as part of scenario 1, were considered. These pages—denoted page 1 and page 2—included a higher number of embedded images and a larger amount of data to be delivered to the learner. Consequently, the subjects perceived long waiting periods when the AHA! system was used. QoEAHA decreased the access time perceived by the students but has also performed some degradation into the quality of the content. Therefore the study time on those pages was analyzed when the two systems were used with the first scenario. *Study time per page* was measured from the moment when the system has received a request for the page until a request for a new page was sent.

The results presented in Table 4 show that on average the students from group 1 (using QoEAHA) spent less time on both page 1 and page 2 for studying the information in these pages than the ones from group 2 (using AHA!). This observation was confirmed by statistical data analysis. By performing a two-sample t-test assuming unequal variances, for each of the two pages it can be said that there is a significant difference between the two groups' means with a confidence level of 95%.

Number of Accesses to a Page

Number of accesses to a page performed by subjects was also measured and analyzed for the same two pages. The average value of this parameter for page 1 was 1.43 when the QoEAHA system was used and 1.76 for the AHA! system, as presented in Table 5. Similar values were obtained for page 2: 1.38 and respectively 1.70. An unpaired two-tailed t-test analysis, with unequal variance assumed, has statistically confirmed with at least 92% confidence that there is a significant difference in the number of visits performed by

Table 5. Study time per person during learning session

Study Time (min)	Page 1					Page 2				
	Avg	Min	Max	St.Dev	σ^2	Avg	Min	Max	St.Dev	σ^2
QoEAHA (Group 1)	4.28	2.09	7.91	1.48	2.2	4.23	2.15	7.7	1.44	2.1
AHA! (Group 2)	5.33	3.69	8.47	1.28	1.7	5.40	3.76	8.23	1.28	1.6

a student to page 1 and page 2 when the two versions of the AHA! are used.

The effect of the version of the AHA! system used by the students had on the number of accesses to a page was investigated by analyzing the variability of the test samples. The results presented in Table 5 show that both standard deviation and variance of group 1 results are lower than the values corresponding to group 2 for both pages. An f-test analysis was performed to determine if variance between the two groups is statistically significant. The results confirm that group 1 and group 2 results do not have the same variance and the difference between the two groups' variances is statistically significant.

It can be noticed that the group 2 results have a higher dispersion than those of group 1. Also a larger number of students (an average of 55%) that used the AHA! system (group 2) required more than one access to page 1 and page 2 for learning. At the same time, a large majority of students (an average of 65%) that used the QoEAHA (group 1) performed only one access to the same pages (See Figures 4 and 5). This shows that QoEAHA users have succeeded to focus better on the studied material. This is due to the fact that the material was delivered faster to the students and the students were constantly focused on their task. Long periods of waiting time for getting access to the material annoy the people and disturb their concentration on the learning task.

Remarks

It can be concluded that important learning performance improvements when the QoEAHA system was used were achieved in comparison when AHA! was used. For example, in tested conditions a 16.3% improvement in the Study Session Time alone was obtained. Since the download time per page provided by QoEAHA does not exceed the user tolerance for delay threshold (12-15 seconds is considered satisfactory for the Web users by the research community; the students were constantly focused on the required task and therefore study time per page decreased by 16.27%. It is noteworthy that most of the QoEAHA group students [71.43%] finished the study in up to 20 minutes whereas only 42.85% of the AHA! students finished in the same period of time. Therefore, the QoE-aware AeLS has ensured a smooth learning process. This observation is also confirmed when assessing Number of Accesses per page [on average 19% decrease with QoEAHA than the result obtained for AHA!]).

Visual Quality Assessment

Results on **visual quality assessment** confirmed that the controlled degradation of the quality of the content performed by the proposed QoE layer did not affect the functionality of the AeLS.

As seen in Figure 6, both groups of students succeeded to complete the "search for information" task presented in scenario 2 in similar periods of time and they answered the questions correctly. The information targeted by the task was

Figure 4. Distribution based on number of visits to Page 1

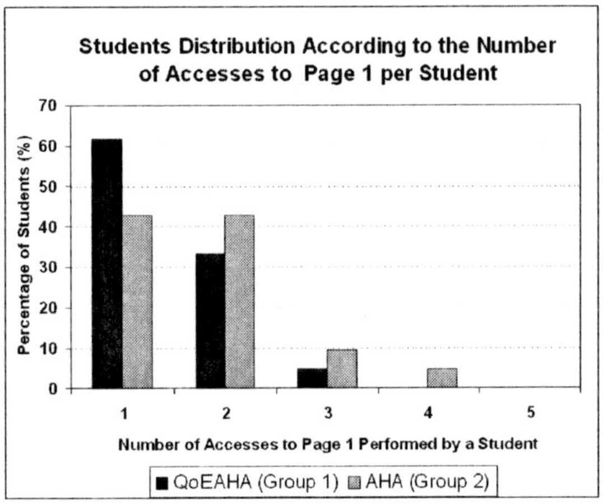

Figure 5. Distribution based on number of visits to Page 2

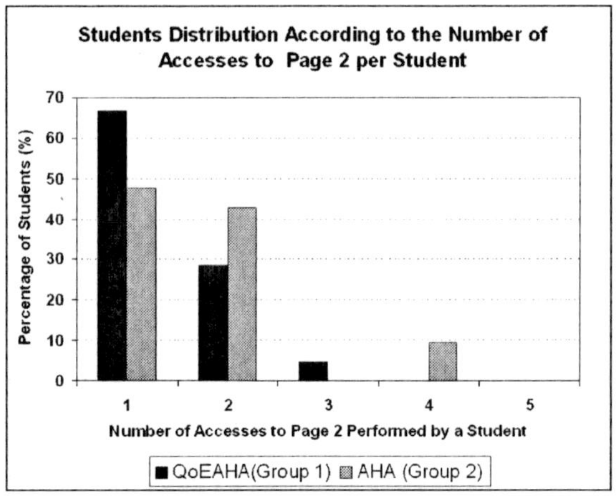

Figure 6. Average completion time for search for a term task (term 1 was located in page 1, whereas term 2 was located in page 2)

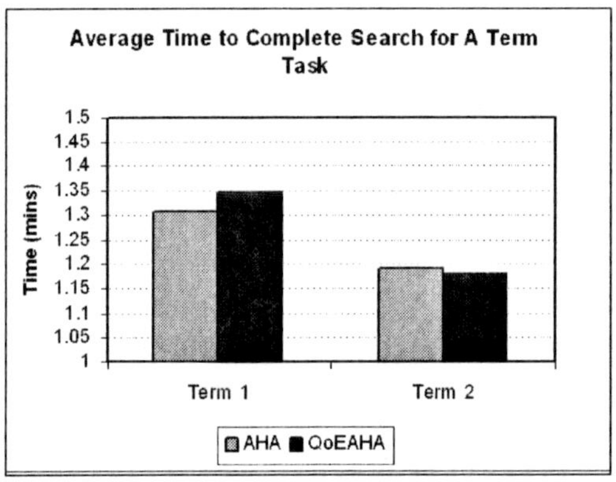

presented in the embedded images of two pages that have the biggest content size and QoEAHA imposed the highest level of image quality degradation as part of its adaptive process. For the worst operational environment case studied (28 kbps connectivity) QoEAHA applied a 57% size reduction to page 1 components and 18% for page 2 items. The subjective-based visual quality assessment investigated through a questionnaire shows that regardless of the high content reduction, the average quality grade given by the subjects to the QoSAHA system was 3.9, very close to "good" perceptual level, and only 4.4% lower than the average quality grade awarded to AHA! (4.3). This suggests that the cost of image quality reduction is not significant as far as user-perceived quality is concerned while at the same time yielding significant improvements in download time and learning performance.

System Usability

The **system usability** investigation was performed using an online questionnaire to which the subjects were asked to respond with grades on a 1-5 Likert scale. It can be noticed from Figure 7 that presents

Figure 7. Average grade comparisons between AHA! and QOEAHA results on the usability questionare

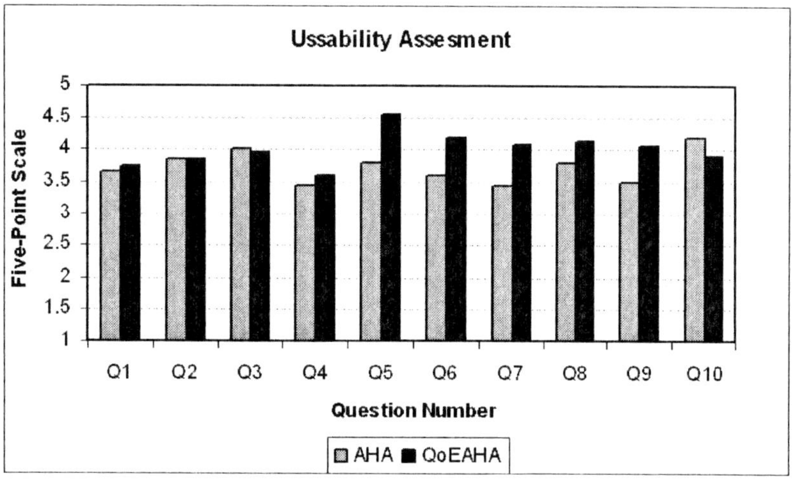

the results to all the questionnaire's questions that the QoEAHA system has provided improving subjects' satisfaction, which was above the "good" level for all QoE-related questions: Q5, Q6, Q7 and Q9. These performance related questions assessed users opinion on the download speed, overall system responsiveness, and performance effect on learning and user satisfaction. The AHA! system scored just above the "average" level on these questions, significantly lower than the QoEAHA! This good performance was obtained in spite of the subjects using slow connection during the study session and not being explicitly informed about this. A two-sample t-test analysis on the results of these four questions confirmed that users' opinion about their QoE is significantly better for QoEAHA than for AHA!, a fact stated with a confidence level above 99%, ($p<0.01$).

Overall, the mean value of QoE usability assessment, assuming that the questions were of equal importance, was 4.22 for QoEAHA and 3.58 for AHA. This leads to an improvement of 17.8% brought by QoEAHA system.

The usability assessment on the other questions related to the navigation and presentation features achieved an average score of 3.83 for AHA! and 3.89 for QoEAHA, demonstrating that these features were not affected by the addition of the QoE enhancements.

Finally, an overall assessment of the all questions from usability questionnaire when all ten questions were considered of equal importance shows that the students considered QoEAHA system (mean value=4.01) significantly more usable then the AHA! system (mean value=3.73). These results were also confirmed by the unpaired two-tailed t-test ($t=2.44, p<0.03$) with a 97% degree of confidence. This increase of 7.5% in the overall QoEAHA usability was mainly achieved due to the higher scores obtained in the questions related to end-user QoE.

By examining in details the provided answers (Figure 7), one can notice that only for the last question (Q10) AHA! has a slight advantage over QoEAHA while in most of the other questions QoEAHA received a higher score. Q10 is related to the user satisfaction with the quality of the provided images. The advantage of the AHA! system is justified by the fact that the QoEAHA system performs controlled image degradation in order to improve the end-user perceived performance. Yet, these image degradations did not disturb the users since they scored this question with an average of 3.9, very close to the "good" level.

CONCLUSION

This chapter describes a novel QoE adaptation layer for AeLS proposed as a solution for increasing end-user QoE. This QoE layer brings significant benefits when the personalized content is delivered to end-users that avail of Web services over various and changeable network conditions, by adapting the content to them.

The QoEAHA evaluation involved a comparison with the AHA! system in home-like low-bite rate operational environments. Different educational-based evaluation techniques such as learner outcome analysis, learning performance assessment, usability survey, and visual quality assessment were used in order to assess QoEAHA in comparison to AHA!. As the students received similar marks on the final evaluation test, regardless of the system used, it can be said that the QoE layer-enhanced system offers similar learning capabilities to the classic one. Results on visual quality assessment confirmed that the controlled degradation of the quality of the content performed by the QoE layer did not affect the functionality of the e-learning system.

At the same time, important learning performance improvements in terms of Study Session Time (16.27% decrease), Study Time per Page (13% decrease) and Number of Accesses to a Page (smaller) were obtained with the QoEAHA system. It is noteworthy that most of the QoEAHA group students (71.43%) finished the study in up to 20 minutes, while only 42.85% of the AHA! group students finished in the same period of time. In terms of system usability, the students thought the QoE layer enhanced system provided much higher user QoE than the classic one. Questions related to the other aspects of the system (e.g., navigation, presentation) achieved similar marks for both systems demonstrating that the QoE layer did not affect them.

In conclusion, the proposed QoE layer brings significant performance benefits to the users that access the adaptive Web content delivered in difficult network conditions.

REFERENCES

AHA! (2006). Retrieved from http://aha.win.tue.nl

Apostolopoulos, T. K. & Kefala, A. (2003). A configurable middleware architecture for deploying e-learning services over diverse communication networks. *IASTED CATE International Conference*. Rhodes, Greece.

Beaumont, I. (1994). User modeling in the Interactive Anatomy Tutoring System ANATOM-TUTOR. *User Models and User Adapted Interaction Journal, 4*(1), 21-45.

Bhatti, N., Bouch, A., & Kuchinsky, A. (2000). Integrating user–perceived quality into Web server design. *Computer Networks Journal, 33*(1-6), 1-16.

Bouch, A., Kuchinsky, A., & Bhatti, N. (2000). Quality is in the eye of the beholder: meeting users' requirements for Internet quality of service. *ACM Conference on Human Factors in Computing Systems* (pp. 297-304). Hague, The Netherlands.

Boyle, C. & Encarnacion, A. O. (1994). MetaDoc: An adaptive hypertext reading system. *User Models and User Adapted Interaction Journal, 4*(1), 1-19.

Brady, A., Conlan, O., & Wade V. (2004). Dynamic composition and personalization of pda-based e-learning—personalized mlearning. *E-Learn'04 Conference* (pp. 234-242). Washington, USA.

Brusilovsky, P. (1996). Methods and techniques of adaptive hypermedia. *User Modeling And User-adapted Interaction Journal, Special Issue: Adaptive Hypertext & Hypermedia, 6*(2-3), 87-129.

Brusilovsky, P., Schwarz, E., & Weber, G. (1996). *ELM-ART: An intelligent tutoring system on World Wide Web.* Intelligent tutoring systems, Berlin, Springer-Verlag, (LNCS 1086, pp. 261-269).

Brusilovsky, P., (2001). Adaptive hypermedia. *User Modeling and User-Adapted Interaction Journal, 11*(1-2), 87-110.

Brusilovsky, P., Karagiannidis, C., & Sampson, D. (2001). The benefits of layered evaluation of adaptive applications and services. *International Conference on User Modeling, Workshop on Empirical Evaluations of Adaptive Systems* (pp. 1-8). Sonthofen, Germany.

Chou, L. D. Wu, C. H., & Lee, S. P. (2004). Position-aware multimedia mobile learning systems in museums. *Web-Based Education Conference* (pp. 416-229). Innsbruck, Austria.

Conati, C., Gertner, A., VanLehn, K., & Druzdzel, M. (1997). On-line student modeling for coached problem solving using Bayesian networks. *International Conference on User Modeling (UM97)* (pp. 231-242). Chia Laguna, Italy.

Conlan, O. & Wade, V. (2004). Evaluation of APeLS—an adaptive e-learning service based on the multi-model, metadata-driven approach. *International Conference on Adaptive Hypermedia & Adaptive Web-based Systems (AH2004)* (pp. 291-295). Eindhoven, Netherlands.

Cristea, A. I. & de Mooij, A. (2003). LAOS: Layered WWW AHS authoring model and their corresponding algebraic operators. *International World Wide Web Conference (WWW'03),* Alternate Track on Education, Budapest, Hungary.

De Bra, P. & Calvi, L. (1998). AHA: A generic adaptive hypermedia system. *ACM HYPERTEXT'98 Conference, Workshop on Adaptive Hypertext and Hypermedia* (pp. 5-12). Pittsburgh, USA.

De Bra, P., Houben, G., & Wu, H. (1999). AHAM: A Dexter-based reference model for adaptive hypermedia. *ACM HYPERTEXT'99 Conference* (pp. 147-156). Germany.

Elissavet, G. & Economides, A. A., (2003). An evaluation instrument for hypermedia courseware. *Journal of International Forum of Educational Technology & Society and IEEE Learning Technology Task Force, 6*(2), 31-44.

Empirix (2003). *Assuring QoE on next generation networks.* White Paper, Retrieved from http://www.empirix.com/Empirix/Corporate/resources/resources+white+papers.html

Grigoriadou, M., Papanikolaou, K., Kornilakis, H., & Magoulas, G. (2001). INSPIRE: An intelligent system for personalized instruction in a remote environment. *Adaptive Hypertext and Hypermedia Workshop.* Sonthofen, Germany.

Heines, J. M. (2000). Evaluating the effect of a course Web site on student performance. *Journal of Computing in Higher Education, 12*(1), 57-83.

IEEE 802.11 (1999). Part 11: Wireless LAN Medium Access Control (MAC) & Physical Layer (PHY) Specifications, IEEE standard 802.11.

IEEE 802.16 (2004). Part 16: Air Interface for Fixed Broadband Wireless Access Systems, IEEE standard 802.16.

Jacobs, L. C. & Chase, C. I., (Eds.). (1992). *Developing and using tests effectively: A guide for faculty.* San Francisco: Jossey-Bass.

Karagiannidis, C., Sampson, D., & Brusilovsky, P. (2001). Layered evaluation of adaptive and personalized educational applications and services. *International Conference on Artificial Intelligence in Education, Workshop on Assessment Methods in Web-based Learning Environments and Adaptive Hypermedia* (pp. 21-29). San Antonio, TX.

Kay, J & Kummerfeld, R. J. (1994). An Individualised course for the C programming language. *International World Wide Web Conference*. Chicago.

Muntean, C. H. & McManis, J. (2006a). Fine grained content-based adaptation mechanism for providing high end-user quality of experience with adaptive hypermedia systems. *W3C International World Wide Web Conference (WWW'06), Hypermedia and Multimedia Track* (pp. 53-62). Edinburgh, UK. New York: ACM Press.

Muntean, C. H. & McManis, J. (2006b). *The value of QoE-based adaptation approach in educational hypermedia: Empirical evaluation*. Springer-Verlag, Berlin (LNCS 4018, pp.121-130).

Muntean, C. H. & McManis, J. (2006c). End-user quality of experience oriented adaptive e-learning system. *Journal of Digital Information, Special Issue on Adaptive Hypermedia, 7*(1). Retrieved from http://journals.tdl.org/jodi/issue/view/29

Muntean, C. H., McManis, J., & Muntean, G.-M. (2006). Improving the performance of content delivery in Web-based information systems.*China-Ireland International Conference on Information and Communications Technology* (pp. 430-435). Hangzhou, China.

Murphy, M. & McTear, M. (1997). Learner modeling for intelligent CALL. In Jameson A., Paris C. & Tasso C. (Eds.) *International Conference on User Modeling (UM97)* (pp. 301-312). Wien Austria: Springer-Verlag.

Nejdl, W. & Wolpers, M. (1999). KBS hyperbook—a data-driven information system on the Web. *W3C International World Wide Web Conference (WWW99)*. Canada.

Ng, M. H., Hall, W., Maier, P., & Armstrong, R. (2002). The application and evaluation of adaptive hypermedia techniques in Web-based medical education. *Association for Learning Technology Journal, 10*(3), 19-40.

Nielsen J. (Ed.). (1994). *Heuristic evaluation. Usability inspection methods*. New York: Wiley.

O'Keeffe, I., Brady, A., Conlan, O., & Wade, V. (2006). Just-in-time generation of pedagogically sound, context sensitive personalized learning experiences. *International Journal on E-Learning (IJeL), Special Issue: Learning Objects in Context, 5*(1), 113-127.

Papanikolaou K. A., Grigoriadou M., Kornilakis H., & Magoulas G. D. (2003). Personalizing the interaction in a Web-based educational hypermedia system: The case of INSPIRE. *User Modeling and User-Adapted Interaction Journal, 13*(3), 213-267.

Pilar da Silva, D., Van Durm, R., Duval, E., & Olivié, H. (1998). Concepts and Documents for Adaptive Educational Hypermedia: A Model and a Prototype. *ACM HYPERTEXT'98 Conference, Workshop on Adaptive Hypertext and Hypermedia* (pp. 35-43). Pittsburgh, USA.

Preece, J. (Ed.). (2000). *Online communities: Designing usability, supporting sociability*. Chichester, UK: John Willey & Sons.

Ramsay, J., Barbasi, A., & Preece, J. (1998). A psychological investigation of long retrieval times on the World Wide Web. *Interacting with Computers Journal*. Elsevier Publishing House.

Servidge, P. (1999). How long is too long to wait for a Web site to load? *Usability News*.

Sevcik, P. J. (2002). Understanding how users view application performance. *Business Communications Review, 32*(7), 8-9.

Smyth, B. & Cotter, P. (2002). Personalized adaptive navigation for mobile portals. *European Conference on Artificial Intelligence* (pp. 608-612). Lyon, France.

Specht, M., Kravcik, M., Klemke, R., Pesin, L., & Huttenhain, R. (2002). Adaptive learning environment in WINDS. *ED-MEDIA'02* (pp. 1846-1851), Denver, CO: AACE Press.

Triantafillou, E., Pomportsis. A, & Georgiadou, E. (2002). AESCS: adaptive educational system base on cognitive styles. *International Conference on Adaptive Hypermedia and Adaptive Web Based Systems (AH'2002), Workshop on Adaptive Systems for Web-Based Education* (pp. 10-20). Malaga, Spain.

Weber, G. & Möllenberg, A. (1995). ELM-Programming-Environment: A tutoring system for LISP beginners, cognition and computer programming. In Wender, K. F., Schmalhofer F. & Böcker, H. D. (Eds.) *Cognition and Computer Programming* (pp. 373-408). Toronto: Ablex Publishing Corporation.

Weber, G., Specht, M. (1997). User modelling and adaptive navigation supporting WWW-based tutoring systems. *International Conference on User Modeling (UM'97)* (pp. 289-300). Sardinia, Italy.

Weber, G. (1999). Adaptive learning systems in the World Wide Web. *International Conference on User Modeling (UM'99)* (pp. 371-378). Banff, Canada.

Weber, G., Kuhl, H. C., & Weibelzahl, S. (2001). Developing adaptive internet based courses with the authoring system NetCoach. *Workshop on Adaptive Hypertext and Hypermedia* (pp. 35-48). Sonthofen, Germany.

Weibelzahl, S. & Weber, G. (2002). Advantages, opportunities, and limits of empirical evaluations: Evaluating adaptive systems. *Künstliche Intelligenz Journal, 3*, 17-20.

This work was previously published in Architecture Solutions for E-Learning Systems, edited by C. Pahl, pp. 154-174, copyright 2008 by Information Science Reference, formerly known as Idea Group Reference (an imprint of IGI Global).

Chapter XIX
High-Tech Meets End-User

Marc Steen
TNO Information & Communication Technology, The Netherlands

ABSTRACT

One challenge within the high-tech sector is to develop products that end users will actually need and will be able to use. One way of trying to match the design of high-tech products to the needs of end users, is to let researchers and designers interact with them via a human-centred design (HCD) approach. One HCD project, in which the author works, is studied. It is shown that the relation between interacting with end users and making design decision is not straightforward or "logical." Gathering knowledge about end users is like making a grasping gesture and reduces their otherness. Making design decisions is not based on rationally applying rules. It is argued that doing HCD is a social process with ethical qualities. A role for management is suggested to organize HCD alternatively to stimulate researchers and designers to explicitly discuss such ethical qualities and to work more reflectively.

HUMAN-CENTRED DESIGN

Many organizations, both private and public, need to or want to innovate. Not for the sake of innovation itself, but in order to create new products, services, or processes that will create added value for their customers, for the end users of their products or services or for citizens. Developing innovations that match end users' needs or wishes is especially (but not exclusively) problematic in the high-tech industry where many innovations are driven by technology push. A risk of technology push is that researchers and designers invent some product or service that nobody needs or nobody can use. One way in which researchers and designers try to match their innovation efforts to end users' needs and wishes is to *interact* with them during an innovation project. They try to learn from them, to be informed or inspired by them. This can be seen as an attempt to narrow the gap between researchers and designers in their high-tech ivory tower vs. end users "out-there."

My current interest is in researchers and designers activities. However, it is acknowledged that their work is only one half of the innovation process: end users also play crucial roles in adoption, domestication, and appropriation processes (Oudshoorn & Pinch, 2003, p. 11-16). Users and technology are "co-constructed" (*ibidem*): people influence technology and technology influences people, both designers *and* end users shape an innovation. However, I am currently more interested in how researchers and designers think and speak about end-users "out-there" (Latour & Woolgar, 1986), rather than being interested in any "real" existence of end-users or their "real" characteristics.

There is a broad variety of methods available for researchers and designers[a] to involve (future, potential, or putative) end users[b] in their projects, for example: participatory design (e.g., Schuler & Namioka, 1993) where people who will be using the system that is being developed are invited to cooperate during development, evaluation, and implementation of that system (such efforts are often related to workers' emancipation); the lead-user approach (e.g., Von Hippel, 2005) where innovative users are seen as a source of innovation and are invited to help develop or improve a product (similar to participatory design, but with less emphasis on emancipation); fieldwork, inspired by ethnography or ethnomethodology, to study the social and cultural aspects of what people do, in order to design applications (often combined with participatory design, for example in the field of computer supported cooperative work) (Crabtree, 2003); contextual design (Beyer & Holzblatt, 1998), a method to observe people doing tasks in their natural context, with attention for their physical surroundings, the artefacts they use as well as their activities, communication, power and culture, and to articulate system requirements based on this; empathic design (e.g., Koskinen, Battarbee, & Mattelmäki, 2003), where researchers or developers try to get closer to end users' lives and experiences, for example by observing their daily life or work, or role-playing some of their activities, and apply what they learn from that in the design process; codesigning (Sanders, 2000), a kind of participatory design where end users make things together with researchers and designers (the focus is on *making* things, and doing that jointly, rather than on *saying* things in interviews, or on being observed *doing* things); and usability engineering, a range of methods to evaluate and improve a product's usability together with end users.

Such approaches are known in more general terms as human-centred design (HCD), which is characterized by four principles: (1) the active involvement of users for a clear understanding of user and task requirements; (2) an appropriate allocation of functions between users and technology; (3) iterations of design and evaluation processes; and (4) a multidisciplinary approach (ISO/IEC, 1999). One goal of HCD is to involve future or potential end users as early as possible: preferably from the start of a project, when problems are articulated, when the design brief is formulated, when directions for searching solutions are chosen so that end users' input can optimally help to steer the project—rather than, for example involving end users at the end of a project for usability testing. Kujala (2003), in a review of various approaches for early user involvement, concludes that early user involvement has "positive effects on both system success and user satisfaction." She is positive about ethnography-inspired field studies as a way to learn about end users' needs and wishes and formulate system requirements (rather than asking them in interviews), and she is critical about whether designers have the necessary skills to conduct and interpret such fieldwork.

STUDYING ONE PROJECT

In many accounts of HCD projects, the relation between interacting with end users and making design decisions is *not* seen as problematic. It seems like taken for granted that such interactions lead to better decisions: "after the observations it was decided to prioritize problem x and make it central in the design process" or "during the workshop solution y was chosen as a basis for further product development." Such phrases seem to cover up what happens in-between interactions with end users and making design decisions. The process of HCD is often presented as a "black box" (Latour, 1987): a container that only shows what goes in (interactions with users) and what goes out (design decisions) and which covers up what happens in-between. However, it is easy to imagine that *something* happens in-between. Some filtering process happens, because researchers and designers do not straightforwardly follow everything that they see and hear from end users, nor do they do nothing with what they see and hear.

My current interest is to open-up the "black box" of HCD. I am interested in how researchers and designers who work in the high-tech ICT industry, for example in research labs, seek interactions with end users, and how this informs or inspires their design process and decision making: how these interactions relate to the process of exploring and articulating of problems and of solutions. Furthermore, I focus on research and design as a *social* process (Bucciarelli, 1994). I focus on the interactions between people, rather than on the technological or economic aspects of research and design.

Such a study requires *access* to researchers and designers doing their work, and therefore I chose to study one project in which I work and of which I coordinate one part, and to conduct the study as "participant observer" (Easterby-Smith, Thorpe, & Lowe, 2002, p. 110-4). This project is sponsored by the Dutch Ministry of Economic Affairs and by the participating organizations, and its goal is to create concepts that will help businesses and organizations to develop and apply innovative information and communication technology (ICT). The project is positioned in the "fuzzy front end" of innovation (Koen et al., 2002). It is about articulating opportunities and creating concepts. This means that the project team members have a relatively large amount of freedom and responsibility concerning what problem to address or what direction to look for solutions—more than in regular product development, implementation, or market introduction projects.

People from different organizations and with various backgrounds work in this project. My study is confined to one part of the project in which we design and evaluate a new kind of telecommunication applications: *we-centric* applications. At the start of the project, there was only this vision of what we wanted to achieve with *we-centric* applications: to stimulate and to facilitate people to communicate and cooperate better with each other, especially among people who currently don't communicate and cooperate enough with each other. During the project we further developed this concept, partly based on interactions with end-users, so that by the end of the project we were able to characterize we-centric telecom applications as follows. A *we-centric* telecom application is typically meant to be used on a mobile phone or *smart phone,* a handheld device that combines phone and basic computer functions, similar to a personal digital assistant. Furthermore, a we-centric telecom application would provide suggestions to communicate or cooperate with others, and has two key functions. These functions were not clear at the start of the project, but they emerged during its course: (1) it composes and presents a dynamic list of potentially relevant people, which is meant as a suggestion to communicate or cooperate with these people—this list is created automatically based upon searching for similarities in data on participants' contexts and interests or preferences; and (2) it presents information about these other people's contexts, which is meant to actually help communicate or cooperate with them—this "context information" may contain their "presence" (one can be "available," "online," "busy," "away," etc., similar to instant messaging) and a short explanation why this person is supposed to be relevant for you currently. Furthermore, we envision that a *we-centric* telecom application is reciprocal: if A receives a suggestion to talk with B, then B receives a similar suggestion about A, and if A can see something of the current activities of B, then B would be able to see something of A's likewise.

The project's goal is to develop and evaluate *we-centric* application together with, and for, two target groups: (1) with/for police officers; and (2) with/for informal carers. These efforts are described in the next two sections.

DESIGNING WITH/FOR POLICE OFFICERS[C]

At the start of this part the project, Albert—who works at an ICT expertise center, which is associated to the police organization and who is a former community police officer—proposed to support and improve the work of community police officers. The work of those kind of police officers consists largely of talking with various people, for example with shop owners, school administrators, people from the municipality, and building corporations as a way to serve citizens and to prevent crime. Their work is rather different from the work of *emergency* police officers, who react to emergencies and who would, for example, drive around with lights flashing. Albert wished to empower community police officers and to create tools for them to better communicate and cooperate with other people for their various tasks. This idea matched the goal to design a *we-centric* application.

Then Mandy, Dirk, and me—we work at two ICT research labs and the three of us are responsible for the design and evaluation of a *we-centric* application—and several fellow project team members accompanied one or more police officers during one working day. We did this to learn hands-on about police work, rather than from documents or from interviews. We made individual notes of their observations and then summarized these in the form of "personas" and "storylines" (e.g., Cooper, 1999). Based on our observations we created three personas, three typical police officers Ad, Bert, and Theo, and described, in the form of storylines, three typical working days of these police officers—see below for an excerpt:

About community police officer Ad
Ad is 45 years old and has been working for 20 years as a community police officer. Ad is married and has two children: a daughter of 18 and a son of 16. Five years ago, Ad and his family moved from Amsterdam to Haretown [in the countryside]. Amsterdam was becoming too hectic for him and he preferred quieter surroundings. [...]

Monday 11:00o'clock.
Ad is walking around in "his" area. He decides to go to the swimming pool to have a quick look how things are. The weather is beautiful and it will probably be very busy in the swimming pool. The last few days a group of teenagers has been causing a lot of trouble. Ad arrives at the swimming pool and looks for his contact person, John. John however appears to have a day off.

We portrayed Ad as relatively old and preferring quiet work, which is typical for some community police officers—so we were told. This is in contrast to a typical emergency police officer, who is younger and enjoys the thrills of emergency police work. Furthermore, we are suggesting that his work may be improved by a *we-centric* telecom application, which would send a notification to Ad that his contact person at the swimming pool is currently absent, so that Ad may decide to visit the swimming pool another time.

We then organized a workshop with the police officers whom we spent a day with, to discuss our observations and interpretations. During that workshop the project's goal shifted from improving communication between police officers and other people *outside* the police organization, to improving communication and cooperation *within* the police organization, mainly because opening-up communication and cooperation towards people outside the police is not desirable from a security perspective. Furthermore, the idea was developed to help community police officers to share their "implicit" knowledge with emergency police officers. Community police officers typically have a lot of knowledge about people and locations in their heads—not everything is made explicit in reports or databases—and such implicit knowledge may be helpful for emergency police officers who often lack such knowledge. A typical "use case" would be this: Emergency police officer Bert is sent to an address because of domestic violence. Without the application, he would enter this situation unprepared, but *with* it he would receive a suggestion to contact community police officer Ad, who was at that address for a similar incident just two days ago and who made some arrangements with these people. If Bert calls Ad, Ad can update Bert, quickly, of course. This idea also builds upon observations of the researchers at the Police Academy that community police officers and emergency police officers rarely communicate with each other on the street, and that their work could improve if they would.

The police officers' manager wanted to participate in the workshop, which had several effects. At the start of the workshop, he became irritated about the storylines that we created. He was irritated because we portrayed his daily practice and problems as "children's stories." During the workshop he probably influenced what other police officers were willing or dared to say. And at the end of the workshop he pointed out that we "learned nothing at all about police work" in one-day observations. He suggested that we should do observations over a longer period.

Mandy, Dirk, and I then made sketches for a telecom application, which we dubbed *PolicePointer.* This application is envisioned to run on a *smart phone.* Its main function is to provide suggestions to contact other police officers who may have information that is useful for you, given your current task or context.

We organized two workshops for which we invited both community police officers and emergency police officers. We discussed the topic of knowledge sharing and our ideas and sketches for receiving suggestions to communicate with other police officers. During these workshops, the *PolicePointer* was further developed. It was discussed that emergency police officers may be reluctant to "waste time" contacting other people; their job is to react quickly. Therefore, a notification function was added: when Bert receives a suggestion to contact Ad, Ad simultaneously receives a notification, so that he may contact Bert proactively. Furthermore, the police officers mentioned that emergency police officers have knowledge what happens outside office hours and what happens in different areas and that this kind of knowledge may be useful for community police officers. We adopted that idea so that the project's focus then became to support *both* kinds of police officers to communicate and share knowledge with each other.

With the help of fellow project team members these sketches were developed into a prototype (see Figure 1). This prototype was evaluated by five police officers during a field trial of two days, during which they used a *smart phone* with the *PolicePointer* on it. During this trial, very few incidents occurred, and, partly because of that, only few situations occurred in which the *PolicePointer* was perceived as having added value. Nevertheless, the participating police officers were positive about the *PolicePointer*, especially when they would receive suggestions for "unstructured" situations, situations for which there are no procedures and in which they must improvise, and when they would receive suggestions when they were in a relatively unknown area in which they don't know many colleagues (yet). Interestingly, both project team members and police officers downplayed the prototype's shortcomings during this trial: the hardware was too bulky, the login

Figure 1. A prototype of PolicePointer, running on a smart phone

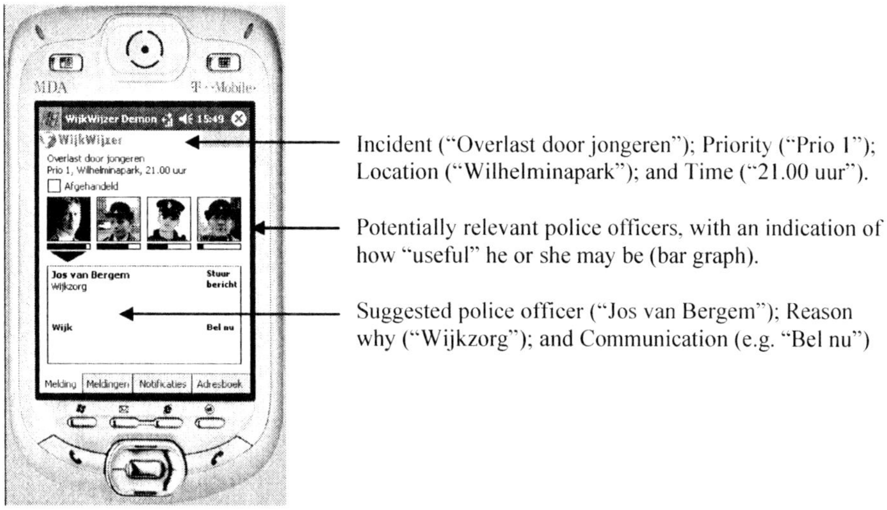

procedure was too long, and the system responded too slow. We explained that these issues would be improved if the prototype were developed into a working product.

Looking at this design process, we see a mixture of participatory design (inviting end users during problem definition and solution finding), of the lead user approach (the police officers who participated in our project were selected based on their innovative behaviour or attitude), and of empathic design (project team members joined a police officer for one day). Furthermore, we can see that each interaction with police officers resulted in some shift of the project's focus. This is in line with the goal of HCD: to let end users have influence upon the research and design and design process. The *PolicePointer* started as a tool for communication between community police officers and people outside the police, then became a tool for community police officers to share their knowledge with emergency police officers, and then became a communication tool between these two sorts of police officers.

However, we may also say that our focus on designing a *we-centric* telecom application gave us eye-flaps. In the process of going from an idea via sketches into a prototype, the *PolicePointer* gradually solidified. But in each step of the research and design process, the end users—the police officers—became less "real" and more "ideal." In each interaction with them, we further modelled real police officers into ideal users of our product. We started with the goal to design a *we-centric* telecom application, a solution "in our heads" looking for a problem "out-there," and we were happy that Albert found community police officers as a target group. We observed some police officers and wrote storylines about them, storylines that irritated one police sergeant, storylines in which we implicitly suggested that they should work differently, and that they should use our *we-centric* application. And we created a trial in which we managed to neglect the prototype practical shortcomings and instead focused on the *PolicePointer*'s theoretical, potential added value.

DESIGNING WITH/FOR INFORMAL CARERS[D]

Project team members Catherine and Edith, who work at a university's medical department and who have years of experience in working with and for people with dementia and their informal carers, articulated this goal: to help informal carers who provide care to people who suffer from dementia and who live at home (*not* in an institution). In many cases, these people provide care to their husband, their wife, or one of their parents and such care is often needed fulltime. The ultimate goal is to improve the quality of life for both the person with dementia and the informal carer, by supporting the informal carer, so that he or she can better provide care and will feel less burdened or will be less likely to suffer from burn-out.

In order to better understand the needs of people with dementia and of their informal carers, Pauline, together with her colleagues Catherine and Edith conducted a large scale survey. Over 300 "dyads"—a person with dementia and his or her "primary" informal carer—were interviewed using several standardized questionnaires. This fieldwork took several months and Pauline presented their preliminary results several times during meetings. Typically, she would show a table with the "most frequently reported (unmet) needs of people with dementia and their informal carer": "memory (40.5%); daily activities (17.9%); information about health and treatment for informal carer (17.3%); company (12.7%); psychic need of informal carer (12.3%) ...."

The other project team members, Annelies and Martin, who work at a new media lab, and Rachel, who works at an ICT research lab, are responsible for the design and evaluations of a *we-centric* application. Dementia and informal care are relatively new topics for them, and therefore they chose to do some additional observations and interviews. This caused friction within the team. Pauline suggested that they first study the literature and the results from their survey, and to do additional research only if they *miss* specific data. Furthermore, there were discussions about method. Pauline and her colleagues do a survey to obtain a statistically representative overview of people's needs, whereas Annelies and Rachel want to get acquainted with a small number of people with dementia and their informal carers and to be inspired by them. This difference is also described by sociologist Haddon and designer Kommonen (2003), who characterize social scientists as being concerned with existing knowledge and studying and documenting reality, and designers as being concerned with originality, imagining alternatives and changing reality. There were several conflicts within the project team about doing a large, statistical representative survey vs. doing "interviews with only one or two people," and about which method can be a basis for conclusions or decisions. Only after project team members had actually cooperated in interpreting each other's data, were they able to cooperate constructively.

Based on the survey and on the additional interviews, a design focus was agreed upon to alleviate the burden of the informal carers, to support them in their "work." In many interviews informal carers told that they experience taking care of a person with dementia as very demanding, especially if he or she is your husband, wife, or parent. Furthermore, the situation of a person who suffers from dementia cannot improve but only become worse. Such informal carers have to do everything alone: there may be nobody who offers help, or they may not dare to ask others for help.

Annelies, Martin, and Rachel, helped by Pauline, then started a user study. Four informal carers were invited to participate in a series of three interviews in their homes. In the first interview the project team members got acquainted with them and learned about their situation. Based on these interviews they created personas and storylines, in the form of "a day in the life of..." – see below for an excerpt:

About Ans and Simon

Ans (73 years) has dementia. Her husband Simon (76 years) is her "primary informal carer." They have been together for 51 years. They have two sons, Johan and Pieter, and four grandchildren. They were born and raised in a rural part of the country. Simon had his own shop where he manufactured and repaired sails. He retired 12 years ago. Ans used to do the financial administration and took care of their two sons. She used to be very active; she did volunteer work and used to go for a bicycle trip once a week with the woman from next door.

Time: 5:30.

Ans is stumbling around in the house, which awakens Simon. He sighs and tries to sleep again. He went to bed late last night, because he wanted to finish reading his book. Ans is calling Simon. Simon gets out of bed and goes to help her.

Time: 06:15.

Simon brushes his teeth and dresses. Then he chooses clothes for Ans. He helps Ans to shower, then he helps her dry herself and dress. Helping her with the sleeves of her dress takes a while. Then Ans is dressed.

Ans is portrayed as a woman who used to be active and social. This shows us the woman she used to be before she got dementia and became passive and isolated. Furthermore, the situations of waking up early and helping Ans shower and dress suggest that Simon "suffers" from Ans suffering from dementia. He has to help her fulltime. This also suggests that we could help Simon, for example with a *we-centric* application, which may help to alleviate informal carer's social and emotional needs.

In a second round of interviews, Rachel and Annelies read these storylines aloud and discussed these. Together with the informal carers, they identified several situations that the informal carers found relevant. After the interviews, Rachel, Annelies, Martin, and Pauline selected three situations to be used as input material for a concept development workshop with six project team members and with six additional "external" creative designers (not project team members). In this workshop, three groups worked in parallel, each on one of the three situations. One group developed an idea for an ICT application that matches people who look for help and people who offer help, which is meant to let informal carers support each other, practically and emotionally. The other two groups came up with a "domotica" application which monitors what the person with dementia does in the home and assists him or her; and a jewelery product with localization technology, which helps the person with dementia to return home if he or she gets lost outside.

The first idea was selected to be further developed because: (1) it is closest to the goal of designing a *we-centric* telecom application, it is about communication and cooperation; (2) it is a "mobile, context-aware and adaptive" application, which fits the project's scope and goal; (3) it is similar to ideas developed previously by project team members and to the *PolicePointer*, which makes the design more comfortable; (4) and it seems most feasible to build a prototype of it within planning and budget. Making this decision was difficult because some project team members thought that the other two ideas were more interesting and creative. Furthermore, it was noted that the people behind the first idea were involved in the research process, they were "experts on informal carers' needs," whereas the people who came up with the other two, more creative ideas were people with little or no experience with dementia or informal carers.

In a third round of interviews, Rachel and Annelies discussed this first concept with the four informal carers who participated in the first two rounds of the interview, and with seven other informal carers. Questions that were discussed were How do you wish to invite informal carers to participate? How would you like to form or control such a group of informal carers? How do you think participants can be motivated to offer and accept help? What kind of help requests would people ask of each other, for example structural or incidental? And should there be mechanisms to monitor how much help a person offers and receives? Based on these interviews, the following functional requirements for a we-centric telecom application—dubbed *WeCare*—were articulated: (1) an informal carer can ask for help by putting a "help request" in her online calendar; (2) an informal carer can offer help by filling in her online calendar (when she is available to help) and her profile (the kind of help she wishes to offer); (3) the system automatically matches the "help requests" in with other participant's calendars and profiles; and (4) messages are sent to the person asking for help and the person offering help—these messages can be sent by e-mail or SMS or as entries in participants' online calendars. Additionally, participants can post on a bulletin board: "help requested" and "help offered."

The current plan is to build a prototype of *WeCare* and to evaluate this together with several informal carers, similar to how the *PolicePointer* prototype was evaluated. The prototype will include a database, which can be accessed via a webpage on their computer and also via their *smart phone*. The idea is that people will be able to do complex tasks, such as scheduling, while sitting at their computer, see Figure 2, and use a subset of the functionalities via their mobile phone or *smart phone*, for example when they are travelling.

In this design process we saw a mixture of participatory design (discussing possible problems and solutions with potential end users) of the lead user approach (the informal carers who participated were selected based on their use of computer, internet and mobile phone), and of usability engineering (discussing concepts and sketches with end users). If we look at how the interactions with informal carers influenced the design process, we see that the initial idea—to support informal carers to do their "work"—remained relatively stable. The interviews with

Figure 2. A mock-up of WeCare: A webpage on a computer

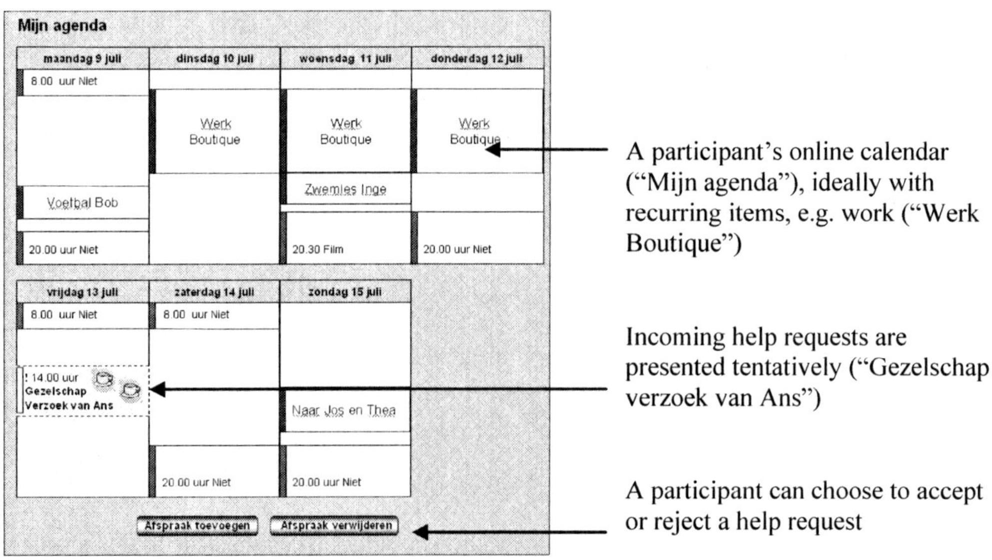

informal carers helped to make design decisions and to articulate functional requirements. However, during the interactions with the informal carers the concept was not critically questioned, and an attempt, in the concept development workshop, to come up with alternative ideas was not welcomed.

Furthermore, we can see that there was friction *within* the (relatively large and heterogeneous) project team: about which research methods to use (a systematic survey, which produces facts vs. informal interviews for inspiration) and about choosing a concept (a conservative concept vs. a more innovative concept). This contrasts with police officers' case, where there was friction between the (relatively small and homogeneous) project team and the police officers, for example when the officers' manager was irritated about "our" storylines about "their" practice. This draws attention to the importance of cooperation *within* the team in a human-centred design project: not only the interactions with end users, but also the interactions between people in a multidisciplinary team are important (cf. principle 4 of HCD).

GATHERING KNOWLEDGE

In the next two sections, I will critically examine two assumptions that seem to be key in human-centred design (HCD): that researchers and designers can gather knowledge about end users' needs, wishes and preferences; and that they can apply this knowledge to make appropriate design decisions.

These assumptions can be criticised from various angles: one can argue that end users are not aware of their needs, cannot articulate these, or do not want to or cannot speak about these (van Kleef, van Trijp, & Luning., 2005); that zooming in too much on a small group of end users will result in an over-customized product that will interest only a few (Stewart & Williams, 2005); or that paying too much attention to end users erodes the role of the designer, whose vision and creativity are essential (Hekkert & Van Dijk, 2001). Such effects can indeed be seen in the cases, for example when some of the police officers seemed to talk less freely with their manager present in the workshop; when the validity of making design decisions based "on interviews with only one or two" informal carers was disputed; and when the experts on the informal carers' needs came up with a relatively conservative idea.

What is probably most striking is that the end users—the police officers and the informal carers—were most often absent in the process. The police officers were invited for several workshops and informal carers were interviewed, but they did *not* participate in design decision making. In project meetings, where design decisions were made, they were not present, but they were *represented.* Project team members portrayed them via statistics, which they constructed out of their survey, and via storylines, which they constructed out of their observations and interviews. The storylines are meant to portray the current situation ("is"), but in it were suggestions that these situations are problematic from a very specific angle, and that these situations can be improved by using our *we-centric* telecom application ("ought").

Furthermore, the project team members acted as end users' spokespersons, for example, when they discussed their needs, wishes, and preferences, and when they made design decisions, decisions which are meant to influence the product, which is meant to influence end users' life and work. Representing end users has resembles the "configuring" (Woolgar, 1991) of end users, and the creation of "scripts" (Akrich, 1992) about end users. Parallel to the product which they create, researchers and designers create an image of what an end-user should look like and how he or she should use their product. Moreover, the gathering knowledge about and from end users and representing them is *not* a neutral activity (Rohracher, 2005, p. 16):

Representing users in design is by no means a simple and straightforward process, but continuously reshaped and negotiated by actors involved in the design process. [...] User representations are constructed and shaped by the interests, specific discourses and traditions of actors involved and often are also entrenched in material infrastructures or methods to investigate demand.

The process of representing end users is a process of mobilising resources to influence others through communication (Latour, 1987). When Pauline talks about her interviews with informal carers or when Mandy quotes what police officers said in a workshop, they mobilize these end users to make their point.

Gathering knowledge about the world around me, including other people, was also examined by Levinas. He wrote that when I gather knowledge I almost automatically reduce everything to concepts that are already familiar for me: I transmute the *other* so that it matches to my *self* (1996a, p. 11-12):

The knowing I is the melting pot of such a transmutation. It is the same par excellence. When the other enters into the horizon of knowledge, it already renounces alterity.

When I gather knowledge about another person, I make a gesture of grasping: I grasp what I see and hear of that other person, and pull him into my own world. Levinas uses words like "the con*cept* or the Be*griff*" (1996b, p. 152, emphasis in original) to exemplify the "concreteness of the grasp." The project team members could not escape this tendency of drawing others into their "melting pot." Their own interests, ambitions, intentions, methods, and their creativity—their "selves"—make them filter what they heard, saw, and understood of "the other" during their observations, workshop and interviews. It is possible to imagine a HCD project in which researchers and designers do their observations, workshops and interviews according to the book, and then follow their own interests, ambitions, intentions and do what they had in mind already. The gesture of the grasp becomes poignant when constructing storylines or personas turns into a way of "involving users by simply excluding them. The users are instead represented by an archetype of a user" (Blomquist & Arvola, 2002).

Researchers and designers who study end users are not engaged in some neutral fact-finding activity, but in negotiating between different people's interests and values. Explicitly or implicitly they are taking sides, they are making decisions all the time.

MAKING DECISIONS

A research and design project is about making decisions. There are many options open, many uncertainties, but at the same time, there are only few criteria or requirements to ground decision making. Nevertheless, one has to make decisions in order to proceed: create ideas, choose between ideas, turn an idea into concepts, choose between concepts, turn a concept into a prototype, and into a product. One proceeds by reducing "design space" gradually until one has one product. Explicitly or implicitly one must make decisions about what problem to address, about criteria to choose between problems, about what direction to search for solutions, about criteria to choose between solutions, etcetera. This makes research and design different from for example a study in social science or an engineering project.

Roozenburg & Eekels (1995) argue that "design thinking" is different from other "logical" ways of thinking: *de*duction starts with premises and then one draws a conclusion (in mathematics); *in*duction starts with several observations and then one speculates about a pattern (in natural science); and *ab*duction starts with observing an effect and a process, and then one reasons backward to a possible cause (in history). Contrastingly, design thinking or *inno*-duction starts with imagining a problem that seems worthwhile to try to solve, one simultaneously imagines possible solutions, and also criteria to choose between solutions, one may reformulate the problem or go looking for solutions in other directions, etcetera. Articulating the design problem and formulating the design brief are part of the design process. Imagining the problem and imaging the solution are intertwined. "The problem and solution co-evolve" (Cross, 2006, p. 80)[e]. A designer will try-out and play with solutions as a way of exploring the problem. This play with problem and solutions can be found especially in the "fuzzy front end" where problems and solutions are discussed iteratively.

This is different from how one (supposedly) conducts a study in social science—one starts with a question, develops a method, does the study, and interprets the data—or how an engineering project is (supposedly) done—one starts with a brief and then proceeds via steps of generating and choosing alternatives until one has an optimal solution. A research and design project starts with a bet: "No innovation, no invention develops without this initial bet" (Akrich et al., 2002, p. 219). Albert betted that designing a telecom application to support community police officers to cooperate better with others would be worthwhile. The project team members betted that designing a telecom application to support informal carers in their "work" by requesting and offering help would be sensible.

Derrida has a special way of looking at the making of decisions. He argued that only in a situation *without* rules, where one cannot use knowledge, logical rules, or moral rules can one make a decision (Derrida, 1995, p. 147-8):

The only decision possible is the impossible decision. It is when it is not possible to know what must be done, when knowledge is not and cannot be determining that a decision is possible as such. Otherwise, the decision is an application: one knows what has to be done, it's clear, there is no more decision possible; what one has here is an effect, an application, a programming.

Making design decisions about what problem to focus upon, what end users to interview, how to interpret what they say, about which direction to look for solutions, and how to choose between alternatives can be called "impossible" decisions. Furthermore, Derrida said that only when one makes such an "impossible" decision, freedom and responsibility become possible (Derrida, 2001, p. 28):

A decision, as its name indicates, must interrupt, cut, rend a continuity, the fabric or the ordinary course of history. To be free and responsible, it must do other and more than deploy or reveal a truth already potentially present, indeed a power or a possibility, an existent force.

This freedom to choose may not be there in every project, but this freedom, which comes together with responsibility was there in the two cases studied. If we organize a research and design project in such a way that logical or moral rules steer the decision making, new possibilities are not possible—no innovation is possible. Innovation can *only* happen if we make "impossible" decisions, without rules. And the making of such "impossible" decisions make it possible to act freely and responsibly.

SELF AND OTHER

A key idea behind HCD, behind interacting with end users and behind multidisciplinary teamwork is that "other people" should be able to influence the project. What researchers and designers do in a HCD project can be seen as making moves towards the *self* or towards the *other*. They move towards the *other* when they listen to end users and try to understand their needs, wishes, and preferences or when they listen to another project team member and try to understand what he or she says. Conversely, they move towards the *self*, when they view end users as raw material for their own creativity, when they study end users with their own methods and represent them within their own argument, or when they remind others about the project goal and focus and mobilize this project goal and focus to make their own point.

Researchers and designers are making movements in both directions, at different moments, in different situations. During an interview or workshop, they try to move toward the other and when they discuss their findings within the project team, they move toward the self. They even make these two movements simultaneously, for example when they decide that other people are in need and must be helped: they move towards the *other* who is (supposedly) in need and they offer help which fits within their own expertise and ambition, within the *self*. This movement is salient when these products are meant to *empower* end users, to help them change their behaviour in a (supposedly) beneficial direction: the *PolicePointer* was developed to empower police officers to become more self-steering, rather than following hierarchical lines of command; and *WeCare* was developed to stimulate informal carers to share their tasks with others, rather than doing everything themselves.

Project team members made these movements in order to accommodate different demands upon them. I see at least three forces at work here.

Firstly, there is the *self* that demands them to use their expertise, their methods, their creativity. This is what they bring into the project. Dirk is clear and positive about team members having their own pet subjects: "Everybody has of course his own ideas which he wants to introduce" (Transcript 31 May 2006, p. 6). He provides examples of Albert who wishes to experiment with location technology and maps and of Barry who likes *smart phones* and who thinks of applying such *smart phones* almost regardless of the target group or the problem at hand.

Due to their ambition to develop a *we-centric* application the people working on the *PolicePointer* learned slowly about what police work is about. It was only after several workshops with police officers that they were able to create something that is both interesting for police officers and for their own project. This focus on telecom brings the risk of missing the larger context of police work. During lunch, after one workshop, the police officers talked heatedly about their uniform's trousers. They currently wear cotton trousers, and their managers want them to wear woollen trousers. But they do not want woollen trousers. They told us how easily you get blood or other stains on your trouser and that they are responsible for keeping their uniform clean. They can easily wash their cotton trousers, but they must bring such woollen trousers to the dry cleaner. Listening to such stories we could have learned more about power and culture and about how innovation works in a police organization. Moreover, there may be interesting parallels to draw between the introduction of woollen trousers and the introduction of something like the *PolicePointer*.

It was only during the field trial of the prototype that we were discussing in detail how the *PolicePointer* may fit in the current police organization and culture. Before that, we had paid attention to communication and information as functional processes, which can be supported by ICT. If the police is currently organized top-down and reacts to incidents, how will police officers think and feel about the *PolicePointer*, which is meant to stimulate working more proactively and more self-steeringly?

The people working on *WeCare* used different methods to study and represent end users. Their focus on their own respective methods and their discussions about methods may have blurred their view on the end users. During one such discussion, there was confusion about *whose* problem we are trying to solve. The table that summarizes the survey's results has four columns: "reported needs" and "reported unmet needs" of people with dementia, and "reported needs" and "reported unmet needs" of their informal carers. Confusingly, many of the informal carers' needs are "caused by" the needs of the people with dementia, for example needs related to "memory." And trying to solve the informal carers' needs is, of course, meant to help solve the needs of the people with dementia. During one such discussion, somebody remarked "Our need is to do something about that problem." This draws attention to the researchers who are concerned with their own methods and interpretations, to designers who are concerned with their own creativity and ideas—and to the risk of altogether forgetting the end users. Another risk is that a focus on technology—often the *self* of researchers and designers—leads to a technological view: as if organizing the exchange of help between informal carers is a technical problem which needs to be solved through a technical solution.

Secondly, there are the end users, the *others*. The project team members tried to move towards the *others*, and the interactions with them influenced the project. Looking critically at the moves made towards the *others*, we saw that the end-uses were often not present, but represented. They are made part of one's argument, and drawn towards the *self*. The project team members—probably unintentionally—reduced the otherness of end users.

In workshops the police officers talked about problems in their work, for example about how they balance conflicting roles or identities, such as being a "spider in the web," being a "go-getter" on an emergency call, and being servant to larger bureaucratic processes simultaneously. Although these utterances appeared in meeting minutes, they were rarely explicitly discussed during decision making within the project team. The police officers' otherness was also reduced when project team members turned their field notes into storylines. The notes described vividly how the different project team members experienced situations like arresting a thief, driving with lights flashing, wearing a bulletproof vest, or interviewing suspects. However, they omitted such descriptions when they made their storylines, which consequently became relatively sterile. They constructed the storylines with the project's focus in mind and focused on situations where a *we-centric* application may be of value.

The informal carers' otherness was reduced when the project team members applied questionnaires with pre-defined concepts to interview people and summarized their findings into one table. Interestingly, Pauline mentioned relatively late in the process of doing the survey fieldwork that almost all the interviewees were crying during the interviews. This fact was not told before nor was it asked about. Crying respondents was maybe not an apparent aspect of conducting and reporting a survey. The informal carers' otherness was also reduced when we constructed storylines based on the interviews, privileging what we are interested in and overlooking other aspects. Rachel talked several times about how it moved her that an elderly lady, who was informal carer, found it hard to call upon others for help, but these emotions of Rachel and of this lady were lost in the translation into storylines.

Human-centred design is an attempt to move toward end users. But, since such projects are often multidisciplinary, and with different organizations participating, it is also an attempt to move towards other team members: people who have different backgrounds, positions and interests. Many attempts were made to move towards the *other*, but often one easily moved (back) to the *self*.

And thirdly, there is the context of the project. There is a project plan, on which the participating organizations had agreed upon, and in it is the goal to develop and evaluate a *we-centric* telecom application. I tried, on various occasions, in my role as coordinator of a part of the project, to influence my fellow project team members to focus on needs or problems that can be related to telecom (and to neglect other problems or needs) and to develop a telecom application (and to disregard other kinds of solutions). This can be illustrated by the observation that, during the development of the *PolicePointer*, we only became interested relatively late in the project in how police officers input and edit reports in their database. We became interested when and because we realised that we needed to understand "their" information process in order to make "our" telecom application work: the *PolicePointer* must access and match these reports in this database in order to make automatically generate its suggestions for the police officers to communicate. The development of *WeCare* was similarly steered by mobilizing the project's scope and goal, for example, when during and after the creative session these criteria were put forward: it must be a *we-centric* application; it must be "mobile, context-aware and adaptive"; it must be similar to current ideas within the project team; and it must be feasible technically.

Although the project plan is text in a paper document, it can be brought alive. It can be mobilized when one team member refers to it, interprets it and makes it into a part of his or her own argument. This mobilizing is similar to how end users are represented and mobilized. The project plan is not part of any one of the project team members' *self*, but it can be drawn towards the *self*.

In summary, project team members are moving in a space which spans between *self* and *other*. And in this space they move towards end users when they interact with them, or they pull them towards their selves when they represent them to make their own point. Similarly, they move towards fellow project team members or pull them into their own argument. And they move toward the project's goal and scope, or pull it towards themselves to make their own point.

SPEAKING ABOUT ETHICS

Talking about *self* and *other* and making references to Levinas and Derrida is my way of arguing that human-centred design (HCD) is a social process which has ethical qualities[f]. Furthermore, I would like to suggest a role for management in bringing these ethical qualities more to the fore.

In the two cases I described how researchers and designers interacted with end users and how they made design decisions, and I described their activities in terms of making "grasping" gestures, of making "impossible" decisions and as moving between *self* and *other*. I would like to argue that such actions have ethical qualities. My argument is *not* concerned with evaluating whether what researchers and designers do is *morally* good or evil, with suggesting that they should behave more "morally" or with suggestions to improve HCD.

The point which I wish to make is that ethics are happening in HCD—all the time, *already*. But these ethical qualities are rarely discussed explicitly. We act as if they do not happen, we marginalize such ethical qualities. My point is also *not* that moves towards the other are better or worse than moves towards the self. The two movements stand for the two "faces" of HCD: it is about meeting another person and trying to create something for this other person and about deploying one's own creativity. Both moves are made, and both moves must be made.

I am interested in the ethics, which happen in interactions between people when researchers and designers interact with end users and with fellow project team members—whom they listen to, how they listen—and when they make design decisions – when they choose to focus on a certain problem or choose to develop a certain solution. I would like to ask questions like How much participation do I allow in "participatory" design? How much empathy do I have in "empathic" design? How central are people in *human-centred* design?

My suggestion can be seen as an attempt to "deconstruct" (Critchley, 1999; Derrida, 1991)

HCD, where I understand deconstruction as a "radical form of questioning text(s)" (Letiche, 1998, p. 125): providing a critical reading of what happens in a project and providing an alternative reading of that project. What would happen if these ethical qualities were brought more to the fore? What if managers, researchers and designers become more aware and explicitly discuss their *selves*, their own respective positions, their own expertise, methods and creativity? What if they become more aware and explicitly discuss how they relate to *others*, to what they hear and see from end users and from fellow project team members? What if they become more aware and explicit about how they move between *self* and *other*?

This suggestion can be positioned in a debate about the relation between ethics and science and technology studies (STS). Many social constructivist studies of technology in the field of STS can be criticized for *not* speaking (enough) about ethics, for their "lack of and, indeed, apparent disdain for anything resembling an evaluative stance or any particular moral or political principles that might help people judge the possibilities that technologies present" (Winner, 1993, p. 371). Recently Poel and Verbeek (2006) stimulated a debate between STS and ethics. They argue that, traditionally, engineering ethics has been concerned with studying the ethical *consequences* of developing or applying technology, and advocate that STS might help to conduct more situated studies of what people involved actually *do*, to "open the black box of technology" (rather than building theoretical arguments), and that, vice versa, engineering ethics "might help STS to overcome its normative sterility."

My suggestion is in line with theirs: I suggest that researchers, designers, their managers, and others involved in a HCD project can be stimulated to reflect upon their practice and upon the ethical qualities of their practice. This could be a way of letting the other put into question my *self*, my spontaneity, my creativity (cf. Critchley, 1999, p. 5; Levinas, 1996a, p. 17), and it could be a way to help researchers and designers to act more freely and more responsibly. Responsibly in the sense that they can question their design decisions more explicitly and ask each other to *respond* to such questioning. And freely in that they can more consciously and explicitly choose between options. I think that many researchers and designers are currently attempting to do HCD, but that they are doing their projects as if they are doing social science or an engineering project. My suggestion is to conduct HCD differently, to conduct it more accordingly to what it *already* is—a process between people, with ethical qualities—to continue doing HCD, but more reflectively.

ACKNOWLEDGMENT

This text was written as part of the Freeband Communication research program (http://www.freeband.nl). I would especially like to thank my fellow project team members for their cooperation while conducting my study and for their permission to write about them.

REFERENCES

Akrich, M. (1992). The description of technical objects. In W. E. Bijker & J. Law (Eds.), *Shaping technology/Building society: Studies in sociotechnical change* (pp. 205-224). Cambridge, MA and London, England: MIT Press.

Akrich, M., Callon, M., & Latour, B. (2002). The key to success in innovation—Part 2: The art of choosing good spokespersons. *International Journal of Innovation Management, 6,* 207-225.

Beyer, H., & Holzblatt, K. (1998). *Contextual design: Defining customer-centred systems.* San Francisco: Morgan Kaufmann.

Blomquist, Å., & Arvola, M. (2002). Personas in action: Ethnography in an interaction design team. In *Proceedings of the 2nd Nordic Conference on Human-Computer Interaction* (pp. 197-200). Aarhus, Denmark: ACM Press.

Bucciarelli, L. (1994). *Designing engineers*. Cambridge, MA and London, England: MIT Press.

Cooper, A. (1999). The inmates are running the asylum: Why high-tech products drive us crazy and how to restore the sanity. Indianapolis, IN: SAMS Publishing.

Crabtree, A. (2003). *Designing collaborative systems: A practical guide to ethnography*. London: Springer-Verlag.

Critchley, S. (1999). *The ethics of deconstruction: Derrida and Levinas* (2nd ed.). Edinburgh University Press.

Cross, N. (2006). *Designerly ways of knowing*. London: Springer-Verlag.

Derrida, J. (1995). *Points: Interviews, 1974-1994*. Stanford, CA: Stanford University Press.

Derrida, J. (2001). Deconstructions: The impossible. In S. Lotringer & S. Cohen (Eds.), *French theory in America* (pp. 12-32). New York and London: Routledge.

Derrida, J. (1991). Letter to a Japanese friend [original 1987]. In P. Kamuf (Ed.), *A Derrida reader: Between the blinds* (pp. 270-276). New York: Columbia University Press.

Easterby-Smith, M., Thorpe, R., & Lowe, A. (2002). *Management research: An introduction* (2nd ed.). London: Sage.

Haddon, L., & Kommonen, K. H. (2003). *Interdisciplinary explorations: A dialogue between a sociologist and a design group*. Helsinki: University of Art and Design.

Hekkert, P., & Van Dijk, M. (2001). Designing from context. In P. Lloyd & H. Christiaans (Eds.), *Designing in context* (pp. 383-394). Delft: Delft University Press Science.

ISO. (1999). ISO 13407: *Human-centred design processes for interactive systems*. Geneva, Switzerland: International Organization for Standardization.

Koen, P., Ajamian, G., Boyce, S., Clamen, A., Fisher, E., Fountoulakis, S. et al. (2002). Fuzzy front end: Effective methods, tools, and techniques. In *PDMA toolbook for new product development* (pp. 2-35). John Wiley & Sons.

Koskinen, I., Battarbee, K., & Mattelmäki, T. (2003). *Empathic design: User experience in product design*. Helsinki: Edita Publishing.

Kujala, S. (2003). User involvement: A review of the benefits and challenges. *Behaviour and Information Technology, 22,* 1-17.

Latour, B. (1987). *Science in action: How to follow scientists and engineers through society*. Milton Keynes: Open University Press.

Latour, B., & Woolgar, S. (1986). *Laboratory life: The construction of scientific facts* (2nd ed.). Princeton, NJ: Princeton University Press.

Lawson, B. (1997). *How designers think: The design process demystified*. Architectural Press (an imprint of Butterworth-Heineman, a division of Reed Educational and Professional Publishing).

Letiche, H. (1998). Business ethics: (In-)justice and (anti-)law. In M. Parker (Ed.), *Ethics and organizations* (pp. 122-149). London: Sage.

Levinas, E. (1996a). Transcendence and height [original 1962]. In A. Peperzak, S. Critchley, & R. Bernasconi (Eds.), *Emmanuel Levinas: Basic philosophical writings* (pp. 11-32). Bloomington and Indianapolis, IN: Indiana University Press.

Levinas, E. (1996b). Transcendence and intelligibility [original 1984]. In A. Peperzak, S. Critchley, & R. Bernasconi (Eds.), *Emmanuel Levinas: Basic philosophical writings* (pp. 149-

159). Bloomington and Indianapolis, IN: Indiana University Press.

Oudshoorn, N., & Pinch, T. (2003). Introduction: How users and non-users matter. In N. Oudshoorn, & T. Pinch (Eds.). *How users matter: The co-construction of users and technology* (pp. 1-25). Cambridge, MA and London, England: MIT Press.

Poel, I. v. d. & Verbeek, P. P. (2006). Ethics and engineering design. *Science Technology & Human Values, 31,* 223-236.

Rohracher, H. (2005). The diverse roles of users in innovation processes. In H. Rohracher (Ed.), *User involvement in innovation processes: Strategies and limitations from a socio-technical perspective* (pp. 9-35). Munchen/Wien: Profil Verlag.

Roozenburg, N. F. M., & Eekels, J. (1995). *Product design, fundamentals and methods.* Chichester, UK: John Wiley and Sons.

Sanders, E. B. N. (2000). Generative tools for codesigning. In Scrivener, Ball, & Woodcock (Eds.), *Collaborative design.* London: Springer-Verlag.

Schuler, D., & Namioka, A. (1993). *Participatory design: Principles and practices.* NJ: Lawrence Erlbaum Associates.

Steen, M. (2006a). *Our need is to do something about that problem—Studying how researchers and developers interact with informal carers during an innovation project.* Presented at EASST 2006 Conference (European Association for Studies of Science and Technology), August 23-26, Lausanne, Switzerland.

Steen, M. (2006b). We don't want woollen trousers—Studying how researchers and developers interact with police officers during an innovation project. In R. ten Bos & R. Kaulingfreks (Eds.), *Proceedings of SCOS 2006 (Standing Conference on Organisational Symbolism)* (pp. 644-666), July 13-15, Nijmegen, The Netherlands.

Stewart, J., & Williams, R. (2005). The wrong trousers? Beyond the design fallacy: Social learning and the user. In H. Rohracher (Ed.), *User involvement in innovation processes: Strategies and limitations from a socio-technical perspective* (pp. 39-71). Munchen/Wien: Profil Verlag.

van Kleef, E., van Trijp, H. C. M., & Luning, P. (2005). Consumer research in the early stages of new product development: A critical review of methods and techniques. *Food Quality and Preference, 16,* 181-201.

Von Hippel, E. (2005). *Democratizing innovation.* Cambridge, MA: MIT Press.

Winner, L. (1993). Upon opening the black box and finding it empty: Social constructivism and the philosophy of technology. *Science Technology & Human Values, 18,* 362-378.

Woolgar, S. (1991). Configuring the user: The case of usability trials. In J. Law (Ed.), *A sociology of monsters: Essays on power, technology, and dominations* (pp. 57-102). London and New York: Routledge.

ENDNOTES

[a] I use "researcher" and "designer" for roles or for activities—not for people or for occupations. One person can at one time do research, "study something carefully and try to discover new facts about it", and at another time do design, "decide how something will look, work, etc. especially by drawing plans or making models" (Oxford Advanced Learner's Dictionary, 7th ed.). I am keen not to say "R&D" because that word is associated with an organizational job, function or department.

[b] I use "end-users" as a more readable alternative for "future, potential or putative end-users." Nevertheless, it is strange to talk about end-users when there is not yet

a product or service that can be used – it is currently being developed. Furthermore, I use "end-users" rather than "users," to refer to people for whom a product or service is meant primarily or ultimately, and not to e.g. a repair-person, who may also use the product, or e.g. a decision-maker who may decide to install or implement the service.

[c] More details about this project are in a conference paper (Steen, 2006b).

[d] More details about this project are in a conference paper (Steen, 2006a).

[e] Similarly, Lawson (1997, p. 121-7) argues that design problems cannot be comprehensively stated and require subjective interpretation, that the number of possible solutions is inexhaustible and no one is optimal, and that the design process is involves finding as well as solving problems and is a prescriptive activity.

[f] I intend to talk about the ethical qualities of research and design. Not about the ethical qualities of applying or using the products that come out of that process.

This work was previously published in Management Practices in High-Tech Environments, edited by D. Jemielniak and J. Kociatkiewicz, pp. 75-93, copyright 2008 by Information Science Reference, formerly known as Idea Group Reference (an imprint of IGI Global).

Compilation of References

Al-Khaldi, M. A., & Wallace, R. S. O. (1999). The influence of attitudes on personal computer utilization among knowledge workers: The case of Saudi Arabia. *Information & Management, 36*(4), 185-204.

Abdinnour-Helm, S.F., Chaparro, B.S., & Farmer, S.M. (2005). Using the end user computing satisfaction instrument to measure satisfaction with a Web site. *Decision Sciences, 36*(2), 341-363.

Adams, D. A., Nelson, R. P., & Todd, P. A. (1992). Perceived usefulness, ease of use, and usage of information technology: A replication. *MIS Quarterly, 16*(2), 227-247.

Adams, S., & Cutshall, C.R. (1987). Refusing to prosecute minor offenses: The relative influence of legal and extralegal factors. *Justice Quarterly, 4,* 595-609.

Agarwal, R., & Prasad, J. (1997). The role of innovation characteristics and perceived voluntariness in the acceptance of information technologies. *Decision Sciences, 28*(3), 557-582.

Agarwal, R., & Prasad, J. (1998). The antecedents and consequents of user perceptions in information technology adoption. *Decision Support Systems, 22*(1), 15-29.

Agarwal, R., & Prasad, J. (1999). Are individual differences germane to the acceptance of new information technologies. *Decision Sciences, 30*(2), 361-391.

Agarwal, R., & Prasad, J. (2000). A field study of the adoption of software process innovations by information systems professionals. *IEEE Transaction on Engineering Management, 47*(3), 295-308.

Agarwal, R., & Venkatesh, V. (2002). Assessing a firm's Web presence: A heuristic evaluation procedure for the measurement of usability. *Information Systems Research, 13*(2), 168-186.

Ajzen, I. (1985). From intentions to action: A theory of planned behavior. In J. Kuhl & J. Beckman (Eds.), *Action control: From cognitions to behaviors*. New York: Springer.

Ajzen, I. (1988). *Attitudes, personality, and behavior.* Chicago: Dorsey.

Ajzen, I. (1991). Theory of planned behavior. *Organizational Behavior and Human Decision Processes, 50,* 179-211.

Ajzen, I., & Fishbein, M. (1980). *Understanding attitudes and predicting social behavior.* Englewood Cliffs, NJ: Prentice Hall.

Akrich, M. (1992). The description of technical objects. In W. E. Bijker & J. Law (Eds.), *Shaping technology/Building society: Studies in sociotechnical change* (pp. 205-224). Cambridge, MA and London, England: MIT Press.

Akrich, M., Callon, M., & Latour, B. (2002). The key to success in innovation—Part 2: The art of choosing good spokespersons. *International Journal of Innovation Management, 6,* 207-225.

Alavi, M. (1994). Computer-mediated collaborative learning: An empirical evaluation. *MIS Quarterly, 18*(2), 159-174.

Alavi, M., & Leidner, D. (1999). Knowledge management systems: Issues, challenges and benefits. *Communications of the Associations for Information Systems, 1*(2), 1-36.

Alavi, M., & Leidner, D. E. (1999). Knowledge management systems: Issues, challenges, and benefits. *Communications of the Association for Information System, 1*(7). Retrieved May 15, 2001, from http://cais.isworld.org./articels/1-7/article.html

Alavi, M., & Leidner, D. E. (2001). Knowledge management and knowledge management systems: Conceptual foundations and research issues. *MIS Quarterly, 25*(1), 107-136.

Alavi, M., & Leidner, D. E. (2002). Knowledge management systems: Issues, challenges and benefits. In S. Barnes (Ed.), Knowledge management systems: Theory and practice. London: Thomson Learning.

Alavi, M., & Leidner, D.E. (2001). Technology mediated learning: A call for greater depth and breadth of research. *Information Systems Research, 12*(1), 1-10.

Alavi, M., Marakas, G.M., & Yoo, Y. (2002) A Comparative study of distributed learning environments on learning outcomes. *Information Systems Research, 13*(4), 404-415.

Albers, J. (2000). Successful speech applications in high noise environments. *SpeechTEK Proceedings* (pp. 147-154).

Al-Gahtani, S. S., & King, M. (1999). Attitudes, satisfaction and usage: Factors contributing to each in the acceptance of information technology. *Behaviour and Information Technology, 18,* 277-297.

Allen, M.J. & Yen, W.M. (1979). *Introduction to measurement theory.* Monterey, CA: Brooks/Cole.

Allwood, C. M. (1984). Error detection processes in statistical problem solving. *Cognitive Science, 8*(4), 413-437.

Al-Mashari, M., & Al-Mudimigh, A. (2003). ERP Implementation: Lessons from a case study. *Information Technology & People, 16*(1), 21-33.

Alvesson, M., & Willmott, H. (1996). *Making sense of management: A critical introduction.* London: Sage.

Ambrosio, J. (2000). Knowledge management mistakes: Experts reveal five pitfalls to avoid when starting down the knowledge management path. *Computerworld, 34*(27), 44.

Anderson, E.W., & Sullivan, M.W. (1993). The antecedents and consequences of customer satisfaction for firms. *Marketing Science, 12*(2), 125-143.

Andrea, D. (2000). Improving the user interface: Digital far-field microphone technology. *SpeechTEK Proceedings*, 155-160.

Anonymous. (2002). Many hack attacks go unreported. Retrieved March, 3, 2008,from http://www.cbsnews.com/stories/2002/04/07/tech/main505563.shtml

Arkes, H. R., Christensen, C., Lai, C., & Blumer, C. (1987). Two methods of reducing overconfidence. *Organizational Behavior and Human Decision Processes, 39*(1), 134-144.

Armitage, C. J., & Conner, M. (2001). Efficacy of the theory of planned behavior: A meta-analytic review. *British Journal of Social Psychology, 40,* 471-499.

Avison, D., & Fitzgerald, G. (2003). *Information systems development: Methodologies, techniques, and tools* (3rd ed.). Berkshire: Mc-Graw-Hill.

Awazu, Y., Baloh, P., Desouza, K. C., Wecht, C. H., Kim, J. Y., & Jha, S. (2006). *Opening up innovation through information-communication technologies* (Technical Report No. #I4I-I3M-InnovICT-1): Institute for Innovation in Information Management, The Information School, University of Washington.

Baars, B. J. (1992). A new ideomotor theory of voluntary control. In B. J. Baars (Ed.), *Experimental slips and human error* (pp. 93-120). New York: Plenum Press.

Bagozzi, R. P. (1981). Attitudes, intentions, and behavior: The test of some key hypothesis. *Journal of Personality and Social Psychology, 41*(4), 607-627.

Bagozzi, R. P. (1992). The self-regulation of attitudes, intentions, and behavior. *Social Psychology Quarterly, 55,* 178-204.

Bagozzi, R. P., & Baumgartner, H. (1994). The evaluation of structural equation models and hypothesis testing. In R.P. Bagozzi. (Ed.), *Principles of Marketing Research* (pp. 386-422). Cambridge: Blackwell.

Bagozzi, R. P., Baumgartner, H. & Yi, Y. (1992). State versus action orientation and the theory of reasoned action: An application to coupon usage, *Journal of Consumer Research, 18*(4), 505-518.

Bailey, J. E., & Pearson, S. W. (1983). Development of a tool for measurement and analysing computer user satisfaction. *Management Science*, *29*(5), 530-545.

Baker, K.R., Powell, S.G., Lawson, B., & Foster-Johns, L. (2006). *A survey of spreadsheet users* (in submission).

Baker, K.R., Powell, S.G., Lawson, B., & Foster-Johns, L. (2006). Comparison of characteristics and practices among spreadsheet users with different levels of experience. Presented at the *European Spreadsheet Risks Interest Group 6th Annual Symposium*, Cambridge.

Baker, T., Miner, A. S., & Eesley, D. T. (2003). Improvising firms: Bricolage, account giving and improvisational competencies in the founding process. *Research Policy, 32,* 255-276.

Bakx, A.W.E.A., Vermetten, Y.J.M., & Van der Sanden, J.M.M. (2003). Self-perceived competence, learning conceptions and preferred learning situations in the domain of communication. *British Journal of Educational Psychology, 73*(2), 223-245.

Baloh, P., Sitar, A. S., & Vasić, V. (2005). Practical approach to KM implementation: Industrial case of Slovenian company Gorenje d.d. In S. Hawamdeh (Ed.), *Knowledge management: Nurturing culture, innovation and technology. Proceedings of the 2005 International Conference on Knowledge Management, Charlotte, North Carolina, USA, October 27-28th, 2005* (pp. 689-690). New Jersey [etc.]: World Scientific.

Baloh, P., Wecht, C. H., & Desouza, K. C. (2006). Managing knowledge in fuzzy front end of open innovation. In *Proceedings of the MIC'06-7th International Conference of the Faculty of Management Koper, University of Primorska, November 23-25, 2006*. Portorož, Slovenia, 2006.

Bandura, A. (1986). *Social foundations of thought and action: A social cognitive theory.* Englewood Cliffs, NJ: Prentice Hall.

Bansler, J. P., & Havn, E. (2002, June 6-8). *Exploring the role of networks in IT implementation—The case of knowledge management systems.* Paper presented at the 10th European Conference on Information Systems, Gdansk, Poland.

Barley, S. R. (1986). Technology as an occasion for structuring: Evidence from observations of CT scanners and the social order of radiology departments. *Adm Sci Q, 31*(1), 78-108.

Barua, A., & Mukhopadhyay, T. (2000). Information technology and business performance: past, present, and future. *Framing the Domains of IT Management: Projecting the Future Through the Past*, 65-84.

Barua, A., Kriebel, C. H., & Mukhopadhyay, T. (1995). Information technologies and business value - An analytic and empirical-investigation. *Information Systems Research, 6*(1), 3-23.

Baskerville, R. (1999). Investigating information systems with action research. *Communications of the Association for Information Systems, 2.*

Baskerville, R., & Wood-Harper, A. T. (1996). A critical perspective on action research as a method for information systems research. *Journal of Information Technology, 11,* 235-246.

Baskerville, R., Pawlowski, S., & McLean, E. (2000). Enterprise resource planning and organizational knowledge: Patterns of convergence and divergence. In S. Ang, H. Krcmar, W. Orlikowski & P. Weill (Eds.), In *Proceedings of the Twenty First International Conference on Information Systems* (pp. 396-406). Atlanta, GA: Association for Information Systems.

Basselilier, G., Benbasat, I., & Reich, B.H. (2003). The Influence of business managers' IT competence on championing IT. *Information Systems Research, 14*(4), 317-336.

Becerra-Fernandez, I., & Sabherwal, R. (2001). Organization knowledge management: A contingency perspective. *Journal of Management Information Systems, 18*(1), 23-55.

Beck, K. (2000). *Extreme programming explained: Embrace change*. Reading, Massachusetts: Addison Wesley.

Becker, G.S. (1968). Crime and punishment: An economic approach. *Journal of Political Economy, 78*(2), 169-127.

Belassi, W., & Fadalla, A. (1998). An integrative framework for FMS diffusion. *OMEGA, 26*(6), 699-713.

Benbasat, I., & Zmud, R. W. (2003). The identity crisis within the IS discipline: Defining and communicating the discipline's core properties. *MIS Quarterly, 27*(2), 183-194.

Benbunan-Fich, R. (2002). Improving education and training with IT. *Communications of the ACM, 45*(6), 94-99.

Benbya, H. (2006). Mechanisms for knowledge management systems effectiveness: Empirical evidence from the silicon valley. In *Proceedings of the Academy of Management Conference* (pp. 1-6).

Bennett, J. (1984). Managing to meet usability requirements. In Bennett, J., Case, D., Sandelin, J. &Smith, M., (eds.), *Visual Display Terminals: Usability issues and Health concerns* Englewood Cliffs, NJ: Prentice-Hall.

Bennett, K. H., & Rajlich, V. T. (2000). Software maintenance and evolution: A roadmap. *The conference on the future of software engineering.* Limerick, Ireland: ACM Press.

Bentler, P. (1995). *EQS structural equations program Manual*, Encino, CA: Multivariate Software.

Bergeron, F. (1986). Factors influencing the use of DP charge back information. *MIS Quarterly, 10*(3), 225-238.

Beyer, H., & Holzblatt, K. (1998). *Contextual design: Defining customer-centred systems.* San Francisco: Morgan Kaufmann.

Beynon-Davies, P. (1995). Information systems "failure": The case of the London ambulance service's computer-aided dispatch project. *European Journal of Information Systems, 4*(3), 171-184.

Beynon-Davies, P. (1999). Human error and information systems failure: The case of the London ambulance service computer-aided dispatch system project. *Interacting with Computers, 11*(6), 699-720.

Bhatt, G. (2001). Knowledge management in organisations: Examining the interaction between technologies, techniques and people. *Journal of Knowledge Management, 5*(1), 68-75.

Bhattacherjee, A. (2001). Understanding information systems continuance: An expectation- confirmation model. *MIS Quarterly, 25*(3), 351-370.

Bijker, W. E. (1994). The social construction of fluorescent lighting, or how and artefact was invented in its diffusion stage. In W. E. Bijker & J. Law (Eds), *Shaping technology/building society: Studies in sociotechnical change* (pp. 75-104). Cambridge, MA: MIT Press.

Bijker, W. E., & Law, J. (1994). *Shaping technology/building society: Studies in sociotechnical change.* Cambridge MA: MIT Press.

Biner, P.M. (1993). The development of an instrument to measure student attitudes toward televised courses. *The American Journal of Distance Education, 7*(1), 63-73.

Blackler, F. (1995). Knowledge, knowledge work and organizations: An overview and interpretation. *Organization Studies, 16*(6), 1021-1046.

Blackwell, B.S. (2000). Perceived sanction threats, sex, and crime: A test and elaboration of power-control theory. *Criminology, 38*(2), 439-488.

Blair, E., & Burton, S. (1987). Cognitive processes used by survey respondents to answer behavioral frequency questions. *Journal of Consumer Research, 14,* 280-288.

Blattberg, R. C., & Hoch, S. J. (1990). Database models and managerial intuition: 50 Model+ 50 Manager. *Management Science, 36*(8), 887-899.

Bleek, W.G. (2004). *Software Infrastruktur. Von analystischer Perspective zu konstruktiver Orientierung.* Hamburg: Hamburg University Press.

Blomberg, J., Suchman, L., & Trigg, R.H. (1996). Reflections on a work-oriented design project. *Human-Computer Interaction, 11*(3), 237-265.

Blomquist, Å., & Arvola, M. (2002). Personas in action: Ethnography in an interaction design team. In *Proceedings of the 2nd Nordic Conference on Human-Computer Interaction* (pp. 197-200). Aarhus, Denmark: ACM Press.

Bloomfield, B. P., & Vurdubakis, T. (1994). Boundary disputes: Negotiating the boundary between the technical and the social in the development of IT systems. *Information Technology and People, 7*(1), 9-24.

Blue, C. L. (1995). The predictive capacity of the theory of reasoned action and the theory of planned behavior in exercise research. *Research in Nursing and Health, 18,* 105-121.

Bokulich, F. (2000). JSF [Joint Strike Fighter] voice recognition. *Aerospace Engineering Online.* Retrieved from http://www.sae.org/aeromag/techupdate_5-00/03.htm

Bose, R. (2004). Knowledge management metrics. *Industrial Management & Data Systems, 6*(104), 457-468.

Boud, D., & Feletti, G. (1991). *The challenge of problem based learning.* New York: St. Martin's Press.

Bowen, T. S. (1999). Making knowledge mean business: Vendors rush to jump on bandwagon. *InfoWorld, 21*(2), 69.

Boyce, C. (2002). Speech recognition market grows. *Research & Development, 44,* 17.

Bradner, S. (2005). The winner so Far: Cardsystems solutions. *Network World, 22*(25), 30.

Brancheau, J. C., & Wetherbe, J. C. (1990). The adoption of spreadsheet software: Testing innovation diffusion theory in the context of end-user computing. *Information Systems Research, 1*(2), 123-139.

Briggs, R. O., de Vreed, G. J., & Kolfschoten, G. L. (2007). *Report of the HICSS-40 Workshop on Collaboration Engineering.*

Briggs, R. O., de Vreed, G. J., & Nunamaker, J. F. (2003). Collaboration engineering with thnkLets to pursue sustained success with group support systems. *Journal of Management Information Systems, 19,* 31-64.

Brooking, A. (1999). *Corporate memory: Strategies for knowledge management.* London: International Thomas Business Press.

Brown, B. (2002). Studying the use of mobile technology. In B. Brown, N. Green, & R. Harper (Eds.), *Wireless world. Social and interactional aspects of the mobile age* (pp. 3-15). London: Springer.

Brown, I. D. (1990). Drivers' margins of safety considered as a focus for research on error. *Ergonomics, 33*(10/11), 1307-1314.

Brown, P. S., & Gould, J. D. (1987). An experimental study of people creating spreadsheets. *ACM Transactions on Office Information Systems, 5*(3), 258-272.

Bruegge, B., & Dutoit, A. H. (2004). *Object-oriented software engineering using UML, Patterns, and Java,* 2nd Ed. Upper Saddle River: Pearson Prentice Hall.

Bucciarelli, L. (1994). *Designing engineers.* Cambridge, MA and London, England: MIT Press.

Buchanan, D., & Badham, R. (1999). Politics and organizational change: The lived experience. *Human Relations, 52*(5), 609-629.

Burke, K., & Chidambaram, L. (1999). How much bandwidth is enough? A longitudinal examination of media characteristics and group outcomes. *MIS Quarterly, 23*(4), 557-579.

Burnett, M., Rothermel, G., & Cook, C. (2003). Software engineering for end-user programmers. *Proceedings of the Conference on Human Factors in Computing Systems (CHI'03),* 12-15.

Business Communications Company, Inc. (2003). RG-268 Voice Recognition: Technologies, Markets and Opportunities. Retreived from http:// www.bccresearch.com

Butler, R. (2000). Is this spreadsheet a tax evader? How H. M. Customs & Excise test spreadsheet applications. In *Proceedings of the 33rd Hawaii International Conference on System Sciences,* CD-ROM. Los Alamitas, CA: IEEE Computer Society Press.

Butler, T. (2003). An institutional perspective on the development and implementation of intranet- and Internet-based IS. *Information Systems Journal, 13*(3), 209-232.

Butler, T., & Fitzgerald, B. (1999). Unpacking the systems development process: An empirical application of the CSF concept in a research context. *Journal of Strategic Information Systems,* 8, 351-371.

Butler, T., & Murphy, C. (2007). Understanding the design of information technologies for knowledge management in organizations: A pragmatic perspective. *Information Systems Journal, 7*(2), 143-163.

Butler, T., Feller, J., Pope, A., Murphy, C., & Emerson, B. (2006, June 12-14). An action research study on the design and development of core it artifacts for knowledge management systems. In *Proceedings of the 14th European Conference on Information Systems*, University of Göteborg, Sweden.

Cale, E. G., Jr. (1994). Quality issues for end-user developed software. *Journal of Systems Management, 45*(1), 36-39.

Calisir, F., & Calisir, F. (2004). The relation of interface usability characteristics, perceived usefulness, and perceived ease of use to end-user satisfaction with enterprise resource planning (ERP) systems. *Computers in Human Behaviour, 20,* 505-515.

Callahan, E. (2005). Cultural similarities and differences in the design of university Web sites. *Journal of Computer-Mediated Communication, 11*(1). Retrieved February 15, 2006, from http://jcmc.indiana.edu/vol11/issue1/callahan.html

Camerer, C. F., & Johnson, E. C. (1991). The process-performance paradox in expert judgment: How can experts know so much and predict so badly? In K. A. Ericsson & J. Smith (Eds.), *Toward a general theory of expertise* (pp. 195-217). Cambridge, England: Cambridge University Press.

Campbell, A. (2000). Cultural identity as a social construct. *Intercultural Education, 11*(1), 31-39.

Campbell, D. T., & Fiske, D. W. (1959). Convergent and discriminant validation by the multitrait-multimethod matrix. *Psychological Bulletin, 56*(1), 81-105.

Carmine, E. G., & Zeller, R.A. (1979). *Reliability and Validity Assessment.* Newbury Park, CA: Sage Publications.

Carter, K., & Henderson, A. (1999). Tailoring culture. *Proceedings of the 13th Information Systems Research Seminar (IRIS'13),* 103-116.

Cash, C.H., & Fox, R. (1992). Elements of successful project management. *Journal of Systems Management, 43*(9), 10-12.

Castells, M. (1997). *The power of identity* (vol. 2). Oxford: Blackwell.

Caulton, D. A. (2001). Relaxing the homogeneity assumption in usability testing. *Behavior and Information Technology*, 20, 1-7.

Cavaye, A. L. M. (1995). User participation in system development revisited. *Information and Management, 28*(5), 311-323.

Cavusoglu, H., Mishra, B., & Raghunathan, S. (2004). The effect of Internet security breach announcements on market value: Capital market reactions for breached firms and Internet security developers. *International Journal of Electronic Commerce, 9*(1), 69-104.

Cedar Crestone (2007). *Cedar Crestone 2007-2008 HR Systems Survey: HR Technologies, Service Delivery Approaches, and Metrics 10th Annual Edition*, Cedar Crestone Inc.

Chait, L. P. (1999). Creating a successful knowledge management system. *Journal of Business Strategy, 20*(2), 23-26.

Chan, Y., & Storey, V. C. (1996). The use of spreadsheets in organizations: Determinants and consequences. *Information & Management, 31*(3), 119-134.

Chandler, A. (1996). The changing definition and image of hackers in popular discourse. *International Journal of the Sociology of Law, 24*, 229-251.

Chantler, N.(1996). Risk: The profile of the computer hacker. Ph.D. dissertation, Curtin University of Technology, Perth, Western Australia.

Charry, M., Pimentel, H., & Camargo, R. (2000). User reactions in continuous speech recognition systems. *AVIOS Proceedings of The Speech Technology & Applications Expo*, (pp. 113-130).

Chau, P. Y. (1996). An empirical assessment of a modified technology acceptance model. *Journal of Management Information Systems, 13*, 185-204.

Chen, L., Soliman, K. S., Mao, E., & Frolick, M. N. (2000). Measuring user satisfaction with data warehouses: An exploratory study. *Information & Management, 37*(3), 103-110.

Chen, Q., & Sharma, V. (2002). Human factors in interface design: An analytical survey and perspective. In Snodgrass, C. R., & Szewczak, E. J. (eds.) *Human Factors in Information Systems*. Hershey, PA: IRM Press.

Chen, W., Boase, J., & Wellman, B. (2002). The global villagers: Comparing Internet users and uses around the world. In B. Wellman & C. Haythornthwaite (Eds.), *The Internet in everyday life* (pp. 74-113). Malden, MA: Blackwell.

Cheney, P. H., Mann, R. I., & Amoroso, D. L. (1986). Organizational factors affecting the success of end user computing. *Journal of Management Information Systems, 3*, 65-80.

Cherns, A. (1976). The principles of sociotechnical design. *Human Relations, 29*(8), 783-792.

Chervany, N. L. (1973). Discussion comments for empirical studies of management information systems. *Data Base, 5*(2-4), 180-181.

Cho, B., Kwan, U., Gentry, J. W., Jun, S., & Kropp, F. (1999). Cultural values reflected in theme and execution: A comparative study of U.S. and Korean television commercials. *Journal of Advertising, 28*(4), 59-73.

Cho, N., & Park, S. (2001). Development of electronic commerce user-consumer satisfaction index (ECUSI) for Internet shopping. *Industrial Management & Data Systems, 101*(8), 400-405.

Choe, J. (1996). The relationships among performance of accounting information systems, influence factors, and evolution level of information systems. *Journal of Management Information Systems, 12*(4), 215-240.

Choi, B., & Lee, H. (2002). Knowledge management strategy and its link to knowledge creation process. *Expert Systems with Applications, 23*(3), 173-187.

Choi, J. H., & Danowski, J. (2002). Making a global community on the Net—global village or global metropolis?: A network analysis of Usenet newsgroups. *Journal of Computer-Mediated Communication, 7*(3). Retrieved December 9, 2005, from http://jcmc.indiana.edu/vol7/issue3/choi.html#theories

Choudhury, V., & Sampler, J. L. (1997). Information specificity and environmental scanning: An economic perspective. *MIS Quarterly, 21*(1), 25-53.

Christ, K. A. (1984). *Literature review of voice recognition and regeneration technology for Army helicopter applications* (Army Report No. HEL-TN-11-84). Aberdeen Proving Ground, MD: Human Engineering Lab.

Christie, B. (1981). Face to file communication: A psychological approach to information systems. New York: John Wiley & Sons.

Chua, A. (2004). Knowledge management system architecture: A bridge between KM consultants and technologies. *International Journal of Information Management*, 24, 87-98.

Chua, A., & Lam, W. (2005). Why KM projects fail: A multi-case analysis. *Journal of Knowledge Management, 9*(3), 6-17.

Chung, K.S., & Peterson, D.K. (2000,2001). Developers' perceptions of information systems success factors. *Journal of Computer Information Systems, 41*(2), 29-35.

Churchill, G. A. (1979). A paradigm for developing better measures of marketing constructs. *Journal of Marketing Research, 19*(4), 491-504.

Ciborra, C., Braa, K., Cordella, A., Dahlbom, B., Failla, A., Hanseth, O., Hepso, V., Ljungberg, J., Monteiro, E., & Simon, K. A. (2000). *From control to drift.* Oxford: Oxford University Press.

Clarke, F. R. (1990). Confidence ratings, second-choice responses, and confusion matrices in intelligibility tests. *Journal of the Acoustical Society of America, 32*(1), 35-46.

Clegg, C. W. (2000). Sociotechnical principles for system design. *Appl Ergon, 31*(5), 463-477.

Clermont, M., Hanin, C. & Mittermeier, R. (2000). A spreadsheet auditing tool evaluated in an industrial context. In *Proceedings of the EuSpRIG 2000 Symposium, Spreadsheet Risks—the Hidden Corporate Gamble* (pp. 35-46). Greenwich, England: Greenwich University Press.

Coldwell, R. (1995). Australian attitudes toward legal intervention into hacking. *Communications of the ACM, 38*(11),115-117.

Collins, H. (1993). The structure of knowledge. *Social Research, 60*(1), 95-116.

Compeau, D., Olfman, L., Sei, M., & Webster, J. (1995). End-user training and learning. *Communications of the ACM, 38*(7), 24-26.

Comrey, A. L., & Lee, H. B. (1992). *A first course in factor analysis.* New Jersey: Lawrence Erlbaum.

Conner, M., & Armitage, C. J. (1998). Extending the theory of planned behavior: A review and avenues for further research. *Journal of Applied Social Psychology, 28,* 1429-1464.

Conner, M., & Sparks, P. (1996). The theory of planned behavior and health behaviors. In M. Conner & P. Norman (Eds.), *Predicating health behavior.* Buckingham: Open University Press.

Constantine, L. L., & Lockwood, L. A. D. (1999). *Software for use: A practical guide to the models and methods of usage-centered design.* New York: ACM Press.

Conti, G. (2005). Why computer scientists should attend hacker conferences. *Communications of the Association for Information Systems, 48,* 23-24.

Cook, J.D., Hepworth, S.J., Wall, T.D., & Warr, P.B. (1981). *The experience of work.* San Diego: Academic Press.

Cooper, A. (1999). The inmates are running the asylum: Why high-tech products drive us crazy and how to restore the sanity. Indianapolis, IN: SAMS Publishing.

Cooper, L. P. (2003). A research agenda to reduce risk in new product development through knowledge management: A practitioner perspective. *Journal of Engineering and Technology Management, 20*(1), 117-140.

Costabile, M. F., Fogli, D., Mussion, P., & Piccinno, A. (2006). End-user development: The software shaping workshop approach. In H. Lieberman, F. Paternò & V. Wulf (Eds.), *End user development* (1st ed., pp. 183-205). Netherlands: Springer.

Costanzo, C. (2003). Consumers still wary of voice recognition. *American Banker, 168,* 10-11.

Couger, J, D., Davis, G.B., Dologite, D G., & Feinstein, D. L. (1995). IS' 95: Guidelines for undergraduate IS curriculum. *MIS Quarterly, 19*(3), 341-359.

Crabtree, A. (2003). *Designing collaborative systems: A practical guide to ethnography.* London: Springer-Verlag.

Cragg, P. G., & King, M. (1993). Spreadsheet modelling abuse: An opportunity for OR? *Journal of the Operational Research Society, 44*(8), 743-752.

Critchley, S. (1999). *The ethics of deconstruction: Derrida and Levinas* (2nd ed.). Edinburgh University Press.

Croll, G.J. (2005). The importance and criticality of spreadsheets in the city of London. Paper presented to the *EuSpRIG Conference*. Available at http://www.eusprig.org/tiacositcol4.pdf.

Cross, N. (2006). *Designerly ways of knowing*. London: Springer-Verlag.

Cyert, R., & March, J. (1963). *A behavioral theory of the firm*. Englewood Cliffs, NJ: Prentice Hall.

Czarniawska, B. (2000). The European capital of the 2000s: On image construction and modeling. *Corporate Reputation Review, 3,* 202-217.

Dabney, D.A., Hollinger, R.C., & Dugan, L. (2004). Who actually steals? A study of covertly observed shoplifters. *Justice Quarterly, 21*(4), 694-728.

Damodaran, L., & Olphert, W. (2000). Barriers and facilitators to the use of knowledge management systems. *Behaviour and information Technology, 19*(6), 405-413.

Das, S. (1999, May 6). E-tag chaos hits tollway opening. *The Age*, p. 1-2.

Davenport, T. H., & Glaser, (2002). Just-in-time delivery comes to knowledge management. *Harvard Business Review,* 107-111.

Davenport, T. H., & Prusak, L. (1998). *Working knowledge: How organizations manage what they know.* Boston: Harvard Business School Press.

Davenport, T. H., De Long, D. W., & Beers, M. C. (1998). Successful knowledge management projects. *Sloan Management Review, 39*(2), 43-57.

Davenport, T., & Glaser, J. (2002). Just-in-time delivery comes to knowledge management. *Harvard Business Review, 80*(7), 5-9.

Davies, N., & Ikin, C. (1987). Auditing spreadsheets. *Australian Accountant*, 54-56.

Davis, F. D. (1986). *A technology acceptance model for empirically testing new end-user information systems: Theory and results.* Unpublished doctoral dissertation, Massachusetts Institute of Technology, Cambridge.

Davis, F. D. (1989). Perceived usefulness, perceived ease of use, and user acceptance of information technology. *MIS Quarterly, 13*(3), 319-340.

Davis, F. D. (1993). User acceptance of information technology: System characteristics, User perceptions, and behavioral impacts. *International Journal of Man-Machine Studies, 38,* 475-487.

Davis, F. D., Bagozzi, R. P., & Warshaw, P. R. (1989). User acceptance of computer technology: A comparison of two theoretical models. *Management Science, 35*(8), 982-1003.

Davis, F.D., Bagozzi, R. P. & Warshaw, P. R. (1992). Extrinsic and intrinsic motivation to use computers in the workplace, *Journal of Applied Social Psychology, 22*(4), 1111-1132.

Davis, L. (1992). Instrument review: Getting the most from your panel of experts. *Applied Nursing Research, 5,* 104-107.

de Mooij, M. (1998). *Global marketing and advertising: Understanding cultural paradoxes.* Thousand Oaks, CA: Sage.

Deans, K. R., & Allmen, S. (2002). **Poo poo portals at your peril.** Proceedings of the *Eighth Australian World Wide Web Conference* (pp. 117-135)**.** Queensland, Australia.

Dearing, J. W., Meyer, G., & Kazmierczak, J. (1994). Portraying the new communication between university innovators and potential users. *Communication Science, 16,* 11-42.

Delone, W. H. & McLean, E. R. (1992) Information systems success: The quest for the dependent variable, *Information Systems Research, 3*(1), 60-95.

DeLone, W. H., & McLean, E. R. (2003). The DeLone and McLean model of information systems success: A ten-year update. *Journal of Management Information Systems, 19*(4), 9-30.

Dennis, A. R., Wixom, B. H., & Vandenberg, R. J. (2001). Understanding fit and appropriation effects in group support systems via meta-analysis. *MIS Quarterly, 25*(2), 167-193.

Dennis, A., & Wixon, B. H. (2000). *Systems analysis and design—An applied approach.* New York: John Wiley & Sons, Inc.

Derrida, J. (1991). Letter to a Japanese friend [original 1987]. In P. Kamuf (Ed.), *A Derrida reader: Between the blinds* (pp. 270-276). New York: Columbia University Press.

Derrida, J. (1995). *Points: Interviews, 1974-1994.* Stanford, CA: Stanford University Press.

Derrida, J. (2001). Deconstructions: The impossible. In S. Lotringer & S. Cohen (Eds.), *French theory in America* (pp. 12-32). New York and London: Routledge.

Desouza, K. C. (2006). *Knowledge management maturity model theoretical development and preliminary empirical testing.* Doctoral Thesis, University of Illinois at Chicago, Chicago, Illinois.

Desouza, K. C., & Awazu, Y. (2005). *Engaged knowledge management: Engagement with new realities.* Basingstoke: Palgrave Macmillan.

Desouza, K. C., Awazu, Y., & Ramaprasad, A. (2007). Modifications and innovations to technology artifacts. *Technovation, 27,* 204-220.

Desouza, K. C., Awazu, Y., & Wan, Y. (2005). Factors governing the consumption of explicit knowledge. *Journal of the American Society for Information Science and Technology, 57*(1), 36-43.

Desouza, K. C., Thomas, D., Yifeng, Z., & Awazu, Y. (2004). Information integrity in healthcare enterprises: strategies for mitigation of medical errors. *International Journal of Healthcare Technology & Management, 6*(2), 1-1.

Dhillon, G., & Backhouse, J. (2000). Information system security management in the new millennium. *Communications of the ACM, 43*(7), 125-128.

Dillon, A. (2001). Usability evaluation, in W. Karwowski (ed.), *Encyclopedia of Human Factors and Ergonomics.* London: Taylor and Francis.

Dishaw, M. T., & Strong, D. M. (1999). Extending the technology acceptance model with task-technology fot constructs. *Information and Management, 36,* 9-21.

Dittrich, Y., & Lindeberg, O. (2002). Designing for changing work and business practices. In N. Patel (Eds.), *Evolutionary and Adaptive Information Systems* (pp. 152-171). USA: IDEA Group Publishing.

Dittrich, Y., Rönkkö, K., Eriksson, J., Hansson, C., & Lindeberg, O. (2007, December). Co-operative method development – combining qualitative empirical research with process improvement. *Empirical Software Engineering Journal.*

Dix, A., Finlay, J., Abword, D., & Beale, R. (2004). *Human-computer interaction* (3rd. ed.). Harlow, UK: Pearson Education Limited.

Doherty, N.F., & King, M. (2001). An investigation of the factors affecting the successful treatment of organizational issues in systems development projects. *European Journal of Information Systems, 10*(3), 147.

Doll, W. J., & Torkzadeh, G. (1988). The measurement of end-user computing satisfaction. *MIS Quarterly, 12*(2), 259-275.

Donald, I., & Cooper, S. R. (2001). A facet approach to extending the normative component of the theory of reasoned action. *British Journal of Social Psychology, 40,* 599-621.

Dosi, G. (1982). Technological paradigms and technological trajectories. *Research Policy, 11,* 147-162.

Drennon, C. (2005). Teaching geographic information systems in a problem-based learning environment. *Journal of Geography in Higher Education, 29*(3), 385-402.

Drommi, A. (2001). Interface design: An embedded process for human computer interactivity. In Chen, Q. (ed.), *Human Computer Interaction: Issues and Challenges*. Hershey, PA: Idea Group Reference.

Druckenmiller, D. A., Acar, W., and Troutt, M. D. 2007. Usability testing of an agent-based modeling tool for comprehensive situation mapping. *International Journal of Technology Intelligence and Planning, 3*(2), 193-212.

Drucker, P. (1959). Long-range planning challenge to management science, *Management Science, 5*(3), 238-249.

Duffy, J. (1999). Harvesting Experience: Reaping the Benefits of Knowledge, ARMA International, Prairie Village, Kansas.

Dunlap, J. (2005). Problem-based learning and self-efficacy: How a capstone course prepares students for a profession. *Educational Technology Research & Development, 53*(1), 65-85.

Dunning, D., Griffin, D. W., Milojkovic, J. D., & Ross, L. (1990). The overconfidence effect in social prediction. *Journal of Personality and Social Psychology, 58*(4), 568-581.

Earl, M. (2001). Knowledge management strategies: Toward a taxonomy. *Journal of Management Information Systems, 18*(1), 215-233.

Easterby-Smith, M., Thorpe, R., & Lowe, A. (2002). *Management research: An introduction* (2nd ed.). London: Sage.

Eisenhardt, K. M., & Martin, J. A. (2000). Dynamic capabilities: What are they? *Strategic Management Journal, 21*(10/11), 1105-1121.

El-Dirghami, A. (1974). Shoplifting among students. *Journal of Retailing, 50*(3), 33-42.

Elen, J., & Lowyck, J. (1998). Students' views on the efficiency of instruction: An exploratory survey of the instructional metacognitive knowledge of university freshmen. *Higher Education, 36*(2), 231-252.

Elen, J., & Lowyck, J. (1999). Metacognitive instructional knowledge: Cognitive mediation and instructional design. *Journal of Structural Learning & Intelligent Systems, 13*(3-4), 145-169.

Elen, J., & Lowyck, J. (2000). Instructional metacognitive knowledge: A qualitative study on conceptions of freshmen about instruction. *Journal of Curriculum Studies, 32*(3), 421-444.

Ellis, P., & Walsh, P. (2004). Hacking the planet: DefCon harbors sex, geeks and rock 'n' roll in one sleepless package, Retrieved March 3, 2008, from http://www.defcon.org/html/links/dc_press/archives/12/citylife_hacking-theplanet.htm.

Emery, F. E. (1993). Characteristics of sociotechnical systems. In E. L. Trist & H. Murray (Eds.), *The Social Engagement of Social Science: A Tavistock Anthology vol. II. The Sociotechnical Perspective.* Philadelphia, PA: University of Pennsylvania Press.

Entwistle, N. J. (1991). Approaches to learning and perceptions of the learning environment. *Higher Education, 22*, 201-204.

Entwistle, N., McCune, V., & Hounsell, J. (2002). *Approaches to studying and perceptions of university teaching-learning environments: Concepts, measures and preliminary findings.* ETL Project Occasional Report.

Entwistle, N.J. (1998). Approaches to learning and forms of understanding. In B. Dart & G. Boulton-Lewis (Eds.), *Teaching and learning in higher education* (pp. 72–101). Melbourne: Australian Council for Educational Research.

Entwistle, N.J. (1998). Improving teaching through research on student learning. In J.J.F. Forest (Ed.), *University teaching: International perspectives* (pp. 73–112). New York: Garland.

Erickson, K., & Howard, P.N. (2007). A case of mistaken identity? News accounts of hacker, consumer, and organizational responsibility for compromised digital records. *Journal of Computer-Mediated Communication, 12*, 1229-1247.

Ericsson, K.A., & Simon, H.A. (1993). *Protocol Analysis: Verbal Reports as Data.* Cambridge, MA: MIT Press.

Eriksson, J. (2004). Can end-users manage system infrastructure? - User-adaptable inter-application communication in a changing business environment. *WSEAS Transactions on Computers*, *6*(3), 2021-2026.

Eriksson, J. (2007). Usability patterns in design of end-user tailorable software, *The Seventh Conference on Software Engineering Research and Practice in Sweden, SERPS 07.* Gothenburg.

Eriksson, J., Warren, P., & Lindeberg, O. (2003). An adaptable rchitecture for continuous development - User perspectives reflected in the architecture. *Proceedings of the 26th Information Systems Research Seminar (IRIS'26)*, Finland.

Erten, G., Paoletti, D., & Salam, F. (2000). Speech recognition accuracy improvement in noisy environments using Clear Voice Capture (CVC) technology. *AVIOS Proceedings of The Speech Technology & Applications Expo* (pp. 193-198).

Fagan, M. E. (1976). Design and code inspections to reduce errors in program development. *IBM Systems Journal, 15*(3), 182-211.

Fahey, L., & Prusak, L. (1998). The eleven deadliest sins of knowledge management. *California Management Review, 40*(3), 265-276.

Fehring, R. J. (1987). Methods to validate nursing diagnoses. *Heart and Lung, 16*(6), 625-629.

Feurestein, R., Rand, Y., Hoffman, M., & Miller, R. (1980). *Instrumental enrichment.* Baltimore, MD: University Park Press.

Ficarrotta, J. C., (2004). Software engineering as a profession: A moral case for licensure. In Brennan, L. L., & Johnson, V. E. (eds.), *Social, Ethical, and Policy Implications of Information Technology.* Hershey PA: Information Science Publishing.

Finlay, P.N., & Wilson, J.M. (2000). A survey of contingency factors affecting the validation of end-use spreadsheet-based decision support systems. *JORS, 51,* 949-958.

Fiol, C. M. (1991). Managing culture as a competitive resource - An identity-based view of sustainable competitive advantage. *Journal of Management, 17*(1), 191-211.

Fiore, S.M., Salas, E., Cuevas, H.M., & Bowers, C.A. (2003). Distributed coordination space: Toward a theory of distributed team process and performance. *Theoretical Issues in Ergonomics Science, 4*(3-4), 340-364.

Fischer, G., & Girgensohn, A. (1990). End-user modifiability in design environments. *The conference on Human Factors in Computing Systems, CHI'90.* Washington, USA.

Fishbein, M., & Ajzen, I. (1975). *Belief, attitude, intention and behavior: An introduction to theory and research.* Reading: Addison-Wesley.

Fleck, J. (1994). Learning by trying: The implementation of configurational technology. *Research Policy, 23*(6), 637-652.

Fleck, J. (1997). Contingent knowledge and technology development. *Technology Analysis and Strategic Management, 9*(4), 383-397.

Flood, J. (2000). Read all about it: Online learning facing 80% attrition rates. *Turkish Online Journal of Distance Education, 3.*

Floyd, B. D., Walls, J., & Marr, K. (1995). Managing spreadsheet model development, *Journal of Systems Management, 46*(1), 38-43, 68.

Franz, C. R., & Robey, D. (1984). An investigation of user-led system design: Rational and political perspectives. *Communications of the Association for Computing Machinery, 27*(12), 1202-1209.

Franz, C. R., & Robey, D. (1986). Organizational context, user involvement, and the usefulness of information systems. *Decision Sciences, 17*(3), 328-357.

Freestone, O., & Mitchell, V-W. (2004). Generation Y attitudes towards E-Ethics and Internet-related misbehavior. *Journal of Business Ethics, 54*, 121-128.

Frenkel, S., Tam, M., Korczynski, M., & Shire, K. (1998). "Beyond bureaucracy" work organisation in call centres. *International Journal of Human Resource Management, 9*(6), 957-979.

Friedman, A. L., & Cornford, D. S. (1989). *Computer systems development: History, organization and implementation.* Chichester, England: John Wiley and Sons.

Fuller, R. (1990). Learning to make errors: Evidence from a driving simulation. *Ergonomics, 33*(10/11), 1241-1250.

Gable, G., Yap, C. S., & Eng, M. N. (1991). Spreadsheet investment, criticality, and control. *Proceedings of the Twenty-Fourth Hawaii International Conference on System Sciences, Vol. III* (pp. 153-162). Los Alamitas, CA: IEEE Computer Society Press.

Gaddy, L. (2000). The future of speech I/O in mobile phones. *SpeechTEK Proceedings* (pp. 249-260).

Galletta, D. F., & Hufnagel, E. M. (1992). A model of end-user computing policy: Context, process, content and compliance. *Information and Management, 22,* 1-28.

Galletta, D. F., Abraham, D., El Louadi, M., Lekse, W., Pollailis, Y. A., & Sampler, J. L. (1993). An empirical study of spreadsheet error-finding performance. *Journal of Accounting, Management, and Information Technology, 3*(2), 79-95.

Galletta, D.F., & Lederer, A.L. (1989). Some cautions on the measurement of user information satisfaction. *Decision Sciences. 20*(3), 419-438.

Gantt, M., & Nardi, B.A. (1992). Gardeners and gurus: Patterns of cooperation among CAD Users. *Proceedings of the SIGCHI Conference on Human Factors in Computing Systems*, (pp. 107-117).

Garrity, E.J., & Sanders, L.G. (1998). *Information systems success measurement.* Hershey, PA: Idea Group Publishing.

Gattiker, U.E., & Kelley, H. (1999). Morality and computers: Attitudes and differences in judgments. *Information Systems Research, 10*(3), 233-254.

Gaudin, S. (1998). Migration plans? Remember: Talk to end users up front. *Computerworld, 32*(43), 1-2.

Gebauer, J., Shaw, M. J., & Gribbins, M. L. (2005). *Task-technology fit for mobile information systems. Working paper.* Retrieved September 1, 2006, from http://www.business.uiuc.edu/Working_Papers/papers/06-0107.pdf. Last Updated July 27, 2005.

Gefen, D. (2000). Lessons learnt from the successful adoption of an erp: The central role of trust. In S. Zanakis, G. Doukidis, & C. Zopounidis (Eds.), *Recent developments and applications in decision making.* Norwood, MA: Kluwer Academic.

Gefen, D., & Straub, D. (2000). The relative importance of perceived ease of use in IS adoption: A study of e-commerce adoption. *Journal of the Association for Information Systems, 1*(8). Retrieved February 14, 2001, from http://jais.aisnet.org/articles/1-8/default.asp?x=95&y=6

Gefen, D., & Straub, D. W. (1997). Gender differences in the perception and use of e-mail: An extension to the technology acceptance model. *MIS Quarterly, 21,* 389-400.

Geiger, M.A., & Cooper, E.A. (1996). Using expectancy to address student motivation. *Issues in Accounting Education, 11*(1), 113-129.

Gerson, M., Chien, I.S., & Raval, V. (1992). Computer assisted decision support systems: Their use in strategic decision making. *Proceedings of the 1992 ACM SIGCPR Conference on Computer Personnel Research* held in Cincinnati, Ohio. New York: ACM Press (pp. 152-160).

Ginzberg, M.J. (1981). Key recurrent issues in the MIS implementation process. *MIS Quarterly, 5*(2), 47-59.

Glimell, H., & Juhlin, O. (Eds.). (2001). *The social production of technology. On the everyday life with things.* Göteborg: BAS.

Godin, G. (1993). The theories of reasoned action and planned behavior: Overview of findings, emergent research problems and usefulness for exercise promotion. *Journal of Applied Sport Psychology, 9,* 491-501.

Godin, G., & Kok, G. (1996). The theory of planned behavior: A review of its application to health-related behaviors. *American Journal of Health Promotion, 11,* 87-98.

Goette, T. (2000). Keys to the adoption and use of voice recognition technology in organizations. *Library Computing, 13,* 67-80.

Gold, A. H., Malhotra, A., & Segars, A. H. (2001). Knowledge management: An organizational capabilities perspective. *Journal of Management Information Systems, 18*(1), 185-214.

Goodhue, D. L., & Thompson, R. L. (1995). Task-technology fit and individual-performance. *Mis Quarterly, 19*(2), 213-236.

Goodliffe, C. (2000). The telephone and the Internet. *AVIOS Proceedings of the Speech Technology & Applications Expo* (pp. 149-151).

Gordon, S. (1996, September).The generic virus writer II. In *Proceedings from the 6th International Virus Bulletin Conference.* Brighton, UK.

Gorham, A., & Graham, J. (2000) Full automation of directory enquiries:A live customer trial in the United Kingdom. *AVIOS Proceedings of the Speech Technology & Applications Expo* (pp. 1-8).

Gottschalk, P. (1999). Use of IT for knowledge management in law firms. *Journal of Information, Law, and Technology.* Retrieved May 15, 2002, from http://elj.warwick.ac.uk/jilt/99~3/gottschalk.html

Grady, R. B. (1995). Successfully applying software metrics. *Communications of the ACM, 38*(3), 18-25.

Graham, P. (2004). *Hackers and painters: Big ideas from the computer age.* O'Reilly.

Grant, J., & Kinney, M. (1992). Using the delphi technique to examine the content validity of nursing diagnoses. *Nursing Diagnosis, 3,* 12-22.

Grant, J.S., & Davis, L.L. (1997). Focus on quantitative methods: Selection and use of content experts for instrument development. *Research in Nursing & Health, 20,* 269-274.

Grant, R. M. (1996). Prospering in dynamically-competitive environments: Organizational capability as knowledge integration. *Organization Science: A Journal of the Institute of Management Sciences,* 7(4), 375-387.

Grant, R. M. (1996). Toward a knowledge-based theory of the firm. *Strategic Management Journal, 17(Winter96),* 109-122.

Grasmick, H., & Bryjak, G. (1980). The deterrent effect of perceived severity of punishment. *Social Forces,* 59(2), 471-491.

Grasmick, H.G., & Scott, W.J. (1982). Tax evasion and mechanisms of social control: A comparison with grand and petty theft. *Journal of Economic Psychology, 2,* 213-230.

Gray, P. H. (2000). The effects of knowledge management systems on emergent teams: Towards a research model. *Journal of Strategic Information Systems, 9*(2-3), 175-191.

Gray, P. H., & Durcikova, A. (2006). The role of knowledge repositories in technical support environments: Speed versus learning in user performance. *Journal of Management Information Systems, 22*(3), 159-190.

Gregor, S. (2006). The nature of theory of information systems. *MIS Quarterly, 30*(3), 611-642.

Grenblad, D. (2003). *Growth area – E-services in the public sector, analyses of the innovation system in 2003.* Vinnova (The Swedish Agency for Innovation Systems).

Gribble, R.K., & Haupt, C. (2005). Quantitative and qualitative differences between handout and mailed patient satisfaction surveys. *Medical Care, 43,* 276-281.

Grossman, T.A. (2002). Spreadsheet engineering: A research framework. *European Spreadsheet Risks Interest Group 3rd Annual Symposium*, Cardiff . Available at: http://www.usfca.edu/sobam/faculty/grossman_t.html

Grossman, T.A. (2003). Accuracy in spreadsheet modeling systems. *European Spreadsheet Risks Interest Group 4th Annual Symposium*, Dublin.

Grossman, T.A., & Özlük, O. (2004). A paradigm for spreadsheet engineering methodologies. *European Spreadsheet Risks Interest Group 5th Annual Symposium*, Klagenfurt, Austria. Available at: http://www.usfca.edu/sobam/publications/AParadigmforSpreadsheetEngineeringMethodologies2004.pdf.

Grover, V. (1993). An empirically derived model for the adoption of customer-based interorganizational systems. *Decision Sciences, 24*(3), 603-640.

Gudykunst, W. B., & Ting-Toomey, S. (1988). *Culture and interpersonal communication.* Newbury Park, CA: Sage.

Gunawardena, C.N. (1995). Social presence theory and implications for interaction and collaborative learning in computer conferences. *International Journal of Educational Telecommunications, 1*(2/3), 147-166.

Habberley, J. S., Shaddick, C. A., & Taylor, D. H. (1986). *A behavioural study of the collision avoidance task in bridge watchkeeping.* Southampton, England: College of Marine Studies. Cited in Reason (1990).

Hackett, B. (2000). *Beyond knowledge management: New ways to work and learn* (Research Rep. No. 1262-00-RR). The Conference Board Inc.

Haddon, L., & Kommonen, K. H. (2003). *Interdisciplinary explorations: A dialogue between a sociologist and a design group.* Helsinki: University of Art and Design.

Hage, J., & Aiken, M. (1970). Social change in complex organizations. New York: Random House.

Hahn, J., & Subramani, M. R. (2000). A framework of knowledge management systems: Issues and challenges for theory and practice. In *Proceedings of the Twenty First International Conference on Information Systems,* Brisbane, Queensland, Australia (pp. 302-312). Atlanta, GA: Association for Information Systems.

Hair, J.E., Anderson, R.E., Tatham, R.L., & Black,W. C. (1998). *Multivariate data analysis with readings.* Englewood Cliffs: Prentice Hall.

Hall, E. T. (1976). *Beyond culture.* Garden City, NY: Doubleday & Company.

Hall, E. T. (1984). *The dance of life.* Garden City, NY: Doubleday/Anchor.

Hall, M. J. J. (1996). A risk and control oriented study of the practices of spreadsheet application developers. In *Proceedings of the 29th Hawaii International Conference on System Sciences* (pp. 364-373).

Hall, M. J. J. (1996). A risk and control oriented study of the practices of spreadsheet application developers. *Proceedings of the Twenty-Ninth Hawaii International Conference on Systems Sciences, Vol. II* (pp. 364-373). Kihei, Hawaii. Los Alamitas, CA: IEEE Computer Society Press.

Hallin, A. (2003). *Mobile technology and social development – Dialogic spaces in msociety.* EGOS Annual Conference, Copenhagen, July 2-5.

Han, S.-P., & Shavitt, S. (1994). Persuasion and culture: Advertising appeals in individualistic and collectivistic cultures. *Journal of Experimental Social Psychology, 30,* 326-350.

Hankins, M., French, D., & Horne, R. (2000). Statistical guidelines for studies of the theory of reasoned action and the theory of planned behaviour. *Psychology and Health, 15,* 151-161.

Hansen, M. T., Nohria, N., & Tierney, T. (1999). What's your strategy for managing knowledge? *Harvard Business Review, 77*(2), 106-116.

Hansen, M., Nohria, N., & Tierney, T. (1999). What's your strategy for managing knowledge? *Harvard Business Review, 77*(2), 106-116.

Hanseth, O., & Braa, K. (1998). In R. Hirschheim, M. Newmann, & J. I. DeGross (Eds.), *Proceedings of the*

19th International Conferenceon Information Systems (pp. 188-196). Helsinki, Finland.

Hardy, C., & Clegg, S. R. (1996). Some dare call it power. In S. R. Clegg, C. Hardy, & W. R. Nord (Eds), *Handbook of organization studies* (pp. 622-641). London: Sage.

Harland, T. (1998). Moving towards problem-based learning. *Teaching in Higher Education, 3*(2), 219-230.

Harrell, A., Caldwell, C., & Doty, E. (1985, October). Expectancy theory predictions of accounting students' academic success motivation. *The Accounting Review*, pp. 724-735.

Harrison, A.W., & Rainer Jr., R.K. (1996). A general measure of user computing satisfaction. *Computers in Human Behavior, 12*(1), 79-92.

Hartwick, J., & Barki, H. (1994). Explaining the role of user participation in information system use. *Management Science, 40,* 440-465.

Harvey, D. (1989). From managerialism to entrepreneurialism: The transformation in urban governance in late capitalism. *Geografiska Annaler, 71B*(1), 3-17.

Hasanali, F. (2002). *Critical success factors of knowledge management.* Retrieved from http://www.kmadvantage.com/docs/km_articles/Critical_Success_Factors_of_KM.pdf

Hatcher, L. (1994). *SAS system for factor analysis and structural equation modeling.* Cary: SAS.

Hausenblas, H. A., Carron, A. V., & Mack, D. E. (1997). Application of the theories of planned behavior and reasoned action to exercise behavior," *Journal of Sport and Exercise Psychology, 19,* 36-51.

Haynes, T. (2000). Conversational IVR: The future of speech recognition for the enterprise. *AVIOS Proceedings of the Speech Technology & Applications Expo,* (pp. 15-32).

Heichler, E. (1995). Move to groupware sparks user resistance. *Computerworld, 29*(11), 12.

Hekkert, P., & Van Dijk, M. (2001). Designing from context. In P. Lloyd & H. Christiaans (Eds.), *Designing in context* (pp. 383-394). Delft: Delft University Press Science.

Henderson, A., & Kyng, M. (1991). There's no place like home: Continuing design in use. In J. Greenbaum, & M. Kyng (Eds.), *Design at Work* (pp. 219-240). Hillsdale, NJ: Lawrence Erlbaum.

Hendry, D.G., & Green, T.R.G. (1994). Creating, comprehending and explaining spreadsheets: A cognitive interpretation of what discretionary users think of the spreadsheet model. *International Journal of Human-Computer Studies, 40,* 1033-1065.

Hermansen, L. A., & Pugh, W. M. (1996). *Conceptual design of an expert system for planning afloat industrial hygiene surveys* (Technical Report No. 96-5E). San Diego, CA: Naval Health Research Center.

Hevner, A. R., March, S. T., Park, J., & Ram, S. (2004). Design science research in information systems. *MIS Quarterly, 28*(1), 75-105.

Hewling, A. (2005). Culture in the online class: Using message analysis to look beyond nationality-based frames of reference. *Journal of Computer-Mediated Communication, 11*(1). Retrieved December 15, 2005, from http://jcmc.indiana.edu/vol11/issue1/hewling.html

Higgins, G.E., Wilson, A.L., & Fell, B. D. (2005). An application of deterrence theory to software piracy. *Journal of Criminal Justice and Popular Culture, 12*(3), 166-184.

Hiltz, S.R., & Wellman, B. (1997). Asynchronous learning networks as a virtual classroom. *Communications of the ACM, 40*(9), 44-49.

Hinkin, T.R., & Schriesheim, C.A. (1989). Development and application of new scales to measure the French and Raven (1959) bases of social power. *Journal of Applied Psychology, 74*(4), 561-567.

Hirschheim, R., Klein, H. K., & Lyytinen, K. (1996). Exploring the intellectual structures of information systems development: A social action theory analysis. *Accounting, Management and Information Technologies, 6,* 1-64.

Hislop, D. (2003). Linking human resource management and knowledge management via commitment: A review and research agenda. *Employee Relations, 25*(2), 182-202.

Hislop, D., Newell, S., Scarbrough, H., & Swan, J. (2000). Networks, knowledge and power: Decision making, politics and the process of innovation. *Technology Analysis and Strategic Management, 12*(3), 399-411.

Hodges, J. (1991). Six (or so) things you can do with a bad model. *Operations Research, 39*, 355-365.

Hofer, B.K. (2004). Epistemological understanding as a metacognitive process: Thinking aloud during online searching. *Educational Psychologist, 39*(1), 43-55.

Hoffer, J., George, J., & Valacich, J. (2002). *Modern systems analysis and design* (3rd ed.). Reading, MA: Addison-Wesley.

Hoffer, J.A., & Straub, D.W. (1989). The 9 to 5 underground: Are you policing computer crimes? *Sloan Management Review, 30*(4), 35-43.

Hoffman, G. M. (2004). Ethical challenges for information system professionals. In Brennan, L. L., & Johnson, V. E. (eds.), *Social, Ethical and Policy Implications of Information Technology*. Hershey PA: Information Science publishing.

Hofstede, G. (1980). *Culture's consequences.* Beverly Hills, CA: Sage.

Hofstede, G. (1983). Dimensions of national cultures in fifty countries and three regions. In J. Deregowski, S. Dzirawiec, & R. Annis (Eds.), *Explications in cross-cultural psychology* (pp. 389-407). Lisse, The Netherlands: Swets and Zeitlinger.

Hofstede, G. (1984). *Culture's consequences: International differences in work-related values.* Beverly Hills, CA: Sage.

Holmström, H. (2001). Virtual communities as platforms for product development: An interpretive case study of customer involvement in online game development. In S. Sarkar, V. C. Storey, & J. I. De Gross (Eds), *Proceedings of the 22nd International Conference on Information Systems Association for Information Systems,* New Orleans, LA (pp. 299-306).

Holsapple, C., & Joshi, K. (2000). An investigation of factors that influence the management of knowledge in organizations. *Journal of Strategic Information Systems, 9*(2-3), 235-261.

Holtzman, T. (2000). Improving patient care through a speech-controlled emergency medical information system," *AVIOS Proceedings of the Speech Technology & Applications Expo*, (pp. 73-81).

Howarth, C. I. The relationship between objective risk, subjective risk, and behavior. *Ergonomics, 31,* 527-535. Cited in Wagenaar & Reason, 1990.

Howcroft, D., & Light, B. (2006). Reflections on issues of power in packaged software selection. *Information Systems Journal, 16*(3), 215-235.

Howe, H., & Simkin, M.G. (2006). Factors affecting the ability to detect spreadsheet errors. *Decision Sciences Journal of Innovative Education, 4*(1), 101-122.

Hoyle, R. H. (Ed.). (1995). *Structural equation modeling: Concepts, issues, and applications.* Thousand Oaks, CA: Sage.

Hu, P. J., Chau, P. Y., Sheng, O. R., & Tam, K. Y. (1999). Examining the technology acceptance model using physician acceptance of telemedicine technology. *Journal of Management Information Systems, 16,* 91-112.

Huang, J. H., Yang, C., Jin, B. H., & Chiu, H. (2004). Measuring satisfaction with business-to-employee systems. *Computers in Human Behaviour, 20*, 17-35.

Huber, G. P. (2001). Transfer of knowledge in knowledge management systems: Unexplored issues and suggested studies. *European Journal of Information Systems, 10*(2), 72-79.

Hummes, J., & Merialdo, B. (2000). Design of extensible component-based groupware. *Computer Supported Cooperative Work (CSCW), 9*(1), 53-74.

Hunton, J.E., & Beeler, J.D. (1997). Effects of user participation in systems development: A longitudinal field experiment. *MIS Quarterly, 21*(4), 359-388.

IDC. (2003). Press release. Retrieved November 30, 2006, from *http://www.idc.com/getdoc.jhtml?containerId=pr2003_01_14_145111*

Igbaria M., Guimaraes, T., & Davis, G. B. (1995). Testing the determinants of microcomputer usage via a structural equation model. *Journal of Management Information Systems, 11*(4), 87-114.

Igbaria, M. (1994). An examination of the factors contributing to microcomputer technology acceptance. *Accounting, Management & Information Technology, 4*(4), 205-224.

Igbaria, M., Pavri, F.N., & Huff, S.L. (1989). Microcomputer applications: An empirical look at usage. *Information & Management, 16*(4), 187-196.

Igbaria, M., Zinatelli, N., Cargg, P., & Cavaye, A. L. M. (1997). Personal computing acceptance factors in small firms: A structural equation model. *MIS Quarterly, 21*(3), 279-302.

Iivari, J. (1990). Implementability of in-house developed v. application package based information systems. *Data Base*, 1-10.

Iivari, J. (1995). Factors affecting perceptions of case effectiveness. *European Journal of Information Systems, 4*(3), 143-158.

Iivari, J., & Ervasti, I. (1994). User information satisfaction: IS implementability and effectiveness. *Information & Management*, 27, 205-220.

Ingram, A. L. (1991). *Report of potential applications of voice technology to armor training* (Final Report: Sep 84-Mar 86). Cambridge: Scientific Systems Inc.

Irani, Z., Sharif, A.M., & Love, P.E.D. (2001). Transforming failure into success through organizational learning: An analysis of a manufacturing information system. *European Journal of Information Systems, 10*, 55-66.

Isakowitz, T., Schocken, S., & Lucas, Jr., H.C. (1995). Toward a logical/physical theory of spreadsheet modeling. *ACM Transactions on Information Systems, 13*(1), 1-37.

ISO. (1999). ISO 13407: *Human-centred design processes for interactive systems.* Geneva, Switzerland: International Organization for Standardization.

Ives, B., Olson, M. H., & Baroudi, J. J. (1983). The measurement of user information satisfaction. *Communications of the ACM, 26*(10), 785-793.

Jackson, C. M., Chow, S., & Leitch, R. A. (1997). Toward an understanding of the behavioral intention to use an information system. *Decision Sciences, 28*(2), 357-389.

Jacobs, B.A., Topalli, V., & Wright, R. (2000). Managing retaliation: Drug robbery and informal sanction threats. *Criminology, 38*(1), 171-197.

James, H.L. (1996). Managing information systems security: A soft approach. *Proceedings of the Information Systems Conference of New Zealand*, 10-20.

Janicki, T., & Steinberg, G. (2003). Evaluation of a computer-supported learning system. *Decision Sciences Journal of Innovative Education, 1*(2), 203-223.

Janvrin, D., & Morrison, J. (2000). Using a structured design approach to reduce risks in end user spreadsheet development. *Information and Management, 37,* 1-12.

Jawahar, I. M., & Elango, B. (2001). The effect of attitudes, goal setting and self efficacy on end user performance. *Journal of End User Computing, 13,* 40-45.

Jazic, A., & Lundevall, K. (2003). *mWatch – A survey on mobile readiness in the Baltic Sea Region.* Presented at the 5th Annual Baltic Development Forum Summit, Riga, Latvia. Retrieved November 20, 2006, from http://www.bdforum.org/download.asp?id=49

Jennex, M. (2005). The issue of system use in knowledge management systems. In *Proceedings of the 38th Hawaii International Conference on System Sciences (HICSS-38 2005)* (pp. 9): IEEE Computer Society.

Jennex, M. E., & Olfman, L. (2004). Assessing knowledge management success/effectiveness models. In *Proceedings of 35th Hawaii International Conference on System, Sciences, HICSS35, IEEE Computer Society* (pp. 1-10).

Jennex, M. E., & Olfman, L. (2006). A model of knowledge management success. *International Journal of Knowledge Management, 2*(3), 51-68.

Jenster, P.V. (1987). Firm performance and monitoring of critical success factors in different strategic contexts. *Journal of Management Information Systems, 3*(3), 17-33.

Jiang, J.J., Klein, G., & Discenza, R. (2002). Pre-project partnering impact on an information system project, project team and project manager. *European Journal of Information Systems, 11*(2), 86-97.

Johnson, D. G. (2001). *Computer ethics* (3rd ed.). Upper Saddle River, NJ: Prentice Hall.

Johnson, E. J. (1988). Expertise and decision making under uncertainty: Performance and process. In M. T. H. Chi, R. Glaser, & M. J. Farr (Eds.), *The nature of expertise* (pp. 209-228). Hillsdale, NJ: Lawrence Erlbaum Associates.

Jonas, K., & Doll, J. (1996). A critical evaluation of the theory of reasoned action and theory of planned behavior. *Zeitschrift fur Sozialpsycgologie*, 27, 19-31.

Jones, M.C., & Beatty, R.C. (2001). User satisfaction with EDI: An empirical investigation. *Information Resources Management Journal, 14*(2), 17-26.

Jones, T. C. (1998). *Estimating software costs.* New York: McGraw-Hill.

Ju, B., & Gluck, M., (2005). User-process model approach to improve user interface usability. *Journal of the American Society for Information Science and Technology, 56*(10), 1098-1112.

Junnarkar, B., & Brown, C. V. (1997). Re-assessing the enabling role of information technology in KM. *Journal of Knowledge Management, 1*(2), 142-148.

Kahan, D. (1997). Between economics and sociology: The new path of deterrence. *Michigan Law Review, 95*(8), 2477-2498.

Kahler, H. (2001). *Supporting collaborative tailoring.* Roskilde: Roskilde University.

Kanellis, P., & Paul, R. J. (2005). User behaving badly: Phenomena and paradoxes from an investigation into information systems fit. *Journal of Organizational and End User Computing, 17*(2), 64-91.

Kanet, J., & Barut, M. (2003). Problem-based learning for production and operations management. *Decision Sciences Journal of Innovative Education, 1*(1), 99-118.

Kangas, K. (1997). A resource-based analysis of perceived core competencies: Applying the VRIO framework in a case of the Russian trade of a Finnish corporation. In R. Galliers, C. Murphy, S. Carlsson, H. R. Hansen, C. Loebbecke, & R. O'Callaghan (Eds.), *Proceedings of the 5th European Conference on Information Systems*, (Vol. 2, pp. 965-980).

Kankanhalli, A., & Tan, B. C. Y. (2004). A review of metrics for knowledge management systems and knowledge management initiatives. In *Proceedings of the 37th Annual Hawaii International Conference on System Sciences (HICSS-37 2004)* (pp. 238-245): IEEE Computer Society.

Kankanhalli, A., Tanudidjaja, F., Sutanto, J., & Tan, B. C. Y. (2003). The role of IT in successful knowledge management initiatives. *Communications of the Acm, 46*(9), 69-73.

Karahanna, E., & Straub, D. W. (1999). The psychological origins of perceived usefulness and ease-of-use. *Information & Management, 35*(4), 237-250.

Karahanna, E., Straub, D. W., & Chervany, N. L. (1999). Information technology adoption across time: A cross-sectional comparison of pre-adoption and post-adoption beliefs. *MIS Quarterly, 23*(2), 183-213.

Kasper, G. M. (1996). A theory of decision support system design for user calibration. *Information Systems Research, 7*(2), 215-232.

Katyal, N. (1997). Deterrence's difficulty, *Michigan Law Review, 95*(8), 2385-2476.

Katz, R. N. (2002). *Web portals and higher education: technologies to make IT personal.* San Francisco: Jossey-Bass.

Keil, M., & Robey, D. (1999) Turning around troubled software projects: An exploratory study of the de-escalation of commitment to failing courses of action, *Journal of Management Information Systems, 15*(4), 63-88.

Keil, M., Beranek, P. M., & Konsynski, B. R. (1995). Usefulness and ease of use: Field study evidence regarding task considerations. *Decisions Support Systems, 13*(1), 75-91.

Kelleher, D., & Levene, S. (2001). *Knowledge management: A guide to good practice*. London: British Standards Institution.

Kemp, R., Schot, J., & Hoogma, R. (1988). Regime shifts to sustainability through process of niche formation: the approach of strategic niche management. *Technology Analysis & Strategic Management, 10*, 175-195.

Keren, G. (1992). Improving decisions and judgments: The desirable versus the feasible. In G. Wright & F. Bolger (Eds.), *Expertise and decision support* (pp. 25-46). New York: Plenum Press.

Ketchell, M. (2003, February 28). RMIT to scrap $47m software system. *The Age*. Retrieved 23 April, 2004, from http://www.theage.com.au/articles/2003/02/27/1046064164879.html

Kim, N., & Srivastava, R. K. (1998). Managing intraorganizational diffusion of technological innovations. *Industrial Marketing Management, 27*(3), 229-246.

Kimberly, J. R., & Evansiko, M. J. (1981). Organizational innovation: The influence of individual, organizational, and contextual factors on hospital adoption of technological and administrative innovation. *Academy of Management Journal, 24*(4), 689-713.

Kincaid, D. L. (1988). The convergence theory and intercultural communication. In Y. Y. Kim & W. B. Gudykunst (Eds.), *Theories in intercultural communication* (pp. 280-298). Newbury Park, CA: Sage.

Kitts, J.A., & Macy, M.W. (1999). Structural learning: Attraction and conformity in task-oriented groups. *Computational & Mathematical Organization Theory, 5*(2), 129-145.

Klein, G., Jiang, J.J., Shelor, R., & Ballon, J.L. (1999). Skill coverage in project teams. *Journal of Computer Information Systems, 40*(1), 69-75.

Klein, H. K., & Myers, M. D. (1999). A set of principles for conducting and evaluating interpretive field studies in information systems. *MIS Quarterley, 23*(1), 67-94.

Knights, D., Murray, F., & Willmott, H. (1993). Networking as knowledge work: A study of strategic inter-organizational development in the financial services industry. *Journal of Management Studies, 30*(6), 975-995.

Koen, P., Ajamian, G., Boyce, S., Clamen, A., Fisher, E., Fountoulakis, S. et al. (2002). Fuzzy front end: Effective methods, tools, and techniques. In *PDMA toolbook for new product development* (pp. 2-35). John Wiley & Sons.

Koenig, M. (2003). Knowledge management, user education and librarianship. *Library Review, 52*(1-2), 10-17.

Koenig, M. E. D. (2004). Knowledge management strategy: Codification Vs. Personalization (A False Dichotomy). In M. E. D. Koenig & T. Srikantaiah (Eds.), *Knowledge management lessons learned : what works and what doesn't* (pp. xxv, 595 p.). Medford, N.J.: Information Today.

Koriat, A., Lichtenstein, S., & Fischoff, B. (1980). Reasons for confidence. *Journal of Experimental Psychology: Human Learning and Memory, 6*, 107-117.

Koskinen, I., Battarbee, K., & Mattelmäki, T. (2003). *Empathic design: User experience in product design.* Helsinki: Edita Publishing.

Kraiger, K., Ford, J.K., & Salas, E. (1993). Application of cognitive, skill-based, and affective theories of learning outcomes to new methods of training evaluation. *Journal of Applied Psychology, 78*(2), 311-328.

Kreytak, S. (2005, June 7). Hacker at UT was harmless, his lawyer says as trial starts. *Austin American-Statesman..*

Kruck, S.E., & Sheetz, S.D. (2001). Spreadsheet accuracy theory. *Journal of Information Systems Education, 12*(2), 93-107.

Kruck, S.E., Maher, J.J., & Barkhi, R. (2003). Framework for cognitive skill acquisition and spreadsheet training. *Journal of End User Computing, 15*(1), 20-37.

Kuhn, T. S. (1962). *The structure of scientific revolutions*. Chicago; London: University of Chicago Press.

Kujala, S. (2003). User involvement: A review of the benefits and challenges. *Behaviour and Information Technology, 22,* 1-17.

Kundupoglu, Y. (2000). Fundamentals for building a successful patent portfolio in the new millennium. *AVIOS Proceedings of the Speech Technology & Applications Expo*, (pp. 229-234).

Kusunoki, K., Nonaka, I., & Nagata, A. (1998). Organizational capabilities in product development of Japanese firms: A conceptual framework and empirical findings. *Organization Science, 9*(6), 699-718.

Kwon, T. H., & Zmud, R. W. (1987). Unifying the fragmented models of information systems implementation. In R. J. Boland & R. A. Hirschheim (Eds.), *Critical issues in information systems research*. New York: John Wiley and Sons.

Lai, J. (2000). Conversational interfaces. *Communications of the ACM, 43,* 24-27.

Lai, J. (2001). When computers speak, hear, and understand. *Communications of the ACM, 44,* 66-67.

Lai, V. S. (1997). Critical factors of ISDN implementation: An exploratory study. *Information & Management, 33*(2), 87-97.

Lallana, E. C. (2004). *eGovernment for development, mgovernment definition on and models page*. Retrieved January 13, 2005, from http://www.e-devexchange.org/eGov/mgovdefn.htm

Lam, D. (2004). Problem-based learning: An integration of theory and field. *Journal of Social Work Education, 40*(3), 371-389.

Lam, W., & Chua, A. (2005). Knowledge management project abandonment: An explanatory examination of root causes. *Communications of the Association for Information Systems*, 16, 723-743.

Lamb, R., & Kling, R. (2003). Reconceptualizing users as social actors in information systems research. *MIS Quarterly, 27*(2), 197-235.

LaPlante, A. (1991). The ups and downs of end-user involvement. *InfoWorld, 13*(47), 45-47.

Larsen, T. J., & Wetherbe, J. C. (1999). An exploratory field study of differences in information technology use between more and less innovative middle manager. *Information & Management, 36*(2), 93-108.

Latour, B. (1987). *Science in action: How to follow scientists and engineers through society.* Milton Keynes: Open University Press.

Latour, B., & Woolgar, S. (1986). *Laboratory life: The construction of scientific facts* (2nd ed.). Princeton, NJ: Princeton University Press.

Laudon, K., & Laudon, J. (1998). *Management Information Systems: New Approaches to Organisation and Technology.* Englewood Cliffs: Prentice-Hall.

Laudon, K., & Laudon, J. (2005). *Essentials of management information systems: Managing the digital firm and student multimedia edition package* (6th ed.). Upper Saddle River, NJ: Prentice Hall.

Law, W.K., & Perez, K. (2005). Crosscultural implementation of information system. *Journal of Cases on Information Technology, 7*(2), 121-130.

Lawrence, R. J., & Lee, J. (2004). *Financial modelling of project financing transactions*. Institute of Actuaries of Australia Financial Services Forum. The Institute of Actuaries of Australia, Level 7 Challis House 4 Martin Place, Sidney, NSW Australia 2000.

Lawson, B. (1997). *How designers think: The design process demystified.* Architectural Press (an imprint of Butterworth-Heineman, a division of Reed Educational and Professional Publishing).

Lazar, J., Adams, J., & Greenridge, K-D. (2005). Web-STAR: Development of survey tools for use with requirements gathering in Web site development. In *Issues of Human Computer Interaction* (Sarmento, A., ed.). Hershey, PA: IRM Press.

Lee, D., Trauth, E., & Farwell, D. (1995). Critical skills and knowledge requirements of IS professionals: A joint academic/industry investigation. *MIS Quarterly, 19*(3), 313-340.

Lee, W., & Choi, S. M. (2005). The role of horizontal and vertical individualism and collectivism in online consumers' response toward persuasive communication on the Web. *Journal of Computer-Mediated Communication, 11*(1). Retrieved December 15, 2005, from http://jcmc.indiana.edu/vol11/issue1/wnlee.html

Lehman, M.M. (1980). Programs, life cycles, and laws of software evolution. *Proceedings of the IEEE, 68*(9), 1060-1076.

Leidner, D.E., & Jarvenpaa, S. L. (1995). The use of information technology to enhance management school education: A theoretical view. *MIS Quarterly, 19*(3), 265-291.

Leitch, D., & Bain, K. (2000). Improving access for persons with disabilities in higher education using speech recognition technology. *AVIOS Proceedings of The Speech Technology & Applications Expo*, (pp. 83-86).

Leonard, D., & Sensiper, S. (1998). The role of tacit knowledge in group innovation. *California Management Review, 40*(3), 112-132.

Letiche, H. (1998). Business ethics: (In-)justice and (anti-)law. In M. Parker (Ed.), *Ethics and organizations* (pp. 122-149). London: Sage.

Letondal, C. (2006). Participatory programming: Developing programmable bioformatics tools for end-users. In H. Lieberman, F. Paternò & V. Wulf (Eds.), *End user development* (1st ed., pp. 207-242). The Netherlands: Springer.

Leuf, B., & Cunningham, W. (2001). *The wiki way: Quick collaboration on the Web.* Addison Wesley Longman.

Levinas, E. (1996). Transcendence and height [original 1962]. In A. Peperzak, S. Critchley, & R. Bernasconi (Eds.), *Emmanuel Levinas: Basic philosophical writings* (pp. 11-32). Bloomington and Indianapolis, IN: Indiana University Press.

Levinas, E. (1996). Transcendence and intelligibility [original 1984]. In A. Peperzak, S. Critchley, & R. Bernasconi (Eds.), *Emmanuel Levinas: Basic philosophical writings* (pp. 149-159). Bloomington and Indianapolis, IN: Indiana University Press.

Levi-Strauss, C. (1966). *The savage mind.* Oxford: Oxford University Press.

Lewis, C., Polson, P.G., Wharton, C., & Rieman, J. (1990). Testing a walkthrough methodology for theory-based design of walk-up-and-use interfaces. *Proceedings of the ACM CHI '90*, (April), 234-242.

Li, Y., & Kettinger, W. J. (2006). An evolutionary information-processing theory of knowledge creation. *Journal of the Association for Information Systems, 7*(9), 593-617.

Lichtenstein, S., & Fischoff, B. (1980). Training for calibration. *Organizational Behavior and Human Performance, 26*(1), 149-171.

Lichtenstein, S., Fischoff, B., & Philips, L. D. (1982). Calibration of probabilities: The state of the art to 1980. In D. Kahneman, P. Slovic, & A. Tversky (Eds.), *Judgment under uncertainty: Heuristics and biases* (pp. 306-334). Cambridge, England: Cambridge University Press.

Liebowitz, J. (1999). Information systems: Success or failure? *Journal of Computer Information Systems, 40*(1), 17-27.

Light, B. (2003). A study of organisational experiences of CRM packaged software. *Business Process Management Journal, 9*(5), 603-616.

Liker, J. K., & Sindi, A. A. (1997). User acceptance of expert systems: A test of the theory of reasoned action. *Journal of Engineering and Technology Management, 14*(2), 147-173.

Limayem, M. Cheung, C. M. K., & Chan, G. W. W. (2003, December 14-17). Explaining information systems adoption and post-adoption: Toward an integrative model. In *Proceedings of the Twenty Fourth International Conference on Information Systems,* Seattle, WA (pp. 720-731). Atlanta, GA: Association for Information Systems.

Linberg, K.R. (1999). Software developer perceptions about software project failure: A case study. *Journal of Systems and Software, 49*(2-3), 177-192.

Lindeberg, O., Eriksson, J., & Dittrich, Y. (2002). Using metaobject protocol to implement tailoring; possibilities and problems. *Proceedings of the 6th World Conference on Integrated Design & Process Technology (IDPT '02),* Pasadena, USA.

Lipshitz, R., Klein, G., Orasanu, J., & Salas, E. (2001). Taking stock of naturalistic decision making. *Journal of Behavioral Decision Making, 14,* 331-352.

Liu, R.X. (2003). The moderating effects of internal and perceived external sanction threats on the relationship between deviant peer associations and criminal offending. *Western Criminology Review, 4*(3), 101-202.

Locke, E., & Latham, G. (1990). *A theory of goal setting and task performance.* Englewood Cliffs, NJ: Prentice Hall.

Loiacono, E. T., Watson, R.T., & Goodhue, D.L. (2002). WebQual: A measure of Web site quality. In Evans, K.K. and Scheer, L.K. (Eds). *Marketing Theory and Applications, 13.* Chicago: American Marketing Association.

Lorge, I., & Solomon, H. (1995). Two models of group behavior in the solution of eureka-type problems. *Psychometrika, 20*(2), 139-148.

Lowyck, J., & Elen, J. (1994). *Student's instructional metacognition in learning environments (SIMILE).* Paper presented at the Leuven: K.U. Leuven, C.I.P. & T.

Lucas, H. C., & Spitler, V. K. (1999). Technology use and performance: A field study of broker workstations. *Decision Sciences, 30*(2), 291-311.

Lucas, H.C., Schultz, R. & Ginzberg, M. (1990). *Information systems implementation: Testing a structural model.* Norwood, NJ, Ablex.

Lukasik, T. (1998). Personal communication with the author via e-mail.

Lund, G. *The Swedish vision of 24-hour public administration and e-government – Speech by Gunnar Lund, Minister for International Economic Affairs and Financial Markets, held December 9th.* Unpublished manuscript.

Lundbergh, A. (2004). *Infra services – A Swedish way to facilitate public e-services development: MEMO.* The Swedish Agency for Public Management.

Lynn, M. (1986). Determination and quantification of content validity. *Nursing Research, 35*(6), 82-385.

Ma, J., & Hemmje, M. (2001, November 8-9). Developing a knowledge management systems step by step. In *Proceedings of the 2nd Europe Conference on Knowledge Management,* Bled, Slovenia (pp. 301-311).

Ma, Q., Pearson, M.J., & Tadisina, S. (2005). An exploratory study into factors of service quality for application service provider. *Information and Management, 43,* 1067-1080

Mackay, W. E. (1990). Patterns of sharing customizable software. *The Conference on Computer Supported Cooperative Work (CSCW '90).* Los Angeles, California, USA: ACM Press.

Mackay, W. E. (1991). Triggers and barriers to customizing software. *The conference on Human factors in Computing Systems (CHI'94).* Boston, Massachusetts, USA.

MacKenzie, D., & Wajcman, J. (1985). *The social shaping of technology.* Open University Press.

Mackenzie, D., & Wajcman, J. (Eds.). (1999). *The social shaping of technology* (2nd ed.). Maidenhead, England: Open University Press..

MacLean, A., Carter, K., Lövstrand, L., & Morgan, T. (1990). User-tailorable systems: Pressing the issues with buttons. *Proceedings of the Conference on Human Factors in Computing Systems (CHI'90)*, 175-182.

Madahar, M., Cleary, P., & Ball, D. (2007). Categorisation of spreadsheet use within organisations incorporating risk: A progress report. In *Proceedings of the European Spreadsheet Risks Interest Group 8th Annual Conference*, University of Greenwich, London, (pp.37-45).

Mahmood, Mo. A., & Medewitz, J.N. (1985). Impact of design methods on decision support systems success: An empirical assessment, *Information & Management*, *9*(3), 137-151.

Malhotra, Y. (2005). Integrating knowledge management technologies in organizational business processes: Getting real time enterprises to deliver real business performance. *Journal of Knowledge Management*, *9*(1), 7-28.

Malhotra, Y., & Galletta, D. F. (2003). The role of commitment and motivation in knowledge management systems: Theory, conceptualisation and measurement of success. In *Proceedings of the 36th Hawaii Conference on Systems Sciences* (pp. 1-10).

Mannheim, K. (2004). The sociology of knowledge and ideology. In C. Lemert (Ed.), *Social theory: The multicultural and classic readings* (3rd ed., pp. 213-216). Oxford, England: Westview Press.

Manstead, A. S., & Parker, D. (1995). Evaluating and extending the theory of planned behavior. In *European Review of Social Psychology*, *6*, 69-95.

Manstead, A. S., Proffitt, C., & Smart, J. L. (1983). Predicting and understanding mother's infant feeding intentions and behavior: Testing the theory of reasoned action. *Journal of Personality and Social Psychology*, *44*(4), 657-671.

March, J.G., & Simon, H.A. (1958). *Organizations*. New York: Wiley.

Margetson, D. (1994). Current educational reform and the significance of problem-based learning. *Studies in Higher Education*, *19*(1), 5-19.

Markowitz, J. (2000). The value of combining technologies. *AVIOS Proceedings of the Speech Technology & Applications Expo*, (pp. 199-206).

Markus, L., & Robey, D. (1988). Information technology and organisational change: Causal structure in theory and research. *Management Science*, *34*(5), 583-598.

Markus, M. L. (1983). Power, politics, and MIS implementation. *Communications of the Association for Computing Machinery*, *26*(6), 430-444.

Markus, M. L. (2001). Toward a theory of knowledge reuse: Types of knowledge reuse situations and factors in reuse success. *Journal of Management Information Systems*, *18*(1), 57-93.

Markus, M. L., & BjØrn-Andersen, N. (1987). Power over users: Its exercise by system professionals. *Communications of the Association for Computing Machinery*, *30*(6), 498-504.

Markus, M. L., & Keil, M. (1994). If we build it, they will come: Designing information systems that people want to use. *Sloan Management Review* (Summer 1994), 11-25.

Markus, M. L., Majchrzak, A., & Gasser, L. (2002). A design theory for systems that support emergent knowledge processes. *MIS Quarterly*, *26*(3), 179-212.

Marshall, N., & Brady, T. (2001). Knowledge management and the politics of knowledge: Illustrations from complex products and systems. *European Journal of Information Systems*, *10*(2), 99-112.

Marton, F. & Säljö, R. (1976). On qualitative differences in learning: 1 - Outcome and process. *British Journal of Educational Psychology*, *46*, 4-11.

Marton, F., Dall'Alba, G. & Beaty, E. (1993). Conceptions of learning. *International Journal of Educational Research, 19*, 277-300.

Martuza, V.R. (1977). *Applying norm-referenced and criterion-referenced measurement in education*. Boston: Allyn and Bacon, Inc.

Maskell, D., & Grabau, P. (1998). A multidisciplinary cooperative problem-based learning approach to embedded systems design. *IEEE Transactions on Education, 41*(2), 101-103.

Mason, D., & Pauleen, D. (2003). Perceptions of knowledge management: A qualitative analysis. *Journal of knowledge Management, 7*(4), 38-48.

Massey, A. P., Montoya-Weiss, M. M., & O'Driscoll, T. M. (2002). Knowledge management in pursuit of performance: Insights from Nortel Networks. *MIS Quarterly, 26*(3), 269-289.

Mather, D. (1999). A framework for building spreadsheet based decision models. *Journals of the Operational Rsearch Society, 50*, 70-74.

Mathieson, K. (1991). Predicting user intentions: Comparing the technology acceptance model with theory of planned behavior. *Information Systems Research, 2*, 173-191.

Mawhinney, C. H., & Lederer, A. L. (1990). A study of personal computer utilization by managers. *Information & Management, 18*(5), 243-253.

Mayor, T. (2001). E-learning: Does it make the grade. *CIO Magazine. 14*(7), 132-144.

Mazzie, M. (2000). Key challenges facing the evolution of knowledge management. In T. K. Srikantaiah & M. E. D. Koenig (Eds.), *Knowledge management for the information professional* (pp. 99-114). Medford, NJ: Information Today, Inc.

McAdam, R., & McCreedy, S. (1999). A critical review of knowledge management models. *The Learning Organization: An International Journal, 6*(3), 91-101.

McCue, A. (2004). *Oxford student "hackers" suspended.* Retrieved March 3, 2008 from http://software.silicon.com/security/0,39024655,39125462,00.htm

McDermott, R. (1999). Why information technology inspired but cannot deliver knowledge management. *California Management Review, 4*(4), 103-117.

McDermott, R., & O'Dell, C. (2001). Overcoming cultural barriers to sharing knowledge. *Journal of Knowledge Management, 5*(1), 76-85.

McGill, T. (2004). The effect of end user development on end user success. *Journal of Organizational and End User Computing, 16*(1), 41-58.

McGowan, M. K., & Madey, G. R. (1998). Adoption and implementation of electronic data interchange. In T. J. Larsen & E. McGuire (Eds.), *Information systems innovation and diffusion: Issues and directions* (pp. 116-140). London: Idea Group.

McGrath, K. (2002). The golden circle: A way of arguing and acting about technology in the London ambulance service. *European Journal of Information Systems, 11*(4), 17-26.

McLean, E. R., Kapperlman, L. A., & Thompson, J. P. (1993). Converging end user and corporate computing. *Communications of the Association for Information Systems, 36*, 76-91.

MedQuist (2003). http://www.rsleads.com/301ht-192

Menzies, H. (1995). *Whose brave new world? The information highway and the new economy*, Toronto, Canada: Between the Lines Press.

Mintzberg, H. (1975, June to August) The manager's job: Folklore and fact. *Harvard Business Review*, 49-61.

Mintzberg, H. (1994). *The rise and fall of strategic planning.* New York: Free Press.

Miracle, G. E., Chang, K. Y., & Taylor, C. R. (1992). Culture and advertising executives: A comparison of selected characteristics of Korean and U.S. television commercials. *International Marketing Review, 9*(4), 5-17.

Moffett, S., McAdam, R., & Parkinson, S. (2003). An empirical analysis of knowledge management applications. *Journal of Knowledge Management, 7*(3), 6-26.

Mogford, R. M., Rosiles, A., Wagner, D., & Allendoerfer, K. R. (1997), *Voice technology study report* (Report No. DOT/FAA/CT-TN97/2). Atlantic City, NJ: FAA Technical Center.

Molich, R., & Nielsen, J. (1990). Improving a human-computer dialogue. *Communications of the ACM, 33*(3), 338-348.

Mondale, J., Scott, S., & Venters, W. (2006). Knowledge management as an image of the organization: Industry standards and processes of knowing in credit risk management. In *Proceedings of the 14. European Conference on Information Systems (ECIS), Gothenburg, Sweden, June 11-14 2006.*

Montealegre, R., & Keil, M. (2000). De-escalating information technology projects: lessons from the Denver International Airport. *MIS Quarterly, 24*(3), 417-447.

Moore, G. C. (1987). End-user computing and office automation: A diffusion of innovation perspectives. *Information Systems and Operational Research, 25*(3), 214-235.

Moore, G. C., & Benbasat, I. (1991). Development of an instrument to measure the perceptions of adopting an information technology innovation. *Information System Research, 2*(3), 192-222.

Moorman, C., & Miner, A. S. (1998). Organizational improvisation and organizational memory. *Academy of Management Review, 23,* 698-723.

Mørch, A. (1995). Three levels of end-user tailoring: Customization, integration, and extension, *the 3rd Decennial Aarhus Conference.* Aarhus, Denmark.

Mørch, A. I., Stevens, G., Won, M., Klann, M., Dittrich, Y., & Wulf, V. (2004). Component-based technologies for end-user development. *Communications of the ACM, 47*(9), 59-62.

Mørch, A., & Mehandjiev, N. (2000). Tailoring as collaboration: The mediating role of multiple representations and application units. *Computer Supported Cooperative Work (CSCW), 9*(1), 75-100.

Morris, M. G., & Venkatesh, M. (2000). Age differences in technology adoption decisions: Implications for a changing workforce. *Personnel Psychology, 53*(2), 375-403.

Morrison, M., Morrison, J., Melrose, J., & Wilson E.V. (2002). A visual code inspection approach to reduce spreadsheet linking errors. *Journal of End User Computing, 14*(3), 51.

Muylle, S., Moenert, R., & Despontin, M.(2004). The conceptualization and empirical validation of Web site user satisfaction. *Information & Management, 41,* 543-560.

Myers, M. D. (1997). *Qualitative research in information systems.* Retrieved November 21, 1997, from http://www.auckland.ac.nz/msis/isworld/

Naatanen, R., & Summala, H. (1976). *Road user behavior and traffic accidents,* Amsterdam: North-Holland. Cited in Wagenaar & Reason, 1990.

Nardi, B. A., & Miller, J. R. (1991). Twinkling lights and nested loops: Distributed problem solving and spreadsheet development. *International Journal of Man-Machine Studies, 34*(2), 161-184.

Nardi, B.A. (1993). *A Small matter of programming: Perspectives on end user computing.* Cambridge, MA: MIT Press.

Nardi, B.A., & Miller, J.R. (1991). Twinkling lights and nested loops: distributed problem solving and spreadsheet development. *International Journal of Man-Machine Studies, 34*(1), 161-184.

Nath, R. (1989). Are frequent users more satisfied. *Information Processing and Management, 25*(5), 557-562.

Nelson, M. R., & Shavitt, S. (2002). Horizontal and vertical individualism and achievement values: A multimethod examination of Denmark and the United States. *Journal of Cross-Cultural Psychology, 33*(5), 439-458.

Nelson, R. R., & Cheney, P. H. (1987). Training end users: An exploratory study. *MIS Quarterly, 11*, 547-559.

Nelson, R. R., & Todd, P. (1999). Strategies for managing EUC on the Web. *Journal of End User Computing, 11*, 24-31.

Neumann, P.G. (1997). Integrity in software development. *Communications of the ACM, 40*(10), 144.

Nielsen, J. (1994). Heuristic evaluation. In J. Nielsen and R. Mack (eds.) *Usability inspection methods*. (pp. 25-62). New York: Wiley.

Nonaka, I., & Takeuchi, H. (1995). *The knowledge-creating company: How Japanese companies create the dynamics of innovation*. New York: Oxford: Oxford University Press.

Nonaka, I., Toyama, R., & Konno, N. (2000). SECI, ba and leadership: A unified model of dynamic knowledge creation. *Long Range Planning, 33*(1), 5-34.

Novak, T. P., Hoffman, D. L., & Yung, Y.-F. (2000). Measuring the customer experience in online environments: A structural modeling approach. *Marketing Science, 19*(1), 22-42.

Nunamaker, J., Chen, M., & Purdin, T. (1991). System development in information systems research. *Journal of Management Information Systems*, 7(3), 89-106.

Nunnally, J. C. (1978). *Psychometric Theory*. (2nd ed.). New York: McGraw-Hill.

O'Leonard, K. (2004). *HP case study: Flexible solutions for multi-cultural learners*. Oakland, CA: Bersin & Associates.

O'Mara, J., Allen, J.L., Long, K.M. & Judd, B. (1996). Communication apprehension, nonverbal immediacy, and negative expectations for learning. *Communication Research Reports, 13*, 109-128.

Offe, C. (1976). *Industry and inequality*. London: Edward Arnold Publishers Limited. (Original work published in 1970)

Oliver, R.L. (1980). A cognitive model for the antecedents and consequences of satisfaction. *Journal of Marketing Research, 17*, 460-469.

Olphert, C.W., & Wilson, J.M. (2004). Validation of decision-aiding spreadsheets: The influence of contingency factors. *Journal of the Operational Research Society, 55*, 12-22.

Olson, J. R., & Nilsen, E. (1997-1998). Analysis of the cognition involved in spreadsheet interaction. *Human-Computer Interaction, 3*(4), 309-349.

Ong, C. S., & Lai, J. Y. (2004). Developing an instrument for measuring user satisfaction with knowledge management systems. In *Proceedings of the 37th Hawaii International Conference on System Sciences*, Hawaii.

Ono, K., & Davis, J. H. (1988). Individual judgment and group interaction: A variable perspective approach. *Organizational Behavior and Human Decision Processes, 41*(1), 211-232.

Orlikowski, W. (2002). Knowing in practice: Enacting a collective capability in distributed organizing. *Organization Science, 13*(3), 249-273.

Orlikowski, W. (2006). Material knowing: The scaffolding of human knowledgeability. *European Journal of Information Systems, 15*(5), 460-466.

Orlikowski, W. J. (1992). The duality of technology: Rethinking the concept of technology in organizations. *Organization Science, 3*(3), 398-427.

Orlikowski, W. J. (1996). Improvising organizational transformation over time: A situated change perspective. *Information Systems Research, 7*(1), 63-92.

Orlikowski, W. J., & Barley, S. R. (2001). Technology and institutions: What can research on information technology and research on organizations learn from each other. *MIS Quarterly, 25*(2), 145-165.

Östberg, O. (2000). *The 24/7 agency. Criteria for 24/7 agencies in the networked public administration*. Stockholm: The Swedish Agency for Administrative Development.

Oudshoorn, N., & Pinch, T. (2003). Introduction: How users and non-users matter. In N. Oudshoorn, & T. Pinch (Eds.). *How users matter: The co-construction of users and technology* (pp. 1-25). Cambridge, MA and London, England: MIT Press.

Ouellette, T. (July 26, 1999). Giving users the keys to their Web accounts. *Computerworld*, 66-67.

Oviatt, S. (2000). Taming recognition errors with a multimodal interface. *Communications of the ACM, 43,* 45-51.

Oz, E. (1994). When professional standards are lax: The CONFIRM failure and its lessons. *Association for Computing Machinery. Communications of the ACM, 37*(10), 29-36.

Oz, E., & Sosik, J.J. (2000). Why information systems projects are abandoned: A leadership and communication theory and exploratory study. *Journal of Computer Information Systems, 41*(1), 66-78.

Pallant, J. (2005). *SPSS Surviving Manual.* (2nd Ed.) New South Wales: Allen & Unwin.

Pan, S. L., & Scarbrough, H. (1998). A socio-technical view of knowledge-sharing at Buckman laboratories. *Journal of Knowledge Management, 2*(1), 55-66.

Panko, R. (2000). Two corpuses of spreadsheet error. IEEE. In *Proceedings of the 33rd Hawaii International Conference on Systems Sciences,* Retreived from http://panko.cba.hawaii.edu/ssr/Mypapers/HICSS33-Panko-Corpuses.pdf

Panko, R. R., & Halverson, R. P., Jr. (1997). Are two heads better than one? (At reducing errors in spreadsheet modeling). *Office Systems Research Journal, 15*(1), 21-32.

Panko, R. R., & Sprague, R. H., Jr. (1998). Hitting the wall: Errors in developing and code inspecting a "simple" spreadsheet model. *Decision Support Systems, 22*(4), 337-353.

Panko, R.R. (1999). Applying code inspection to spreadsheet testing. *Journal of Information Management Systems, 16*(2), 159-176.

Panko, R.R. (2000). Spreadsheet errors: What we know. What we think we can do. In *Proceedings of the European Spreadsheet Risks Interest Group Conference* , University of Greenwich, London, (pp. 7-17). Retrieved from www.arxiv.org

Panko, R.R. (2005). What we know about spreadsheet errors. Available at: http://panko.cba.hawaii.edu/ssr/Mypapers/whatknow.htm, updated 2005. Previously published in Journal of End User Computing 10(2).

Panko, R.R. (2007) Thinking is bad: Implications of human error research for spreadsheet research and practice. *Proceedings of the European Spreadsheet Risks Interest Group 8th Annual Conference*, pp. 69-80, available at www.arxiv.org.

Paper, D., Rodger, J. A., & Simon, S. J. (in press). Voice says it all in the Navy. *Communications of the ACM.*

Parker, D. B. (2007). The dark side of computing: SRI International and the study of computer crime. *IEEE Annals of the History of Computing, 29*(1), 3-15

Parthasarathy, M., & Bhattacharjee, A. (1998). Understanding post-adoption behavior in the context of online services. *Information Systems Research, 9*(4), 362-379.

Paternoster, R., & Iovanni, L. (1986). The deterrent effort of perceived severity: a reexamination. *Social Forces, 64*, 751-777.

Patton, M. Q. (1990). *Qualitative evaluation and research methods* (2nd ed.). Newbury Park, CA: Sage.

Peace, A.G., Galletta, D., & Thong, J.Y.L (2003). Software piracy in the workplace: A model and empirical test. *Journal of Management Information Systems,* 20(1), 153-177

Perry, W. (1968). *Patterns of development in thought and values of students in a liberal arts college: A validation of a scheme* (No. 5-0825, Contract No. SAE-8973). Washington, DC: Department of Health, Education and Welfare.

Pew Internet & American Life Project. (2005). *Who's online: Demographics of Internet users* (February-March 2005 tracking survey). Retrieved July 29, 2005, from http://www.pewinternet.org/trends/User_Demo_05.18.05.htm

Phillips Fox. (1998). *Leading the way in knowledge management.* Retrieved July 20, 2001, from http://www.phillipsfox.com.au/phillipsfox/publications/pbo98006.htm

Piccoli, G., Ahmad, R. & Ives, B. (2001). Web-based virtual learning environments: A research framework and a preliminary assessment of effectiveness in basic IT skills training. *MIS Quarterly, 25*(4), 401-426.

Pinch, T., & Bijker, W. E. (1987). The social construction of facts and artifacts: Or how the sociology of science and the sociology of technology might benefit each other. In W. E. Bijker, T. P. Hughes, & T. Pinch (Eds), *The social construction of technological systems* (pp. 17-50). London: MIT Press.

Pinker, E. J., & Van Horn, R. L. (2000). Worker incentives to learn in gatekeeper systems: Lessons for the implementation of knowledge management systems. In *Proceedings of the 33rd Hawaii International Conference on System Science.* Retrieved March 21, 2002, from http://computer.org/proceedings/hicss/0493/04933/04933027.pdf

Pisano, G. P. (1994). Knowledge, integration, and the locus of learning - An empirical-analysis of process-development. *Strategic Management Journal, 15*, 85-100.

Plous, S. (1993). *The psychology of judgment and decision making.* Philadelphia: Temple University Press.

Plous, S. (1995). A comparison of strategies for reducing interval overconfidence in group judgments. *Journal of Applied Psychology, 80*(4), 443-454.

Poel, I. v. d. & Verbeek, P. P. (2006). Ethics and engineering design. *Science Technology & Human Values, 31,* 223-236.

Polanyi, M. (1966). *The tacit dimension*: London: Routledge & Kegan Paul.

Popper, K. R. S. (1972). *Objective knowledge: An evolutionary approach*: Oxford: Clarendon Press.

Powell, S.G., Lawson, B., & Baker, K.R. (2007). Errors in operational spreadsheets. Under review at *Journal of Organizational and End User Computing.*

Powell, S.G., Lawson, B., & Baker, K.R. (2007). Impact of errors in operational spreadsheets. In *Proceedings at the European Spreadsheet Risks Int. Grp.*, (pp. 57-68).

Powell, S.G., Lawson, B., & Baker, K.R. (2008). A critical review of the literature on spreadsheet errors. *Forthcoming in Decision Support Systems.*

Powell, S.G., Lawson, B., & Baker, K.R. (2008). An auditing protocol for spreadsheet models. Forthcoming in *Information & Management.*

Prizer, B., Thomas, D., & Suhm, B. (2000). The business case for speech. *SpeechTEK Proceedings,* 15-26.

Pulford, B. D., & Coleman, A. M. (1996). Overconfidence, base rates and outcome positivity/negativity of predicted events. *British Journal of Psychology, 87*(3), 431-445.

Purser, M., & Chadwick, D. (2006). Does an awareness of differing types of spreadsheet errors aid end-users in identifying spreadsheet errors. *Proceedings of the European Spreadsheet Risk Interest Group Annual Conference,* Cambridge, UK, (pp. 185-204).

Putnam, L. H. & Myers, W. (1992). *Measures for excellence: Reliable software on time, on budget.* Englewood Cliffs, NJ: Yourdon.

Rajalingham, K., Chadwick, D., & Knight, B. (2000). Classification of spreadsheet errors. *British Computer Society (BCS) Computer Audit Specialist Group (CASG) Journal, 10*(4), 5-10.

Rajalingham, K., Chadwick, D., Knight, B., & Edwards, D. (2000, January). Quality control in spreadsheets: A software engineering-based approach to spreadsheet development. *Proceedings of the Thirty-Third Hawaii International Conference on System Sciences*, Maui, Hawaii.

Rasmussen, J. (1990). The role of perception in organizing behaviour. *Ergonomics, 33*(10/11), 1185-1199.

Ray, G., Muhanna, W. A., & Barney, J. B. (2005). Information technology and the performance of the customer service process: A resource-based analysis. *MIS Quarterly, 29*(4), 625-651.

Reason, J. (1990). *Human error.* Cambridge, England: Cambridge University Press.

Reithel, B. J., Nichols, D. L., & Robinson, R. K. (1996). An experimental investigation of the effects of size, format, and errors on spreadsheet reliability perception. *Journal of Computer Information Systems, 36*(3),54-64.

Riempp, G. (2004). *Integrierte wissensmanagement-systeme: Architektur und praktische anwendung; Mit 26 Tabellen.* Berlin [u.a.]: Springer.

Rifkin, G. (1991). End-user training: Needs improvement. *Computerworld, 25*(15), 73-74.

Rigby, D., Reichheld, F., & Schefter, P. (2002). Avoid the four perils of CRM. *Harvard Bus Review, 80*(2), 101-109.

Roberts, T.L., Leigh, W., & Purvis, R.L. (2000). Perceptions on stakeholder involvement in the implementation of system development methodologies. *The Journal of Computer Information Systems, 10*(1), 78-83.

Robson, C. (2002). *Real world research.* Oxford, UK: Blackwell Publishers Ltd.

Robson, J. (2000). Evaluating on-line teaching. *Open Learning, 15*(2), 151-172.

Rockart, J. F. (1979). Chief executives define their own data needs. *Harvard Business Review, 57*(2), 81-93.

Rogers, E. M. (1995). *Diffusion of innovations* (4th ed.). New York: The Free Press.

Rogers, E. M., & Kincaid, D. L. (1981). *Communication networks: Toward a new paradigm for research.* New York: The Free Press.

Rogers, M. (2001). *A social learning theory and moral disengagement analysis of criminal behavior: An exploratory study.* Ph.D. Thesis, Dept. of Psychology, University of Manitoba, Winnipeg.

Rohracher, H. (2005). From passive consumers to active participants: The diverse roles of users in innovation processes. In H. Rohracher (Ed.), *User involvement in innovation processes: Strategies and limitations form a socio-technical perspective* (pp. 9-35). Wien, Austria: Profil.

Rohracher, H. (2005). The diverse roles of users in innovation processes. In H. Rohracher (Ed.), *User involvement in innovation processes: Strategies and limitations from a socio-technical perspective* (pp. 9-35). Munchen/Wien: Profil Verlag.

Rolandi, W. (2000). Speech recognition applications and user satisfaction in the imperfect world. *AVIOS Proceedings of the Speech Technology & Applications Expo*, (pp. 153-158).

Rolland, K. H., & Monteiro, E. (2002). Balancing the local and the global in infrastructural information systems. *The Information Society, 18,* 87-100.

Ronen, B., Palley, M. A., & Lucas, H. C. (1989). Spreadsheet analysis and design. *Communications of the ACM, 32*(1), 84-93.

Roozenburg, N. F. M., & Eekels, J. (1995). *Product design, fundamentals and methods.* Chichester, UK: John Wiley and Sons.

Roth, J. (2003). Enabling knowledge creation: Learning from an R&D organization. *Journal of Knowledge Management, 7*(1), 32-48.

Rovai, A.P. (2002). A preliminary look at the structural differences of higher education classroom communities in traditional and ALN courses. *Journal of Asynchronous Learning Networks, 6*(1), 41-56.

Rubin, J. (1994). *Handbook of usability testing: How to plan, design, and conduct effective tests.* New York: Wiley.

Ruggles, R. (1998). The State of the notion: Knowledge management in practice. *California Management Review, 40*(3), 80-89.

Saga, V., & Zmud, R. W. (1994). The nature and determinants of IT acceptance, routinization, and infusion.

In L. Levine (Ed.), *Diffusion, transfer, and implementation of information technology* (pp. 67-86). Amsterdam: North-Holland.

Salas, E., DeRouin, R., & Littrell, L. (2005). Research-based guidelines for designing distance learning: What we know so far. In H.G. Geutal & D.L. Stone (Eds.), *The brave new world of e-HR: human resources management in the digital age*. San Francisco: Jossey-Bass.

Sambamurthy, V. (2001). Editor's comments-research in information systems: What we haven't learned. *MIS Quarterly, 21*(4), v-xv.

Sanders, E. B. N. (2000). Generative tools for codesigning. In Scrivener, Ball, & Woodcock (Eds.), *Collaborative design*. London: Springer-Verlag.

Sarvary, M. (1999). Knowledge management and competition in the consulting industry. *California Management Review, 41*(2), 95-107.

Sauer, C. (1993). *Why information systems fail: A case study approach*. Oxfordshire: Alfred Waller Ltd.

Sauer, C., & Burn, M. J. (1997). The Pathology of Alignment. In C. Sauer & P. Yetton (Eds.), *Steps to the Future*. San Francisco: Jossey Bass.

Schach, S. R. (2005). *Object-oriented & classical software engineering*, 6th Ed. Boston: McGraw-Hill.

Schalk, T. (2000). Design considerations for ASR telephony applications. *AVIOS Proceedings of the Speech Technology & Applications Expo*, (pp. 103-112).

Schlosser, A. E., Shavitt, S., & Kanfer, A. (1999). Survey of Internet users' attitudes toward Internet advertising. *Journal of Interactive Marketing, 13*(3), 34-54.

Schmidt, R., Lyytinen, K., Keil, M., & Cule, P. (2001). Identifying software project risks: An international Delphi study. *Journal of Management Information Systems, 17*(4), 5-37.

Schommer, M. (1994). An emerging conceptualization of epistemological beliefs and their role in learning. In R. Garner & P.A. Alexander (Eds.), *Beliefs about text and instruction with text* (pp. 25-40). Hillsdale, NJ: Lawrence Erlbaum.

Schommer, M., Crouse, A. & Rhodes, N. (1992). Epistemological beliefs and mathematical text comprehension: Believing it is simple does not make it so. *Journal of Educational Psychology, 84*(4), 435-443.

Schommer-Aikins, M. (2004). Explaining the epistemological belief system: Introducing the embedded systemic model and coordinated research approach. *Educational Psychologist, 39*(1), 19-29.

Schuler, D., & Namioka, A. (1993). *Participatory design: Principles and practices*. NJ: Lawrence Erlbaum Associates.

Schultheis, R., & Sumner, M. (1994). The relationship of application risks to application controls: A study of microcomputer-based spreadsheet applications. *Journal of End User Computing, 6*(2), 11-18.

Scollon, R., & Wong-Scollon, S. (2001). *Intercultural communication* (2nd ed.). Oxford: Blackwell.

Seal, K., Przasnyski, Z., & Leon, L. (2000, October). A literature survey of spreadsheet based MR/OR applications:1985-1999. *OR Insight, 13*(4), 21-31.

Seddon, P.B., Staples, S., Patnayakuni, R., & Bowtell, M. (1999). Dimensions of information systems success. *Communications of AIS, 2*(3), 2-24.

Seeley, C. (2002). Igniting knowledge in your business processes. *KM Review, 5*(4), 12-15.

Segars, A. H., & Grover, V. (1993). Re-examining perceived ease of use and usefulness: A confirmatory factor analysis. *MIS Quarterly, 17*, 517-525.

SEI/CMU (2008). http://www.sei.cmu.edu/. (Accessed 7 February 08.) SEI/CMU is the Systems Engineering Institute at Carnegie Mellon University.

Shanteau, J. (1992). The psychology of experts: An alternative view. In G. Wright & F. Bolger (Eds.), *Experts and decision support* (pp. 11-23). New York: Plenum Press.

Shanteau, J., & Phelps, R. H. (1997). Judgment and swine: Approaches and issues in applied judgement analysis.

In M. F. Kaplan & S. Schwartz (Eds.), *Human judgment and decision processes in applied settings.* New York: Academic Press. Cited in Reason (1990).

Shaw, N., DeLone, W., & Niederman, F. (2002). Sources of dissatisfaction in end-user support: An empirical study. *Database for Advances in Information Systems, 33*(2), 41-56.

Shepard, D. (2000). Human user interface—HUI. *SpeechTEK Proceedings*, 262-270.

Shepherd, A., & Cosgriff, B. (1998). Problem-based learning: A bridge between planning education and planning practice. *Journal of Planning Education and Research, 17,* 348-357.

Sheppard, B. H., Hartwick, J., & Warshaw, P. R. (1988). The theory of reasoned action: A meta-analysis of past research with recommendations for modifications and future research. *Journal of Consumer Research, 15,* 325-343.

Sheppard, B. H., Hartwick, J., & Warshaw, P. R. (1988). The theory of reasoned action: A meta-aanlysis of past research with recommendations for modifications and future research. *Journal of Consumer Research, 15,* 325-343.

Shneiderman, B. (2000). The limits of speech recognition. *Communications of the ACM, 43,* 63-65.

Short, J., Williams, E., & Christie, B. (1976). *The social psychology of telecommunciations.* New York: John Wiley & Sons.

Simon, H.A. (1960). *The new science of management decision.* New York: Harper & Row.

Singlis, T. M., & Brown, W. J. (1995). Culture, self, and collectivist communication: Linking culture to individual behavior. *Human Communication Research, 21,* 354-389.

Singlis, T. M., Triandis, H. C., Bhawuk, D., & Gelfand, M. J. (1995). Horizontal and vertical dimensions of individualism and collectivism: A theoretical and measurement refinement. *Cross-Cultural Research, 29,* 240-275.

Sinrod, E.J., & Reilly, W.P. (2000). Hacking your way to hard time: Application of computer crime laws to specific types of hacking attacks. *Journal of Internet Laws, 4*(3), 1-14.

Skibell, R. (2003). Cybercrimes & Misdemeanors: A reevaluation of the computer fraud and abuse act. *Berkeley Technology Law Journal,* 18(3), 909-944

Skorodumova, O. (2004). Hackers as information space phenomenon. *Social Sciences, 35*(4), 105-113.

Skyrme, D., & Amidon, D. (1997). The knowledge agenda. *Journal of Knowledge Management, 1*(1), 27-37.

Smith, H. A., & McKeen, J. D. (2004). Developments in practice XII: Knowledge-enabling business processes. *Communications of AIS, 2004*(13), 25-38.

Sniezek, J. A., & Henry, R. A. (1989). Accuracy and confidence in group judgment. *Organizational Behavior and Human Decision Processes, 43*(1), 1-28.

Sones, R. (2000). Improving voice application performance in real-world environments. *SpeechTEK Proceedings*, 179-210.

SØrensen, K. H. (2002). Social shaping on the move? On the policy relevance of the social shaping of technology perspective. In K. H. Sørensen & R. Williams (Eds.), *Shaping technology, guiding policy: Concepts, spaces and tools* (pp. 19-35). Cheltenham, England: Edward Elgar.

Soule, E. (2000). Selecting the best embedded speech recognition solution. *SpeechTEK* Proceedings, 239-248.

Sparks, P. (1994). Attitudes towards food: Applying, assessing, and extending the theory of planned behavior. In D. Rutter & L. Quine (Eds.), *Social psychology and health: European perspectives.* Aldershot: Avery.

Speier, C. & Brown, C. V. (1996). Perceived risks and management actions: Differences in end-user application development across functional groups. In *Proceedings of the Twenty-Ninth Hawaii International Conference on System Sciences, Vol. II* (pp. 374-384). Kihei, Hawaii. Los Alamitas, CA: IEEE Computer Society Press.

Spring, C. (2003). Productivity gains of speech-recognition technology. *Health Management Technology, 24,* 54-55.

Srinivasan, A. (1985). Alternative measures of system effectiveness: associations and implications. *MIS Quarterly, 9*(3), 243-253.

Stamati, T., Kanellis, P., Stamati, K., & Martakos, D., (2005). Migration of legacy information systems. In M. Khosrow-Pour (Ed.), *Encyclopedia of information science and technology,* second edition. Hershey, PA: IGI Global.

Standish Group (1994). *The CHAOS report (1994).* Retrieved November 30, 2006, from http://www.standishgroup.com/sample_research/chaos_1994_1.php

Standish Group (1999). *Project resolution:* The 5-year view. Retrieved November 30, 2006, from *http://www.standishgroup.com/sample_research/PDFpages/chaos1999.pdf*

Standish Group (2001) *Extreme chaos.* Retrieved November 30, 2006, from http://www.standishgroup.com/sample_research/PDFpages/extreme_chaos.pdf

Stanton, J.M, Guzman, I., Stam, K.R., & Caldera, C. (2003). Examining the linkage between organizational commitment and information security. *International Conference Systems, Man, and Cybernetics, 3,* 2501-2506.

Steen, M. (2006). *Our need is to do something about that problem—Studying how researchers and developers interact with informal carers during an innovation project.* Presented at EASST 2006 Conference (European Association for Studies of Science and Technology), August 23-26, Lausanne, Switzerland.

Steen, M. (2006). We don't want woollen trousers—Studying how researchers and developers interact with police officers during an innovation project. In R. ten Bos & R. Kaulingfreks (Eds.), *Proceedings of SCOS 2006 (Standing Conference on Organisational Symbolism)* (pp. 644-666), July 13-15, Nijmegen, The Netherlands.

Steinemann, A. (2003). Implementing sustainable development through problem-based learning: Pedagogy and practice. *Journal of Professional Issues in Engineering and Education and Practice, 129*(4), 216-224.

Stenmark, D. (2002). Information vs. knowledge: The role of intranets in knowledge management. In *Proceedings of the 35th Hawaii International Conference on System Sciences* (pp. 1-10).

Stenmark, D., & Lindgren, R. (2004). Integrating knowledge management systems with everyday work: Design principles leveraging user practice. In *Proceedings of the 37th Hawaii International Conference on System Sciences (HICSS-37 2004)* (pp. 9): IEEE Computer Society.

Stewart, J., & Williams, R. (2005). The wrong trousers? Beyond the design fallacy: Social learning and the user. In H. Rohracher (Ed.), *User involvement in innovation processes: Strategies and limitations from a socio-technical perspective* (pp. 39-71). Wien, Austria: Profil.

Stewart, T. A. (2002). The case against knowledge management. *Business 2.0, 3*(1), 80.

Stiemerling, O. (2000). *Component-based tailorability.* Bonn: Bonn University.

Stiemerling, O., & Cremers, A.B. (1998). Tailorable component architectures for CSCW-systems. *Proceedings of the 6th Euromicro Workshop on Parallel and Distributed Programming,* 302-308.

Storey, J., & Barnett, E. (2000). Knowledge management initiates: Learning from failure. *Journal of Knowledge Management, 4*(2), 145-156.

Straub, D, Loch, K, Evaristo, R, Karahanna, E., & Strite, M. (2002). Towards a theory-based measurement of culture. *Journal of Global Information Management, 10*(1), 13-23.

Straub, D. W. (1989). Validating instruments in MIS research. *MIS Quarterly, 13,* 147-169.

Straub, D. W., Limayem, M., & Karahanna, E. (1995). Measuring system usage: Implications for is theory testing. *Management Science, 41*(8), 1328-1342.

Stromberg, A. (2000). Professional markets for speech recognition. *SpeechTEK Proceedings*, 101-124.

Sugianto, L., & Tojib, D.R. (2006). Information at your fingertips: B2E portal as a strategic tool for todays workforce. In A Cox (Ed.) *Portals: people, processes, and technology*, London: Facet Publishing.

Sugianto, L., Tojib, D.R., & Burstein, F. (2007). A practical measure of employee satisfaction with B2E portals. In *Proceedings of the 28th International Conference on Information Systems*, Montreal, Canada.

Sultan, F., & Chan, L. (2000). The adoption of new technology: The case of object-oriented computing in software companies. *IEEE Transactions on Engineering Management, 47*(1), 106-125.

Sunassee, N. N., & Sewry, D. A. (2002). A theoretical framework for knowledge management implementation. In *Proceedings of SAICSIT 2002* (pp. 235-245).

Sutton, D. C. (2001). What is knowledge and can it be managed? *European Journal of Information Systems, 10*(2), 80-88.

Sutton, S. (1998). Predicting and explaining intentions and behavior: How well are we doing? *Journal of Applied Social Psychology, 28,* 1317-1338.

Svensen, O. (1997). *Risks of road transportation from a psychological perspective: A pilot study, Report 3-77,* Project Risk Generation and Risk Assessment in a Social Perspective, Committee for Future-Oriented Research, Stockholm, Sweden. Cited in Fuller, 1990.

Swan, J., & Newell, S. (2000). Linking knowledge management and innovation. In B. H. Hansen, M. Bichler & H. Mahrer (Eds.), *Proceedings of the 8th European Conference on Information Systems* (pp. 591-598). Vienna, Austria.

Swanson, E. B. (1982). Measuring user attitudes in MIS research: A review. *Omega, 10*(2), 157-165.

Swanson, E.B. (1988). *Information systems implementation: Bridging the gap between design and utilization,* IRWIN, Homewood, IL.

Szajna, B. (1996). Empirical evaluation of the revised technology acceptance model. *Management Science, 42,* 85-92.

Tabachnick, B. G., & Fidell, L. S. (2001). *Using multivariate statistics* (4th ed.). Boston: Allyn & Bacon.

Tallon, P. P., Kraemer, K. L., & Gurbaxani, V. (2000). Executives' perceptions of the business value of information technology: A process-oriented approach. *Journal of Management Information Systems, 16*(4), 145-173.

Taylor, P. (2000). IT projects: Sink or swim. *Computer Bulletin, 42*(1), 24-26.

Taylor, P., & Bain, P. (1999). An assembly line in the head: Work and employee relations in the call centre. *Industrial Relations Journal, 30*(2), 101-117.

Taylor, P., Mulvey, G., Hymann, J., & Bain, P. (2002). Work organization, control and the experience of work in call centres. *Work, Employment and Society, 16*(1), 133-150.

Taylor, S., & Todd, P. A. (1995). Understanding information technology usage: A test of competing models. *Information Systems Research, 6*(2), 145-177.

Taylor, S., & Todd, P. A. (1995). Assessing IT usage: The role of prior experience. *MIS Quarterly, 19,* 561-591.

Teng, J. T. C., & Galletta, D. F. (1991). MIS research directions: A survey of researcher's views. *Data Base,* 53-62.

Teo, T.S. H., & Tan, M. (1999). Spreadsheet development and what if analysis: Quantitative versus qualitative errors. *Accounting, Management and Information Technology, 9,* 141-160.

The 24/7 Delegation. (No. Dir 2003:81). Retrieved November 19, 2006, from http://www.sou.gov.se/24timmarsdel/PDF/Eng%20version.pdf

The City of Stockholm's E-Strategy. (2001, February 19). Available through the City of Stockholm +46 (0)8-508 00 000. The Swedish version can be retrieved (last retrieval, November 20, 2006) from http://www.stockholm.se/files/16100-16199/file_16185.pdf

Thierauf, R. J. (1999). *Knowledge management systems for business*. London: Quorum Books.

Thomas, D. M., & Bostrom, R. P. (2007). The role of a shared mental model of collaboration technology in facilitating knowledge work in virtual teams. *Proceeding of the 40th Hawaii International Conference on Systems Sciences*. http://csdl2.computer.org/comp/proceedings/hicss/2007/2755/00/27550037a.pdf

Thompson, R. L., Higgins, C. A., & Howell, J. M. (1991). Personal computing: Toward a conceptual model of utilization. *MIS Quarterly, 15*(1), 125-143.

Thompson, R. L., Higgins, C. A., & Howell, J. M. (1994). Influence of experience on personal computer utilization: Testing a conceptual model. *Journal of Management Information Systems, 11*(1), 167-187.

Thong, J. Y. L. (1999). An integrated model of information systems adoption in small business. *Journal of Management Information Systems, 15*(4), 187-214.

Thong, J. Y. L., & Yap, C. S. (1995). CEO characteristics, organizational characteristics and information technology adoption in small businesses. *Omega, 23*(4), 429-442.

Todd, P. & I. Benbasat (1991). An experimental investigation of the impact of computer based decision aids on decision making strategies. *Information Systems Research*, 2(2), 87-115.

Tojib, D. R. (2003). *The use of business to employee (B2E) portal in workplace: A preliminary investigation*. Unpublished honours thesis, Monash University, Victoria, Australia.

Trafimow, D., & Finlay, K. A. (1996). The importance of subject norms for a minority of people. *Personality and Social Psychology Bulletin, 22,* 820-828.

Triandis, H. C. (1995). *Individualism and collectivism*. Boulder, CO: Westview.

Triandis, H. C. (2001). Individualism-collectivism and personality. *Journal of Personality, 69,* 907-924.

Triandis, H. C., & Gelfand, M. J. (1998). Converging measurements of horizontal and vertical individualism and collectivism. *Journal of Personality and Social Psychology, 74,* 118-128.

Triandis, H. C., & Suh, E. M. (2002). Cultural influences on personality. *Annual Review of Psychology, 53,* 133-160.

Triandis, H. C., McCusker, C., & Hui, C. H. (1990). Multimethod probes of individualism and collectivism. *Journal of Personality and Social Psychology, 59*(5), 1006-1020.

Trigg, R., & Bødker, S. (1994). From implementation to design: Tailoring and the emergence of systematization in CSCW. *Proceedings of the Conference of Computer Supported Cooperative Work (CSCW 94)*, 45-54.

Troutt, M. D. (2002). IT security issues: The need for end user oriented research. *Journal of Organizational and End User Computing, 14*(2), 48-49.

Troutt, M. D. (2007). Some usability and ethical issues for captive end-user systems. *Journal of Organizational and End User Computing, 19*(3), i-vii.

Truex III, D. P., Baskerville, R., & Klein, H. (1999). Growing systems in emergent organizations. *Communications of the ACM, 42*(8), 117-123.

Tsui, E. (2005). The role of IT in KM: Where are we now and where are we heading? *Journal of Knowledge Management, 9*(1), 3-6.

Tu, C.-H. & McIsaac, M. (2002). The relationship of social presence and interaction in online classes. *American Journal of Distance Education, 16*(3), 131-150.

Tu, C.H. (2000). On-line learning migration: From social learning theory to social presence theory in a CMC environment. *Journal of Network and Computer Applications, 23*(1), 27-37.

Turban, E., & Aronson, J. E. (2001). *Decision Support Systems and Intelligent Systems* (6th ed.): Prentice Hall.

Turban, E., Leidner, D., McLean, E., & Wetherbe, J. (2006). *Information technology for management*. Hoboken, NJ: John Wiley & Sons.

United States Bureau of the Census (2003). *Computer use in the United States, August 2003*. Department of Commerce, Washington, D.C.

Urry, J. (2000). *Sociology beyond societies, mobilities for the twenty-first century*. London & New York: Routledge.

UsabilityNet (2008). http://www.usabilitynet.org/tools/r_international.htm. (Accessed 7 February 08.) UsabilityNet is a project funded by the European Union (EU) to provide resources and networking for usability practitioners, managers and EU projects.

van Bruggen, G. H., Smidts, A., & Wierenga, B. (2001). The powerful triangle of marketing data, managerial judgment, and marketing management support systems. *European Journal of Marketing, 35*(7/8), 796-816.

Van Den Broek, D., Callaghan, G., & Thompson, P. (2002). Call centres: Exploring the team paradox. In *Proceedings of the 6th International Workshop on Teamworking*, Malmo, Sweden.

Van den Putte, B. (1991). *20 years of the theory of reasoned action: A meta-analysis*. Unpublished working paper, University of Amsterdam, The Netherlands.

Van der Sanden, J.M.M., Terwel, J. & Vosniadou, S. (2000). New learning in science and technology: A competency perspective. In P.R.J. Simons, S.J.L. van der Linden & T.M. Duffy (Eds.), *New learning*. Dordrecht, The Netherlands: Kluwer Academic Publishers.

van Kleef, E., van Trijp, H. C. M., & Luning, P. (2005). Consumer research in the early stages of new product development: A critical review of methods and techniques. *Food Quality and Preference, 16*, 181-201.

Van Maanen, J. (1988). *Tales of the field: On writing ethnography*: University of Chicago Press.

Vance, D., & Eynon, J. (1998). On the requirements of knowledge-transfer using IS: A schema whereby such transfer is enhanced. In E. Hoadley & I. Benbasat (Eds.), *Proceedings of the 4th Americas Conference on Information Systems, Baltimore, 1998* (pp. 632-634).

Venkatesh, V. & Davis, F. D. (1996). A model of the antecedents of perceived ease of use: development and test, *Decision Sciences, 27*(3), 451-481.

Venkatesh, V., & Davis, F. D. (1994). Modelling the determinants of perceived ease of use. In J. I. DeGross, S. L. Huff, & M. C. Munro (Eds.), *Proceedings of the Fifteenth International Conference on Information Systems*, Vancouver, British Columbia (pp. 213-227).

Venkatesh, V., & Davis, F. D. (2000). A theoretical extension of the technology acceptance model: Four longitudinal field studies. *Management Science, 45*(2), 186-204.

Venkatesh, V., & Morris, M. G. (2000). Why don't men ever stop to ask for directions? Gender, social influence, and their role in technology acceptance and usage behavior. *MIS Quarterly, 24*(1), 115-139.

Venkatesh, V., & Ramesh, V. (2006). Web and wireless site usability: Understanding differences and modeling use. *MIS Quarterly*.

Venkatesh, V., Morris, M. G., Hall, M., Davis, G. B., & Davis, F. D. (2003). User acceptance of information technology: Toward a unified view. *MIS Quarterly, 27*(3), 425-478.

Vermetten, Y.J., Vermunt, J.D., & Lodewijks, H.G. (2002). Powerful learning environments? How university students differ in their response to instructional measures. *Learning & Instruction, 12*(3), 263-284.

Vermunt, J.D. & Verloop, N. (1999). Congruence and friction between learning and teaching. *Learning & Instruction, 9*, 257-280.

Vermunt, J.D. & Verloop, N. (2000). Dissonance in students'regulation of learning processes. *European Journal of Psychology of Education, 15*(1), 75-89.

Vermunt, J.D. (1996). Metacognitive, cognitive and affective aspects of learning styles and strategies: Aphenomenographic analysis. *Higher Education, 31*, 25-50.

Vermunt, J.D. (1998). The regulation of constructive learning processes. *British Journal of Educational Psychology, 68*, 149-171.

Verton, D. (2001). Hacker conferences highlight security threats. *PC World.*

Vlahos, G. E., & Ferratt, T. W. (1995). Information technology use by managers in Greece to support decision making: amount, perceived value, and satisfaction. *Information & Management, 29*, 305-315.

Von Hippel, E. (2005). *Democratizing innovation.* Cambridge, MA: MIT Press.

Von Hirsch, A., Bottoms, A.E., Burney, E., & Wickstrom, P.O. (1999). *Criminal deterrence and sentence severity.* Hart, Oxford.

von Krogh, G., & Roos, J. (1995). A perspective on knowledge, competence and strategy. *Personnel Review, 24*(3), 56.

Vygotsky, L.S. (1978). *Mind in society.* Cambridge, MA: Harvard University Press.

Wagenaar, W. A., & Keren, G. B. (1996). Does the expert know? The reliability of predictions and confidence ratings of experts. In E. Hollnagel, G. Manici, & D. D. Woods (Eds.), *Intelligent decision support in process environments* (pp. 87-103). Berlin: Springer-Verlag.

Wagenaar, W. A., & Reason, J. T. (1990). Types and tokens in road accident causation. *Ergonomics, 33*(10/11), 1365-1375.

Wallace, L., Keil, M., & Rai, A. (2004). How software project risk affects project performance: An investigation of the dimensions of risk and an exploratory model. *Decision Sciences, 35*(2), 289-321.

Walsham, G. (1993). *Interpreting information systems in organizations.* Chichester, England: John Wiley & Sons.

Walsham, G. (1995). Interpretive case studies in IS research: Nature and method. *European Journal of Information Systems, 4*(2), 74-81.

Walsham, G. (2001). *Making a world of difference: IT in a global context.* Chichester, England: John Wiley and Sons Ltd.

Walther, J.B. (1995). Relational aspects of computer-mediated communication: Experimental observations over time. *Organization Science, 6*(2), 186-203.

Wang, Y. S. (2003). Assessment of learner satisfaction with asynchronous electronic learning systems. *Information & Management, 41*, 75-86.

Wang, Y-S., Tang, T-I., & Tang, J.E. (2001). An instrument for measuring customer satisfaction toward websites that market digital products and services. *Journal of Electronic Commerce Research, 2*(3), 89-102.

Ward, S. V. (1998). *Selling places. The marketing and promotion of towns and cities 1850-2000.* New York: E & Fn Spon.

Webopedia (2005). Retrieved September 14, 2006, from www.webopedia.com

Webster, J. & Hackley, P. (1997). Teaching effectiveness in technology-mediated distance learning. *Academy of Management Journal, 40*(6), 1282-1309.

Wecht, C. H., & Baloh, P. (2006). Open Innovation: Can Modern ICT-Tools Facilitate Customer Integration Into the Fuzzy Front End of the Innovation Process? In *Proceedings of the 2nd European Conference on Management of Technology (EuroMOT), Birmingham, UK, September 10-12, 2006*: International Association for Management of Technology (IAMOT).

Weick, K. E. (1979). *The social psychology of organizing* (2nd ed. ed.). Reading, MA; London: Addison-Wesley.

Wenger, M. (2000). "Noise rejection—The essence of good speech recognition. *AVIOS Proceedings of the Speech Technology & Applications Expo*, (pp. 51-63).

Wheeler, B. C. (2002). NEBIC: A dynamic capabilities theory for assessing net-enablement. *Information Systems Research, 13*(2), 125-146.

Whittaker, D. (1999). Spreadsheet errors and techniques for finding them. *Management Accounting, 77*(9), 50-51.

Whitten, J. L., Bentley, L. D., & Dittman, K. C. (2001). *Systems analysis and design methods.* 5th Ed. Boston: McGraw-Hill Irwin.

Whyte, W. F., Greenwood, D. J., & Lazes, P. (1991). In W. F. Whyte (Ed.), *Participatory action research* (pp. 19-55). Newbury Park, CA: Sage.

Wickstrom, T. (2000). Microphone voice recognition performance in noise: A proposed testing standard. *SpeechTEK Proceedings*, (pp. 211-219).

Willemain, T.R., Wallace, W.A., Fleischmann, K.R., Waisel, L.B., & Ganaway, S.N. (2003). Bad numbers: Coping with flawed decision support. *Journal of the Operational Research Society, 54*, 949-957.

Williamson, S., & Öst, F. (2004). *The Swedish telecommunications market 2003.* (No. PTS-ER-2004-24), The Swedish National Post and Telecom Agency.

Wilson, M., & Howcroft, D. (2002). Reconstructing failure: Social shaping meets IS research. *European Journal of Information Systems, 11*(4), 236.

Winden, W. v. (2003). *Essays on Urban ICT Policies.* Rotterdam: Erasmus University Rotterdam.

Wing, L., & Chua, A. (2005). Knowledge management project abandonment: An exploratory examination of root causes. *Communications of AIS, 2005*(16), 723-743.

Winner, L. (1993). Upon opening the black box and finding it empty: Social constructivism and the philosophy of technology. *Science Technology & Human Values, 18,* 362-378.

Wirt, J. & Livingston, A. (2004). *Condition of education 2002 in brief* (No. NCES 2002011): National Center for Education Statistics.

Wolin, L. D., Korgaonkar, P., & Lund, D. (2002). Beliefs, attitudes and behavior towards Web advertising. *International Journal of Advertising, 21*(1), 87-113.

Wong, K. Y. (2005). *Industrial Management and Data Systems, 105*(3), 261-279.

Woo, D. (2000). Desktop speech technology: A Macos perspective. *AVIOS Proceedings of the Speech Technology & Applications Expo*, (pp. 39-50).

Woolgar, S. (1991). Configuring the user: The case of usability trials. In J. Law (Ed.), *A sociology of monsters: Essays on power, technology, and dominations* (pp. 57-102). London and New York: Routledge.

Workman, M., & Gathegl, J. (2007). Punishment and ethics deterrents: A study of insider security contravention. *Journal of the American Society and Technology, 58*(2), 212-222.

Worthen, B. (2008) Laws go soft on hackers. *Wall Street Journal,* Feb 22nd, Retrieved February 25, 2008, from http://blogs.wsj.com/biztech/2008/02/22/laws-go-soft-on-hackers/

Wurtz, E. (2005). A cross-cultural analysis of Websites from high-context cultures and low-context cultures. *Journal of Computer-Mediated Communication, 11*(1). Retrieved December 15, 2005, from http://jcmc.indiana.edu/vol11/issue1/wuertz.html

Xiao, L., & Dasgupta, S. (2002). Measurement of user satisfaction with Web-based IS: An empirical study. *Proceedings of the Eight Americas Conference on Information Systems* (pp. 1149-1156).

Yamaguchi, S., Kuhlman, D. M., & Sugimori, S. (1995). Personality correlates of allocentric tendencies in individualist and collectivist cultures. *Journal of Cross-Cultural Psychology, 26,* 658-672.

Yang, Z., Cai, S., Zhou, Z., & Zhou, N. (2005). Development and validation of an instrument to measure user perceived service quality of information presenting Web portals. *Information & Management, 42*, 575-589.

Yap, C. S., Soh, C. P. P, & Raman, K. S. (1992). Information systems success factors in small business. *OMEGA International Journal of Management Science, 20*(5/6), 597-609.

Yi, M.Y. & Davis, F.D. (2003). Developing and validating an observational learning model of computer software training and skill acquisition. *Information Systems Research, 14*(2), 146-149.

Yoo, B. & Donthu, N. (2001). Developing a scale to measure the perceived quality of an Internet shopping site (SiteQual). *Quarterly Journal of Electronic Commerce, 2*(1), 31-45.

Young, R., Zhang, L., & Prybutok, V.R. (2007). Hacking into the minds of hackers. *Information Systems Management, 24*(4), 281-287.

Yourdon, E. (1986). *Managing the structured techniques: Strategies for software development in the 1990s* (3rd ed.). Englewood Cliffs, NJ: Yourdon Press.

Zack, M. H. (1999). Managing codified knowledge. *Sloan Management Review, 40*(4), 45-58.

Zack, M. H. (1999a). Developing a knowledge strategy. *California Management Review, 41*(3), 125-145.

Zack, M. H. (1999b). Managing codified knowledge. *Sloan Management Review, 40*(4), 45-58.

Zigurs, I., & Buckland, B. K. (1998). A theory of task/technology fit and group support systems effectiveness. *MIS Quarterly, 22*(3), 313-334.

Zimet, D., Dahlem, N.W., Zimet, S.G., & Farley, G.K. (1988). The multidimensional scale of perceived social support. *Journal of Personality Assessment, 52*(1), 30-41.

Zmud, R. W. (1979). Individual differences and MIS success: A review of the empirical literature. *Management Science, 25*(10), 966-979.

Zmud, R. W., & Cox, J. F. (1979). The implementation process: A change approach. *MIS Quarterly, 3*(2), 35-43.

Zviran, M., & Erlich, Z. (2003). Measuring IS user satisfaction: Review & implications. *Communications of the Association for Information Systems, 12*, 81-103.

About the Contributors

Steve Clarke received a BSc in economics from The University of Kingston Upon Hull, an MBA from the Putteridge Bury Management Centre, The University of Luton, and a PhD in human centred approaches to information systems development from Brunel University—all in the United Kingdom. He is professor of information systems in the University of Hull Business School. Steve has extensive experience in management systems and information systems consultancy and research, focusing primarily on the identification and satisfaction of user needs and issues connected with knowledge management. His research interests include: social theory and information systems practice; strategic planning; and the impact of user involvement in the development of management systems. Major current research is focused on approaches informed by critical social theory.

* * *

William Acar (Dipl. Ing.; MASc Waterloo, CA; PhD Wharton-Upenn, USA) teaches management and information systems at Kent State University. The author of numerous published articles, Dr. Acar has developed a causal mapping method for the analysis of complex business situations and coauthored a book entitled *Scenario-Driven Planning.* Currently, his method is being used to develop a computerized GSS for solving strategic and organizational learning problems. He has also developed measures of diversity better calibrated than the Herfindahl index and the Entropy function. He has published in journals such as: *Strategic Management Journal, Journal of Management Studies, INFOR, Decision Sciences, OMEGA, Journal of Management, Information Systems, European Journal of Operational Research, International Journal of Operational Research, Systems Research, Behavioral Science, FUTURES, International Journal of Organizational Analysis, International Journal of Technology Management, Production & Inventory Management, Strategic Change, INTERFACES* and the *International Journal of Commerce & Management.*

Peter Baloh is an assistant lecturer at Faculty of Economics Ljubljana University. His primary research focus lies in the areas of information systems, technological innovation and knowledge management, which are considered through the lens of successful implementation in various organizational settings. He has authored or co-authored three books and over forty articles, which were presented at international conferences, and were featured, published or are forthcoming in journals such as *MIT Sloan Management Review, IEEE Software, Research-Technology Management, Journal of Organizational and End User Computing, Knowledge and Process Management*, and *Manager,* among others. He serves on the editorial review board of *International Journal of Knowledge Management.* He has traveled the world and loved it.

Jonathan P. Caulkins (PhD, operations research, MIT, 1990) does research and teaches quantitative analysis and decision modeling at Carnegie Mellon University's Qatar campus and its Heinz School of Public Policy and Management in Pittsburgh.

Yvonne Dittrich researches and teaches as associate professor at the IT University of Copenhagen, Denmark. In 1997, after completing her PhD in computer science from the University of Hamburg, Germany, she built up research in cooperation with Industry at Blekinge Institute of Technology in Sweden. Her research combines software development as cooperative work, end-user development, and use oriented design and development of software. Her current work addresses the development and evolution of software products.

Douglas A. Druckenmiller is an assistant professor in the Department of Information Systems and Decision Sciences at Western Illinois University. Dr Druckenmiller received his PhD in management information systems from Kent State University. His research focuses on development and testing of software to support facilitation processes involving causal mapping, agent-based modeling and simulation tools, systems dynamics modeling and thinking, group support systems (GSS) and virtual meeting technologies. He is currently principle investigator of a major international dual-degree project in business and technology involving multiple United States and European Union partner universities. He has more than 35 years of collaboration engineering experience, and has published several journal articles in the area of group decision support and problem formulation in journals such as *The International Journal of Intellegence, Technology and Planning, FUTURES*, and *Journal of Information Systems Education*.

Jeanette Eriksson is a lecturer and researcher in software engineering at Blekinge Institute of Technology in Sweden. Her research area is within tailorable and flexible systems. Her research interests include exploring how to reflect different user aspects in software architectures to achieve software that adapt to altered conditions and requirements in the end user environment. She is also interested in how to involve end users in the technical design process of adaptable software.

Jeremy Fowler has a Bachelor of Computing (Honours) degree from La Trobe University and is currently completing his Master of Science (by research) degree at the same institution. His current research involves the investigation of the interaction between the cultural and known success and failure factors within an information system that went from failure to success during its development. It is hoped that this research will help to improve our understanding of information systems development failure through better understanding of the interactions that occur between the cultural and known success and failure factors.

Pat Horan is a senior lecturer in the Department of Computer Science and Computer Engineering, Bendigo Section, of La Trobe University. She holds a PhD in education from Monash University. Her research interests include information systems education, information systems failure and success, and soft systems approaches to systems development.

Erica Layne Morrison (MSPPM, CMU, 2005) received her MSPPM degree from Carnegie Mellon's Heinz School and is now a consultant in public sector practice for IBM's Public Sector Supply Chain practice.

Victor R. Prybutok is a regents professor of decision sciences in the Information Technology and Decision Sciences Department and director of the Center for Quality and Productivity in the College of Business Administration at the University of North Texas. He received, from Drexel University, his BS with High Honors in 1974, a MS in bio-mathematics in 1976, a MS in environmental health in 1980, and a PhD in environmental analysis and applied statistics in 1984. He is a senior member of the American Society for Quality (ASQ) and active in the American Statistical Association, Decision Sciences Institute, Institute of Electrical and Electronic Engineers, and Operations Research Society of America. Dr. Prybutok is an ASQ certified quality engineer, certified quality auditor, certified quality manager, and served as a Texas Quality Award Examiner in 1993. Journals where his published articles have appeared include *The American Statistician, Communications of the ACM, Communications in Statistics, Data Base, Decision Sciences, European Journal of Operational Research, IEEE Transactions on Engineering Management, MIS Quarterly, OMEGA: The International Journal of Management Science,* and *Operations Research.* In addition, he is in *Who's Who in American Education* and *Who's Who in America*, and *Who's Who in the South and Southwest.*

Ly Fie Sugianto holds Bachelor of Engineering (H1) degree from Curtin University and Doctor of Philosophy from Monash University. She has been appointed as an expert of international standing by the Australian Research Council College of Experts. At Monash, Dr. Sugianto lectures and supervises doctoral students in e-commerce. She has researched and published extensively (70+ refereed articles) in the fields of e-commerce, DSS, B2E portal and deregulated electricity market. She has also received several grants for deregulated markets and information systems research. Her research reflects her ongoing interests in the study and development of support tools and techniques for intelligent decision making.

Dewi Rooslani Tojib holds Bachelor of Business Systems (H1) and Doctor of Philosophy (PhD) from Monash University, Australia. Her PhD research has focused on the development of a scale to measure user satisfaction with business-to-employee (B2E) portals. She is currently a research fellow in the Department of Marketing at Monash University. While she is still continuing her research on B2E portals, she has extended her research exposure to consumer behavior research and experimental designs. Her research has been published in a number of academic and practitioner journals as well as presented in several international conferences. Her research interests include electronic business, mobile commerce, consumer shopping behavior, user satisfaction measurement, web-based Information Systems, scale development and validation.

Marvin D. Troutt is a professor in the Department of Management & Information Systems and in the Graduate School of Management at Kent State University, Kent, Ohio. He is a fellow of the Decision Sciences Institute. He received the PhD in mathematical statistics from The University of Illinois at Chicago in 1975. His publications have appeared in *Management Science, Decision Sciences, Journal of the Operational Research Society, European Journal of Operational Research, Operations Research, Decision Support Systems, Naval Research Logistics, Statistics,* and others. He received the 2005 Distinguished Scholar Award at Kent State University. He has served as the director of the Center for Information Systems at Kent State University and the Rehn Research professor in Management at Southern Illinois University, Carbondale, Illinois. He served as visiting scholar in the Department of

Applied Mathematics at the Hong Kong Polytechnic University during 1994-95. His current interests include supply chain management, applied probability, and applied optimization.

Timothy Weidemann (MSPPM, CMU, 2004) received his MSPPM degree from Carnegie Mellon's Heinz School and is now a consultant in public sector practice for Fairweather Consulting.

Randall Young is an assistant professor in the Accounting and Business Law Department at the University of the University of Texas – Pan American. He earned a PhD in business computer information systems from the University of North Texas, a master's of accountancy from Abilene Christian University and a BBA in finance from the University of Texas at Arlington. His research interests include IT infrastructure, IT auditing and information security management. He has published in *Information Resource Management Journal*, *Information Systems Management*, and *Journal of Organizational and End User Computing.*

Lixuan Zhang is an assistant professor in the Hull School of Business at Augusta State University. She holds a MS in management information systems and Master of Business Administration from University of Oklahoma and a PhD in business computer information systems from University of North Texas. Her current research interests include IT personnel management, IT ethics and security and human computer interaction. She has published in *Journal of Information Technology and Management*, *Information Systems Management*, *Journal of Organizational and End User Computing*, and *Information Resource Management Journal.*

Index

E

F

G

H

I

J

K

T

U

V

W

X